Orbit of Discovery

To Jim —

Watch out for the woodpeckers and keep your eyes on the stars!

[signature]

STS-65, 70, 83, 94

To Tim—

Watch out for the
woodpeckers and keep
your eyes on the stars!

Owen
STS-65, 70, 83, 94

Orbit of Discovery
The All-Ohio Space Shuttle Mission

Don Thomas
with Mike Bartell

Ringtaw Books
Akron, Ohio

Copyright © 2014 by Don Thomas
All rights reserved • First Edition 2014 • Manufactured in the United States of America. All inquiries and permission requests should be addressed to the Publisher, the University of Akron Press, Akron, Ohio 44325–1703.

18 17 16 15 14   5 4 3 2 1

ISBN: 978-1-937378-72-1 (cloth)
ISBN: 978-1-937378-77-6 (ePDF)
ISBN: 978-1-937378-76-9 (ePub)

LIBRARY OF CONGRESS CATALOGING-IN-PUBLICATION DATA
Thomas, Don, 1955–
  Orbit of Discovery : the all-Ohio space shuttle mission / Don Thomas ; with Mike Bartell. — First edition.
    pages cm
  ISBN 978-1-937378-72-1 (hardcover : alk. paper)
  ISBN 978-1-937378-77-6 (ePDF)
  ISBN 978-1-937378-76-9 (ePub)
  1. Discovery (Spacecraft) 2. Space flights. 3. Thomas, Don, 1955– —Travel. 4. Outer space—Exploration—United States. 5. Astronauts—Ohio. I. Bartell, Mike. II. Title.
  TL795.5.T47 2014
  629.44'1—dc23
                    2013033877

∞ The paper used in this publication meets the minimum requirements of ANSI / NISO z 39.48–1992 (Permanence of Paper).

Cover: Photo courtesy of NASA. Cover design by Lauren McAndrews.
Shuttle graphic designed by Michael Clarke.

The information and conversations within this volume are the direct result of the author's recollections, his notes, and details gathered from his viewing of NASA videos.

Orbit of Discovery was designed and typeset in Minion with Helvetica display by Amy Freels, with assistance from Lauren McAndrews, and printed on sixty-pound natural and bound by Bookmasters of Ashland, Ohio.

For Kai—

*While my trips to space were absolutely incredible,
I treasure the time we have spent together even more.*

Contents

Foreword by John Glenn ... ix
Acknowledgments ... xiii
Introduction by George Voinovich ... 1

1 *Discovery* Attacked! ... 7
2 In the Footsteps of Glenn and Armstrong ... 42
3 The All-Ohio Shuttle Crew ... 92
4 *Discovery* to Orbit ... 141
5 Open Season on Woodpeckers ... 176
6 Scientific *Discovery* ... 206
7 Columbus was Right ... 227
8 Close Encounter with Space Junk ... 246
9 Fireball to Earth ... 262
10 Hail, Hail, Rock 'n' Roll ... 306
11 Onward and Upward ... 327
12 Those Clang Jangin' Woodpeckers ... 348

Appendix ... 357
Index ... 397

# Foreword

Ohio has always had a long and rich aeronautics tradition since the Wright Brothers first accomplished powered flight in 1903. From the great National Air Races that took place in Cleveland in the 1930s and 1940s, to the establishment of Wright-Patterson Air Force Base, the NACA Lewis Research Center, and the many universities and colleges across the state, Ohio has always been and continues to be a leader in aeronautics research and operations.

It was only natural then for Ohio to take a similar leading role in the new area of astronautics with the dawning of the Space Age in the mid-1950s. Aerodynamic design, wind tunnel testing, rocket propulsion, and advanced materials development for space applications are just some of the contributions Ohio has made in the field of aerospace. The NASA Glenn Research Center continues to be a world leader in air-breathing propulsion, communications technology, power generation, energy storage and conversion, space propulsion, and cryogenic fluids management, all critical technologies for our future.

Ohio's excellent educational institutions have also played a vital role in our country's aeronautic and aerospace advancements. Numerous colleges and universities across our state have advanced degree and research programs in aerospace-related fields, including Case Western Reserve University, Cleveland State University, Air Force Institute of Technology, Ohio State University, Ohio University, University of Akron, University of Cincinnati, University of Dayton, University of Toledo, and Wright State University, to name a few.

Besides engineering and scientific research in the aerospace field, Ohio has also played a vital role in the human exploration of space. Since 1961, when the first astronauts and cosmonauts ventured into space, over five hundred people have made this journey. Included in that group are twenty-four Ohioans. These Ohio astronauts have completed a total of sixty-one missions to space, participating in the Mercury, Gemini, Apollo, Space Shuttle, and International Space Station Programs. Since my own solo Mercury flight aboard *Friendship 7* on February 20, 1962, Ohio astronauts have completed over sixteen thousand orbits of the Earth. Ohioans have also participated in many of NASA's exploration missions beyond the Earth, completing three trips to the Moon and back and orbiting the Moon a total of twenty-eight times. An Ohioan is also among the elite group of twelve humans to have set foot on the Moon, being the first to ever do so.

Today Ohio astronauts are among the crews working on the International Space Station (ISS). On November 18, 2012, Ohio astronaut Sunita Williams returned back to Earth after her second long duration mission aboard the ISS. With a total time in space of nearly 322 days, Sunita holds the record for most time spent spacewalking for a female astronaut. In addition, she now ranks sixth all-time for US time in space and ranks second among all female astronauts. Some of the work she and our other astronauts continue to perform on the ISS include numerous experiments in fluids and combustion science, using facilities that were developed and built at the NASA Glenn Research Center. Through these experiments it is hoped that we can help improve life here on Earth, gaining a better understanding of combustion processes and potentially minimizing pollution from automobiles and power plants that burn fossil fuels. These astronauts are also participating in valuable life science research to better understand the effects of weightlessness on the human body. From gaining a better understanding of the vestibular system to minimizing bone loss to prevent osteoporosis, our astronauts and researchers are helping to pave the way for our future space explorers while benefiting people back on Earth as well.

In total, our Ohio astronauts have spent over 1,050 days in space and have traveled nearly 428 million miles, a distance roughly equivalent to three round trips to Mars!

With NASA's focus on sending humans out to explore beyond the Earth once again, with missions to asteroids, Mars, and other distant destinations currently under discussion, our country will continue to need new astronauts for these flights. When the human race sends astronauts to land on the surface of Mars in the decades ahead, there is no reason why we shouldn't expect an Ohioan to be among those explorers, carrying on the rich tradition of the Ohio astronauts. After all, Ohio astronauts represent the first American to orbit the Earth, the first to orbit the Moon, and the first to set foot on the Moon. Clearly Ohioans have a strong heritage of discovery, and exploration is in their blood!

Senator John Glenn

Acknowledgments

This is a story I wanted to tell for many years and it took me more than a decade to put down on paper. Getting from my draft manuscript to a final book was a monumental effort, which involved nearly another decade of work compressed into about six months. Leading that effort was Mike Bartell, an experienced reporter from the *Toledo Blade* who enthusiastically covered our STS-70 mission, as well as many other shuttle flights. Without Mike's ideas, suggestions, comments, and frequent encouragement, this project would never have been completed. His help creating a vision for the book was instrumental and his patience and tireless efforts in helping craft the final manuscript are very appreciated. My thanks as well to Ereck Wheeler for his outstanding technical support during the numerous rewrites and the editing process.

My thanks to Tom Bacher from The University of Akron Press for believing in this project and overseeing its publication. Tom was extremely valuable guiding this first-time author through the publishing process. My thanks to Amy Freels at The University of Akron Press for her outstanding work in the design and editing of the book. My appreciation as well to Carol Slatter at The University of Akron Press for her help on the print production of the book.

My gratitude to Mike Gentry and Jodie Russell at the NASA Johnson Space Center who helped me obtain copies of interviews and photographs that were used in the book. Mike always provided great support to me during my twenty years at NASA and continued to do so during the writing of this book.

My thanks to Ted Sobchak in the Space Network Office at the NASA Goddard Space Flight Center for updates on the TDRS-G satellite. Thanks to Ron Zaleski (NASA GSFC), and Mike Bielucki, Don Shinners, Lynn Vaught, and Steve Gollery, all from the NASA White Sands complex, for sharing their experiences with the TDRS-G satellite.

I am grateful to David DeFelice at the NASA Glenn Research Center for his help pulling together information on the Ohio astronauts and for providing photographs from the 2008 Ohio Astronaut Reunion. My thanks as well to many of the other friends and coworkers at NASA Glenn who have supported me over the years and who invited many of my friends and family members, those unable to make it to Florida, to watch my launches from their facility in Cleveland.

I am indebted to the incredible team of folks at the NASA Johnson Space Center, from the trainers who prepared me for my missions to the flight controllers in Mission Control and everyone else who was involved in supporting my flights. I was so proud to be part of your team.

My thanks to Valerie Neal and Frank McNally from the Udvar-Hazy Center of the National Air and Space Museum who arranged for me to photograph *Discovery* in her new home. Special thanks to my friend, Kanji Takeno, who patiently worked with me during the photo shoots.

My gratitude to Jimmy Pritchett, a friend and fellow space enthusiast in Houston who provided me with hours of video recordings from the STS-70 mission that I used in the writing of this book.

To Malcolm Denemark for his help locating the original negatives from photos he had taken of our crew with Woody Woodpecker after the flight. Thanks to *Florida Today* and the *West Kentucky News* for permission to use their photographs in this book.

To Charles Atkeison IV and Chuck Morely for sharing their memories, photographs, and videos taken at the Space Day Rally in Cape Canaveral a few weeks after we landed following the STS-70 mission.

To Dr. Phil P. Lamarre, a great friend for many decades, who early on helped review my manuscript and provided valuable perspectives, suggestions, and encouragement.

To my wife, Simone, and son, Kai, who patiently supported me during my years at NASA and then continued to do so as I endlessly worked on this book. Thanks for enduring the many nights away from home, the birthdays and anniversaries that went on without me, and the times when my mind and attention were focused more on the mission than on you. It was always so great to get back home to you.

To all my friends and family members who supported me through the years and attended the launches. I could feel your hands gently pushing me to orbit and I appreciate that you were always there for me when I needed you.

A special thanks to my twin brother and best friend, Dennis, who taught me about sacrifice and dedication to your work, which enabled me to go on and achieve my lifelong dream of flying in space. I never would have made it without you.

And to my mom, 'The Mother of the Astronaut,' whose love and support helped keep me on track to follow my dream.

And last but not least, to the workers at all of the NASA and contractor facilities across the United States who participated in the space shuttle program. I am indebted to you for your tremendous dedication and for all you put into the program, making sure that I made it to orbit and returned safely back to Earth.

Thanks to you all. It was such an incredible ride!

 # Introduction

On July 13, 1995, space shuttle *Discovery* and her crew of five astronauts launched from Pad 39B at the Kennedy Space Center in Florida. It was *Discovery*'s twenty-first trip to space. The main objective for this mission was the delivery of a huge communications satellite known as the Tracking and Data Relay Satellite, which is still used today to communicate with the astronauts aboard the International Space Station and to transmit scientific data from the Hubble Space Telescope and other earth-orbiting satellites. Weighing forty thousand pounds, the equivalent of nearly ten automobiles, the satellite and its upper stage filled just about the entire payload bay of *Discovery*.

The crew members, all seasoned astronauts assigned by NASA to this mission, were an unusually unique group. Four of the five astronauts, as it turned out, were from the Great State of Ohio. Tom Henricks, a native of Woodville, was assigned to command the mission. Don Thomas, from Cleveland, would be mission specialist one. Nancy Currie, from Troy, would be mission specialist two, and Mary Ellen Weber, also from Cleveland, would be mission specialist three. Only Kevin Kregel, the pilot assigned to the STS-70 mission, was not from Ohio. He had grown up in Amityville, New York.

To this day I don't know if NASA was fully aware that four of the astronauts were from Ohio when they assigned the crew to this mission. For sure the four Ohioans were aware of it and thought it would be a neat idea to make it an All-Ohio shuttle crew and asked me, as the governor of Ohio, if I would make Kevin an honorary Ohioan.

I thought it was a great idea and a way to underscore to the nation and to my fellow Ohioans the Buckeye State's contribution to the space program and our strong aeronautical and aerospace heritage. After all, Ohio had produced more astronauts than any other state at the time and was home to NASA's Lewis Research Center. We would gladly welcome Kevin as an honorary Ohioan!

On May 4, 1995, I proudly signed the decree officially recognizing Kevin as an honorary Ohioan—"thus extending to him all rights and privileges held dear by all citizens of our state." In addition, I issued a resolution acknowledging our pride in the achievements and contributions that the STS-70 mission would be making to our space program and highlighting what exemplary role models the crew were for the youth of Ohio. I ended the resolution by officially proclaiming STS-70 as "The Ohio Shuttle Flight."

To further underscore the Ohio involvement in the mission, I sent the crew an Ohio flag that flew over the state capitol on Statehood Day that year and, of course, some Ohio buckeyes and a few other items. Tom Henricks asked if the crew could fly my watch with them on the mission, which I gladly sent along as well. The watch was given to me in 1992 for my service as president of the Midwest Governors Conference. The watch face featured a map of the world, highlighting the Midwest. I thought this a great choice to fly on the mission, as our astronauts would be looking at the Midwest and Ohio from a totally different perspective up in space.

NASA arranged for my wife Janet and I to attend the launch and we arrived at the Kennedy Space Center the evening before the scheduled liftoff. We toured the space center's huge area and then to our great surprise and excitement, were driven out to the launch pad to see *Discovery* up close—a rare privilege extended to very few people. Janet and I stood there in awe, staring up at the towering shuttle—the approximate height of a twenty-one-story building—as it rested on the launch pad. As we stood there at the base of the shuttle, it was difficult to comprehend that this magnificent machine would soon be leaving Earth, carrying our Ohio crew to space.

When a NASA official handed me one of the STS-70 crew patches, I couldn't help but notice the patch was shaped in the form of the familiar Ohio State University Block 'O' and that the colors were scarlet and gray. I figured Nancy Currie, a proud Ohio State alumna, must have had a strong hand in its design. Several years later, she was rewarded by having the honor of dotting the 'i' as part of the Ohio State University's marching band signature "Script Ohio" formation at one of their home football games.

The next morning we were up at 4:45 and enjoyed a continental breakfast on the bus ride to Spaceport USA at the Kennedy Space Center. They began the briefing with the singing of the national anthem, performed by a children's choir from California, which gave everyone goose bumps. After this we watched on a huge screen the final preparations as space shuttle *Discovery* was readied for launch.

Soon we were bused to the Banana River viewing site. We sat at the top of the stands set up overlooking Launch Pad 39B, only three miles away. We kept a keen eye on the countdown clock, which kept moving ever closer to the moment of liftoff. With great excitement but with a hush of expectation in the air, STS-70 launched at 9:41 a.m., just as scheduled. With my eyes riveted on the shuttle during its climb to orbit, tears formed as I prayed that we'd get our crew up there safely. It was an extremely emotional event and such a textbook launch!

My wife Janet called the launch "just remarkable" and in her journal wrote, "I can't describe the thrill of being there. God is truly in control. I couldn't speak and had tears of joy. I pray all continues to go well." Afterwards she told a reporter covering the launch "It was almost like a religious experience. No one would ever doubt there was a God in Heaven if they sat through one of those launches like we did today. To have given human beings the capacity to do what we just saw, I think is absolutely right." She went on to add that if the reporter had interviewed her right at launch she would have been unable to speak. She never imagined she would have reacted quite like that, as shuttle launches had become quite common and were nearly everyday occurrences. "But to see it firsthand, it was just unforgettable."

Several weeks after the successful flight and keeping a promise they made to me to help open the Ohio State Fair, Janet and I had the privilege of hosting the crew at the governor's residence for dinner and an overnight stay. The dinner menu featured salmon, our homegrown tomatoes and zucchini, Spanish Costa, and peach cobbler for dessert. After dinner the crew shared with us thrilling pictures and stories from their flight. What an exciting evening with such lovely, down-to-earth people.

The next morning after breakfast we were all taken in a van to the fairgrounds, where we were about to board a tram for a tour when the clouds let go. After a short delay we finally got in that tour, which included a stop at the dairy barn for some chocolate milk. From there we went to the George V. Voinovich Livestock and Trade Center for the opening ceremonies, held indoors because of the rain.

The governor's color guard was there, along with the All-Ohio State Fair Youth Choir, who beautifully sang the national anthem. After a short introduction, Tom Henricks spoke on behalf of the STS-70 crew. He presented me with my watch, the one that had flown with them in space, along with the Ohio flag I had sent them. I still treasure the watch and enjoy wearing it at special events, especially when children are present—as I proudly brag the watch has orbited the Earth 143 times and has traveled nearly four million miles!

I believe that the opening of the 1995 Ohio State Fair was the most impressive one in the 162-year history of the fair, made special because of the visit of the STS-70 'All-Ohio' shuttle crew. It was a memorable day for everyone in attendance.

All in all, the flight of STS-70 is an important part of our Ohio history and Janet and I feel extremely fortunate to have been a part of it. This is the story of that mission aboard space shuttle *Discovery* and the five astronauts that made the State of Ohio and our nation proud.

Senator George Voinovich

Chapter One
Discovery Attacked!

The already grueling pace began to quicken even more. In just hours, the five-member STS-70 crew would be entering quarantine. In a few days, we'd be heading to the Kennedy Space Center (KSC) in Florida. And in little over a week—on June 8, 1995, if all went as planned—we'd be floating in the weightlessness of space.

It was a pretty heady time for me and the other members of the so-called All-Ohio Crew—my fellow astronauts Tom Henricks, the mission's commander; Kevin Kregel, the pilot; and Nancy Currie and Mary Ellen Weber, who like myself were mission specialists.

Shortly after being assigned to the mission, we realized that four of the five of us were from Ohio. Kevin was from the state of New York. At the Buckeye crew members' request, Ohio Governor George Voinovich issued a proclamation making Kevin an 'Honorary Ohioan'—thus creating the 'All-Ohio Space Shuttle Mission.' Soon we 'Ohioans' and the orbiter *Discovery* would be part of NASA's milestone one hundredth US manned spaceflight.

But first, there'd be one more day of prequarantine training. I scheduled a final session to go through the Tracking and Data Relay Satellite deployment a few more times. This was going to be the mission's primary objective and my number one responsibility. I wanted to be 100 percent certain that I would be totally prepared for a flawless deploy or any possible problems that might arise.

I completed the minimum number of classes long ago, but scheduled more as time permitted. I went over the deployment checklist in the simulator many, many times. Since Mary Ellen would be assisting me, we worked on our crew coordination one last time. We would go through the checklist and perform the procedures together just to make sure we nailed it.

We had previously organized the activities and split up the responsibilities. There were a few computer displays on her side of shuttle *Discovery*'s aft cockpit where she could easily monitor things, while most of the switches and displays for the deployment were on my side. In addition, I trained to do the deployment by myself because I wanted to know the entire procedure and wanted to be able to handle the deploy on my own, just in case. Still, it was important for the two of us to train together as a team and coordinate our actions, so that things would go smoothly when we conducted the real deploy on orbit. This final training session went extremely well and, after two hours, we were done.

It was 10:00 a.m. and we declared victory. We were fully trained and ready to fly. In another seven hours, we were to report to the Astronaut Crew Quarters at the Johnson Space Center (JSC) in Houston to begin our quarantine period. But just as we were leaving the training session, we got word of a possible mission delay. The reason, someone told us, was because some woodpeckers had poked holes in the orange foam insulation covering the shuttle's massive external tank (ET). I laughed. It had to be a joke—it couldn't be true. It was just too ridiculous, I thought. So I didn't really take it seriously.

Mary Ellen and I headed to our crew office. As we entered, I saw Tom sitting at his desk and he turned to greet us. Nancy and Kevin were working at their desks. "Is it true about the woodpeckers, Tom? Are they really going to delay our flight?" I asked. His answer was quick and to the point: "I'm afraid so."

KSC's launch team was looking at the possibility of delaying our launch because of the woodpecker damage. I felt totally deflated.

Nancy and Kevin had time that morning to take in the news and were already laughing about how ridiculous it all was. The big bad space shuttle taken down by a woodpecker! But for Mary Ellen and me, the news was just beginning to sink in. My initial reaction was of deep disappointment; no astronaut wants to hear about delays for their mission. That was especially true for Kevin and Mary Ellen, who were making their first flights; you just want to get up there so that you can finally become a 'real' astronaut. But delays were very common in the space shuttle program. They were a fact of life and missions without any delays were a bit of a rarity. While some of the delays were weather-related, most were due to technical problems. In either case, it was always best not to launch until everything was 100 percent ready. We all knew and appreciated that fact. Rationally speaking, nearly all the astronauts would prefer delaying missions instead of launching with unsafe conditions. We were already taking a huge risk riding on this rocket and didn't need any more just because of a hurry to launch. Delays were a necessary part of the program, but the news was always hard to hear.

I think most astronauts had underlying fears of their missions being delayed; I know I did for each of my flights. Every day that a mission was delayed presented an opportunity for some other issue to crop up to further delay things. Whether technical, political, or budgetary, there was a minefield of at least a million things that could delay your flight. Even with all the possibilities, nobody had ever considered that a woodpecker could be the root cause for a delay. In standard NASA lingo we would have said, "This is not a credible scenario."

So, after getting the confirmation from Tom, the reality of the situation sank in pretty quickly. There was little we could do, since the decision to delay or not to delay was not ours to make. Mary Ellen and I accepted our fate and joined the rest of the crew in laughing about the oddity of our situation. Then I called my wife Simone to tell her about the change in plans. While we didn't know the full extent of the damage or the repairs that would be needed, we did know one thing for certain—we were not going to be going into quarantine that night and we were not going to launch on June 8. We suspected that we would probably have at least a few days' delay. All we could do was wait for the repair assessment; that would tell us how long the delay would be.

After first feeling disappointment and then laughing about the situation, we all started to realize that this news had a bright side—that we would spend at least one more night at home with our families. I was particularly excited about this possibility because my son Kai was only 106 days old, and the opportunity to spend at least one more day with him and Simone was a bonus. We felt a little like kids who had just been told that classes were cancelled. We were 100 percent trained and ready to go, so we were all but guaranteed at least one day off from training. It was an exciting possibility.

Meanwhile, NASA was evaluating whether patching the holes could be done while the shuttle was on the launch pad. If so, the delay might only be a couple of days, perhaps a week at most. If not, then they would have to roll the shuttle back to the Vehicle Assembly Building (VAB) where more work-access platforms were available and the repairs could be done appropriately and safely. If that was the case, the delay would be at least a month.

We had a few more training sessions scheduled that day, and then I went home to sleep in my own bed instead of reporting to crew quarters. The next morning, we learned there would not be a quick fix. There was talk of delaying the mission until early August.

Of course, we were disappointed; the crew was ready and wanted to launch. But at the same time, I had a sense of relief. We had trained extremely hard the past year; I realized a delay would mean that we could take a few days off. Our training had been so rushed and hectic and at one point, NASA even accelerated our launch date, which compressed our training schedule further. We had been working long hours and a few days off sounded pretty good, so the news was not all bad.

A day later, word of the woodpecker attack and our launch delay was making news. While not earth-shattering, it's peculiarity caught the public's attention. "Lovesick Woodpeckers Ground a Space Shuttle," one newspaper headline declared. "Birds Poke Holes in Scientists' Plan," another advised. "Wings Clipped by Woodpecker?" and "Pesky Woodpeckers Rule Roost at NASA" were other entries. The *Miami Herald* headlines on Saturday June 3 summed up the situation best: "Woodpeckers 1, *Discovery* 0."

Initial repair attempts were made at the launch pad, with NASA bringing in a special 250-foot crane, plus numerous smaller cranes. Scaffolding also was erected atop the payload change-out room of the rotating service structure. A special team of workers from Lockheed Martin, with experience working high up on the shuttle made valiant efforts to access and repair all the holes. But the more they looked at the situation, the more they realized it would be impossible to make the necessary repairs at the launch pad. Many of them were just not accessible or it was deemed too dangerous for workers to access some of them while dangling in baskets from a crane high above the launch pad.

So after a week of patching attempts and after carefully evaluating the situation, NASA management decided to roll the shuttle back to the VAB. The launch delay would be at least a month.

But the shuttle program would not come to a stop just because of a woodpecker. NASA decided to launch the next shuttle mission ahead of STS-70. The STS-71 flight of the orbiter *Atlantis* would be the first shuttle to dock with the Russian space station *Mir*. STS-70 would lose its claim of being the hundredth manned mission in US space program history. NASA had made some limited fanfare about our mission being number one hundred, but this fleeting fame would be passed to STS-71.

It was a little sad to lose that distinction. STS-70 was only a vanilla shuttle mission, performing the last TDRS satellite deploy from a shuttle. It was not a sexy flight by most standards and considered outright boring by some. We took solace in the fact that we would still be known as 'The All-Ohio Space Shuttle Mission,' but cringed somewhat at the notion that STS-70 would forever be known as 'The Woodpecker Flight.' That was our destiny!

Holey ET, Woodpeckers in Love!

Within a very short period of time, I learned much more about woodpeckers than I had ever learned before... or since! Evidently, the damage was caused exclusively by a single northern flicker. Some of the other more uncommon names for these woodpeckers include yellowhammer, clape, gaffer woodpecker, harry-wicket, heigh-ho, wake-up, walk-up, wick-up, yarrup, and gawker bird, to name a few. Before all was said and done, we and our NASA colleagues would have a few additional—and much more colorful names—for the pesky pecker!

Woodpeckers drill holes to make nests. Normally, male and female northern flickers conduct the excavation work jointly. The entrance hole is about three inches across in an ideal home and the cavity they form is normally thirteen to sixteen inches deep. The cavity widens at the bottom to make room for the eggs and the incubating adult. With the foam insulation on the external tank only three to four inches thick, the woodpecker in question hit the aluminum metal tank underneath before it could excavate an appropriately sized nest. After banging on the aluminum tank a few times, he would simply move to a new location on the tank and try again.

Launching a thorough investigation as only NASA could, bird specialists from the US Fish and Wildlife Service were called to the scene, since the Kennedy Space Center also is a national wildlife refuge. They were able to tell NASA that late May to early June is the woodpeckers' normal nesting season and that the male woodpecker was undoubtedly trying to attract the attention of females by taking on something far more monumental than a dead tree limb.

Years later, I learned of reports that workers at NASA had disturbed some of the woodpeckers' natural habitat areas while clearing brush near our launch pad, which possibly led the birds to seek new nesting territory. Whether the woodpecker attack was only a one time bad-luck event or whether it was caused by the NASA workers didn't much matter—a single woodpecker had grounded our flight!

Ever so slowly, the details of the situation began to emerge. Initial press reports on the number of holes created by the woodpecker started at 78 and ranged in size from between a half-inch to four inches in diameter. A closer examination of the external tank revealed a dramatic increase in the final count to 205.

Two things amazed me at the time. First, that the damage was thought to have been caused by a single woodpecker—one woodpecker, 205 holes! The other was that NASA had hours of video of the woodpecker drilling the holes; there were dozens of cameras mounted all over the rotating service structure to monitor the shuttle. Why didn't someone react while the damage was being done? Why didn't someone say after the first 67 holes, "Gee, look at that little woodpecker." Or 30 holes later, "Think we should do something?" Or another 38 holes later, "I definitely think we ought to report this thing!" I couldn't believe that someone watching wouldn't have brought it to someone else's attention and stopped it. I never heard a satisfactory explanation of why nothing was done sooner.

The external tank was essentially a giant fuel tank. It held slightly more than a half-million gallons of fuel, which fed into the space shuttle orbiters' three main engines. With a length of 154 feet and diameter of 28 feet, the ET is the largest component of the space shuttle and is the structural backbone of the three-part system that also includes an orbiter and the solid rocket boosters. To put it in perspective, the ET was 34 feet longer than the distance flown by Ohioan Wilber Wright on his historic first flight aboard the Wright Flyer at Kitty Hawk, North Carolina, on December 17, 1903.

Because of the extremely cold temperatures of the liquid oxygen (-297°F) and liquid hydrogen (-423°F) used for fuel, the tank was covered with insulation to prevent ice from forming while waiting for liftoff. With the warm humid air typical for the Florida launch location, moisture would otherwise condense on the tank (much like it does on a glass of ice-cold lemonade on a hot summer day) and then freeze into ice. Any buildup of ice would add unnecessary weight to the space shuttle and, equally important, any accumulated ice could potentially break free and strike the fragile surfaces of the space shuttle orbiter due to the vibrations and aerodynamic forces experienced during launch.

Additionally, during launch there was significant aerodynamic heating on the forward end of the external tank, where temperatures could reach nearly 1,800 degrees Fahrenheit due to friction as the space shuttle accelerated through Earth's atmosphere. That heat could have resulted in much of the super-cold propellants boiling off and being wasted during launch. For these reasons, the outside of the tank was covered with a special thermal protection system.

Holes in the external tank's foam—whether created by woodpeckers, hail, or by some other means—were thought to be dangerous. Gaps could permit a buildup of ice, which could break free during the 8½-minute ascent to orbit. In areas where ice buildup was not a concern, gaps in the insulation could expose the external tank and its cryogenic propellants to excessive aerodynamic heating. This could lead to overpressurization and possibly a tank rupture or explosion, which could result in loss of the vehicle and crew. In addition, such holes could cause the tank to heat up and disintegrate earlier than desired during its reentry over the Pacific Ocean. This could result in the debris falling outside the targeted footprint in the ocean, possibly raining down on populated areas, jetliners, or ships.

On June 8 (our originally planned launch date), *Discovery* made her way back to the VAB for the needed repairs. NASA announced our new launch date as July 13, 1995.

Through the end of the shuttle program, there were twenty rollbacks—some for mechanical or payload problems and others due to weather (hurricanes, tropical storms, or hail damage). STS-70 remains the lone instance where the rollback was caused by a woodpecker.

With word that our mission was being delayed a little more than thirty days, we started talking about the possibility of taking some time off. Tom knew we were all trained, had been working very hard, and that everyone would benefit from a short vacation, so he authorized all of us to take the next ten days off. We were thrilled to be getting that much time so close to launch; it was an extremely rare scenario. So as we left the Johnson Space Center that Friday afternoon, we were all excited about the vacation windfall.

All of us wanted to spend the time with our families. It was early June in Houston, which meant you were guaranteed plenty of hot, humid weather, with temperatures that time of year in the upper nineties. The kids were already out of school for summer vacation, so we could all leave town if we wanted to. But with the expense of buying airline tickets at the last minute (especially for Kevin with four kids), some of us stayed in town and hung out with the kids at swimming pools, while others took advantage of the opportunity to leave the area.

Simone, Kai, and I stayed home for a few days, spent a day or two in Galveston, and then took a road trip to the Austin area just to get away for awhile before returning home the following weekend.

Upon our return, it was back to business and training. The vacation was wonderful for everybody. Instead of launching exhausted and a bit burned out, we were all well rested. We knew how lucky we were to have been given that opportunity.

While we vacationed, NASA workers had been busy repairing our ET. For each hole, a layer of adhesive was applied, followed by a manual application of the spray-on urethane foam material. Once hardened, the patch was sanded to minimize the roughness on the surface of the tank. The entire process took about five hours per section of the tank, with most of the time spent waiting for the foam to fully cure and harden.

After all the holes were patched and the quality inspectors had signed off on the repairs, space shuttle *Discovery* was ready to roll back to Launch Pad 39B. NASA management once again cleared *Discovery* for flight, but was still struggling with what to do about the woodpeckers and how to prevent future attacks.

The woodpecker attack had brought more attention to our mission than anything else. David Letterman, alluding to the situation, remarked that "You get the feeling these days that NASA couldn't plan a mission to Planet Hollywood." Even the Space Frontier Foundation, an organization started in 1988 dedicated to opening the space frontier to human settlement as rapidly as possible, reportedly put out a statement congratulating the woodpecker for navigating around NASA's bureaucracy to establish the first private sector use of the external tank.

A Team for the Birds

When word got out about the woodpecker attack, help quickly materialized from around the world. It seemed almost everyone had a solution. NASA received hundreds of calls and letters from people with woodpecker knowledge and experience offering to help solve the problem.

One of the more unusual ideas involved draping the external tank with strips of white bed sheets, wire mesh, or toilet paper. Another person suggested placing a net suspended by hot air balloons around the space shuttle. One individual suggested painting the external tank blue because woodpeckers were said to hate that color. NASA ruled out that idea because of weight considerations. For the first two space shuttle flights, the ETs were painted white—until NASA figured out it could lighten the weight of the tanks by six hundred pounds by eliminating the paint.

Another novel concept involved spraying a solution of boiled skunk cabbage on the tanks. While I have never smelled boiled skunk cabbage, I have a feeling it would repel the woodpeckers as well as the astronauts. Someone else suggested suspending a jet engine over the external tank, hoping that the downward flow of exhaust would keep the birds away. There was a suggestion that a sharpshooter kill the woodpeckers. KSC's wildlife refuge status ruled out that one.

While some of the ideas were impractical and even a bit farfetched, other suggestions were a bit more feasible. When John M.

Tepoorten, of Little Canada, Minnesota, heard of NASA's problem, he recalled that while on a 1984 business trip to Japan he saw beautiful rice fields that had strange, brightly colored balloons with large eyes positioned around them. He brought the idea back to the United States, formed a company called Bird Scare, and developed a product called Predator Eyes. The bright yellow balloons—much like an inflatable beach ball—are about sixteen inches across and have a ring of six big predator-looking eyes around them, any two of which are staring at you from whatever angle you view the balloon. At the center of each eye is a small circle of reflective silver Mylar and hanging from the bottom of the balloon are four streamers of red Mylar. With a gentle breeze blowing, the Mylar strips move and help scare away birds. So Mr. Tepoorten sent of few of his Predator Eyes for NASA to try out.

The key feature of the balloons are the large predator eyes, which mimic those of hawks, owls, and eagles, the main predators of the northern flicker. Any woodpecker seeing such a large set of eyes, it was surmised, would be frightened enough to stay away from the space shuttle without harming the birds in any way.

Other individuals and companies sent in similar products and suggestions. Other balloon designs, like Terror Eyes, manufactured by a Chicago-based company named Bird-X, were sent to NASA for consideration. The Terror Eyes balloon featured a big bright orange balloon with two large eyes and a face to represent a not-so-friendly owl.

Along these same lines, another company suggested placing life-sized plastic owls around the space shuttle to scare the woodpeckers away. Even a company from Australia got involved, sending along cardboard placards of a black cat silhouette which were hoped to scare away the woodpeckers. It was an international effort to save the space shuttles from future woodpecker attack!

The space shuttle management team formed a group to look into the problem and sift through these and other proposed solutions. This group was called the Bird Investigation Review and Deterrent

(BIRD) team and their task was to come up with a plan for keeping the woodpeckers away from the launch pads and prevent them from inflicting any further damage to the space shuttles.

After much study and deliberation, the BIRD team came up with a three-phase approach to solve the problem. The first involved establishing "an aggressive wildlife management program to make the launch pads less attractive to the woodpeckers." Phase two implemented "scare and deterrent tactics at the pads," using plastic owls, balloons, water sprays, and loud air horns. NASA would try frightening them away without harming them. Phase three involved the "implementation of bird sighting response procedures," where definite plans would be implemented in the event that a woodpecker was sighted.

The first thing they did was clear the areas around the launch pads of old trees and brush in an effort to eliminate habitats desirable to the woodpecker. They also let the grass grow longer around the pads, which made it more difficult for the woodpeckers to feed on ants and other insects on the ground. They loaded up the rotating service structure, which surrounds the shuttle, with various devices to scare away the woodpeckers—plastic replicas of horned owls and reflective silver and red Mylar ribbons. Numerous predator balloons of various designs were installed, but NASA didn't stop there. They also relied on human involvement, with sentinels armed with clipboards and air horns that they would blast from time to time. For even more fire power, there were water cannons that they could use to shoo away any woodpeckers that may have gotten within range. And finally, NASA continuously blared high-pitched sounds, using mechanical chirpers to frighten away the woodpeckers. While the shrill chirping sounds mimicking the sharp cries of a hawk or eagle or the screech of an owl were intended to drive the woodpeckers away, their main impact was reportedly driving the workers at Pad 39B crazy. In any event, with the full complement of woodpecker deterrents, no additional woodpecker ever showed its face around *Discovery* again, or any other launch pad for that matter, through the end of the space shuttle program in 2011.

Rolling Toward History

I had always wanted to go to KSC and watch one of the rollouts of the shuttle as it headed from the VAB to the pad about six weeks before launch, but the last few months of training before launch were some of the busiest, making it nearly impossible for astronauts to witness the rollout of the shuttle for their own mission. For my first flight, STS-65, I was training in Huntsville during the rollout of *Columbia*; for STS-70, we had a simulation at the Johnson Space Center that prevented our crew from making it to KSC to watch the event.

By the time we got back from our week of vacation, the woodpecker holes had been patched and space shuttle *Discovery* was ready to return to the launch pad. I told Tom and Kevin that I was interested in seeing a rollout and hoped that we could schedule something. I had seen many pictures of the Saturn V rockets and space shuttles being rolled to their launch pads, but never the real thing in person. So I was really excited to learn that Tom had arranged for some of us to fly to KSC for two days to watch the second STS-70 *Discovery* rollout, scheduled for June 15.

Tom, with Mary Ellen in the backseat of his T-38, and Kevin, with me in his backseat, departed Ellington Field for KSC on June 14, arriving at the Shuttle Landing Facility shortly after 9:00 p.m. We went to crew quarters for a late dinner, hung out there a bit, and then about 11:15 p.m. drove over to the VAB.

Because of the normal sea breezes and the daily threat of afternoon thunderstorms, space shuttles were normally rolled out to the pads in the middle of the night, when the wind and chance of rain was minimized. For this second rollout, *Discovery* was scheduled to leave the VAB about midnight for her eight-hour journey to Pad 39B.

As we approached the VAB about 11:30 p.m., I spotted *Discovery* already a quarter mile or so from the VAB, slowly inching its way to the pad. It had probably started its journey a half hour earlier. We initially tried to drive up close to the crawler transporter carrying the shuttle, but were stopped by security and told we needed special clearance to get any closer.

Buddy Blackman, one of the Vehicle Integration and Test Team members and the lead for STS-70, scrambled to get us proper clearance and about midnight we were allowed to approach. We parked our cars along the side of the road and climbed out.

It was an exciting sight to see—*Discovery*, sandwiched between the two white solid rocket boosters and bolted to the orange external tank, standing high above us on the mobile launch platform. The entire stack was brilliantly illuminated by bright lights that followed the stack as it moved along the crawler/transporter track leading to the pad. It was enough to take your breath away. My heart raced and the butterflies in my stomach took flight as it sank in that this was *our* shuttle and it wouldn't be long before we would be headed to orbit inside of it!

Besides this visual sight, my senses were bombarded with the noise (more like a roar) of the diesel engines of the crawler/transporter as it strained under the weight of the shuttle stack balanced precariously on top. At this point, as it would be for launch, the shuttle was bolted to the mobile launch platform with eight rather large bolts—four bolts each along the base of each of the two solid rocket boosters (SRB).

The bolts were impressive, each slightly larger than a baseball bat. The nuts used to bolt the shuttle down were equally impressive. They were roughly as big as a sandwich plate, about seven inches across and nearly five inches thick. At the moment of launch, precisely at SRB ignition, explosive charges were set off on each bolt, severing it into two pieces that fell away and allowed the shuttle to fly free of the launch platform. It was a long-standing tradition to present each crew member with a set of bookends made from two halves of one of the nuts.

The black smoke and oily smell reminded me that the crawler/transporters were developed in the mid-1960s for the Saturn V rockets. Here we were, nearly thirty years later, using the same vehicles to move the space shuttles from the VAB to the launch pads.

We all climbed aboard the crawler/transporter. Kevin and I walked around a bit before we headed up to the main control cabin located in the front right corner of the crawler. There was a separate control cabin for each of the four sets of treads, one located at each corner of the crawler.

The crawler crew welcomed Kevin and me and one of them gave us a quick tour of the control room and pointed out the crawler speedometer. Here we were taking a vehicle capable of flying 17,500 mph to the launch pad at the maxed-out speed of 0.9 mph.

After about five minutes, we headed outside for a tour of the rest of the transporter. We climbed up a series of stairs and made our way to the shuttle. As we stood at the base of the solid rocket boosters, I looked toward the top of the external tank and started to laugh. Tied to the very top of the tank was a large round balloon, maybe three feet across, that had two giant owl eyes painted on it, one of the predator-eye balloons that NASA installed at the recommendation of the BIRD team.

With the noise, smoke, lights, and other commotion, my guess was that there wasn't a single woodpecker within a few miles on that particular night. I cannot say with any certainty whether the balloon was effective. But I can assure you that on that evening and out on the pad on launch morning, I never saw a single woodpecker. Not one. And just for the record, I also did not see any lions, tigers, or bears.

The patchwork on the external tank was pretty impressive. They had patched 197 of the holes, the largest one about four inches across. On the upper part of the oxygen dome, on the back side away from the orbiter, the surface had been totally peppered with holes. Near the base of the ET, there were about 10 holes that I could see, some of which they were still working on. There was still a lot of scaffolding around the base of the ET that they were using to complete the patching of the final holes.

Kevin brought four eight-inch plastic woodpeckers, one for each of us, and we wore them velcroed to our shoulders as we walked around. The workers got a big kick out of it.

It was a real thrill for me to ride along with *Discovery* as it headed to the pad that night. We stayed out there about half an hour or so.

When it was time to leave we climbed down off the crawler and headed back to our cars, now parked a few hundred yards away. We got back to crew quarters about 1:30 a.m. and got a few hours sleep as *Discovery* continued its slow march toward Pad 39B. I wanted to see the shuttle make the last quarter mile of the trip as it gently climbed to the pad, so I set my alarm for 5:00 a.m. and drove back out to Pad 39B by myself. When I got there about thirty minutes later, I was disappointed to see the shuttle had already completed its journey. Workers were in the process of performing the final stages of leveling and alignments before lowering the launch platform onto the launch pad pedestals.

It was a beautiful sight to see *Discovery* sitting on the launch pad, just as it would look on launch morning. I brought along one of the NASA Nikon F4 cameras that we used in space and spent the next hour or so taking pictures of *Discovery* from all possible angles. While I was out there, the sky began to lighten as sunrise approached and for a few minutes the sky was a beautiful purple, which made for some spectacular pictures with the shuttle still brilliantly illuminated by the lights at the pad. After enjoying the solitude and sights and using up endless rolls of film, I headed back to crew quarters.

That was the one and only opportunity I ever had to see a shuttle rollout, and it was a very special night for me—made even more special because this was my shuttle on the way to the pad for my mission.

About 7:30 a.m. I rejoined my crew mates and we drove to the VAB to meet with some of the workers who had set up the scaffolding and done some of the patching work on our ET. Once again, there were plenty of jokes about woodpeckers. Everywhere we went at KSC, people seemed to be having fun with woodpecker stories and jokes. It might have been the shear silliness of the incident or because it broke up the routine of processing space shuttles a bit, but it was clearly fun for everyone involved. We spent about thirty minutes talking with the workers and signing autographs. There is a long-standing tradition at KSC for the workers to collect dollar bills signed by astronauts. I signed twenty to thirty of them that day.

From the VAB we headed directly to the Shuttle Landing Facility, where we checked the weather for our flight back to Houston, filed our flight plan, and shortly after 8:30 a.m., were airborne. After our takeoff on Runway 33, we made a slow turn to the right over the Atlantic Ocean and then back to the west. Down on the right I could see *Discovery* on Pad 39B and as we continued our climb to our assigned altitude, I thought to myself, *Take care, Discovery. We'll be back soon and we'll see you then.* We were once again a month from launch, and everything was looking good.

We landed at Ellington Field about 10:30 a.m. A few hours later, I drove Simone and Kai to Houston Intercontinental Airport. Simone wanted to take advantage of the month delay to fly to her native Germany so her parents and grandparents could meet their new (and first) grandchild and great-grandchild. I took advantage of the quiet time during the two weeks they were gone to relax a bit and mentally prepare myself for the flight. A month seemed like a long time away, but our launch was coming up quickly.

The next few weeks of training were relatively uneventful. Since we had completed all the necessary training just before the woodpecker attacked, the focus now was just refresher training. I signed up for as many TDRS deploy classes as I could to make sure I was going to be ready. I also scheduled a few additional sessions with Mary Ellen so we could practice working together and fine-tune our coordination during the deploy.

I asked the training team at these refresher sessions to explain each system on a deeper level and also asked that they throw in some of the more obscure malfunctions, along with ones that were considered more likely to occur. Since we were given the luxury of more training time, I wanted to be ready for anything, even malfunctions that probably would never take place. In short, I wanted to be prepared for anything and everything because I took the responsibility of being the TDRS deploy lead crew member very seriously.

As a crew, we each needed to learn our jobs and learn them well. Each crew member relies on the others to do the same. We each had specialized tasks that we would be performing during the mission. For some of the more critical parts of the mission, like launch, landing, and the deployment of our satellite, we had backups in case the primary person became incapacitated. The astronaut crew is a team and we all relied on one another to accomplish the goals of the mission. Our No. 1 duty was to be well trained and proficient in our assigned tasks. We also continued our launch simulation training, with a few sessions each week to go through various malfunctions that might occur. Our ascent team was a good one and we always had fun during these sessions.

Simone and Kai arrived back home and the peace and quiet I had enjoyed while they were gone soon gave way to the usual pattern of waking up in the middle of the night to get Kai a bottle. This was my normal routine at home. Nearly every night, Kai would wake up crying for a bottle. And every night, I groggily walked downstairs to get him one to quiet him so I could get back to sleep. That routine left me dead tired, but I enjoyed every moment at home with my family because very soon it would be time for me to enter quarantine and then head for space.

We were scheduled to enter quarantine July 6, one week before launch. We were given most of that day off to take care of last-minute personal business. It's amazing how many things you need to deal with before leaving the planet. We were scheduled to report to the Astronaut Crew Quarters at the Johnson Space Center at 5:00 p.m. It was time for me to start saying my goodbyes and get ready for launch.

Quarantine? Really?

Normally, crews enter quarantine in the late afternoon so they can share a meal prepared by a team of NASA food specialists and dieticians. The overriding concept of quarantine was to remove the astronauts from coworkers, young family members, and the general public to prevent any exposure to the flu, colds, or any other bugs that may be going around. The last thing you needed in space was someone coming down with something. The Apollo 7 crew got bad head colds during their eleven-day mission, which made all of them miserable and caused problems and tension to develop between the crew and Mission Control. Since the incubation time for most of these germs is less than a week, NASA figured that if you were exposed to anything before entering quarantine, you would show symptoms before you launched and they could consider delaying the mission a few days until everyone was well.

Once we entered quarantine, we were permitted to come into contact only with people who had been checked by NASA doctors and certified as not possessing any germs that could be transmitted to the crew. Cleared individuals were given yellow Prime Crew Contact (PCC) badges that they were required to wear along with their normal employee badges. All the astronauts were issued these as well as members of the training team, flight directors, NASA management team, and key members of Mission Control who may have a need to interact with the crew during the week before launch.

Astronauts in quarantine were reminded to stay away from any individual not wearing a PCC badge. "Assume they are sick and avoid any contact," we were told, and stay at least ten feet away from them (apparently so no germs can jump from them to you like a flea exchange between dogs).

Anyone else needing to have contact with the crew was required to undergo a quick physical from one of the NASA doctors. Once cleared, they were given temporary PCC badges that allowed them to have contact with the astronauts for periods of time from a day to a few days depending on the circumstance. Astronaut spouses routinely received PCC badges so that they could visit in quarantine. Special guests or friends invited to lunch or dinner while we were in quarantine also underwent physicals to get a PCC badge.

What was strictly forbidden in the Health Stabilization Program was exposure or any contact with children under the age of sixteen, whether healthy or not. As well-known germ factories, they provided the greatest risk to astronaut health. Once we entered quarantine, we were no longer permitted any contact with our children. We could talk to them on the phone or send them an email, but we were not permitted to hug them or get within ten feet of them at any time, an extremely emotionally difficult thing for both the children and the astronauts.

Kai was only five months old when I entered quarantine for STS-70. He was a relatively new addition to my life, but it didn't make it any easier to say goodbye. He was too young to understand what was going on as I held him for the last time, about fifteen minutes before entering quarantine. I gave him a big kiss and hugged him tightly for a few minutes. Then exhaling deeply, I handed him over to Simone, gave her a quick kiss, and got in the car and drove over to NASA, a two-mile trip that only took ten minutes door to door. I pulled up to the Astronaut Crew Quarters within a few minutes of our 5:00 p.m. deadline.

The JSC's crew quarters are located in a far back corner of the facility, in a nondescript single-story metal structure designated as Building 259. It had corrugated metal siding, painted white, but hadn't seen a fresh coat of paint in years. The building was the typical architecture for many small businesses or storage sheds that pop up in Texas seemingly overnight. It had a front door, a small side window, and three long windows along the side where the dining area was located.

I punched in the well-known code into the cipher-locked door and officially entered quarantine. There was no force field around the place and no high-tech security alarm system, but it was always clear to me that I was entering the different and amazing world of astronaut quarantine.

Tom, Kevin, and Nancy were already there, and a few minutes later Mary Ellen arrived. We enjoyed our meal together, full of anticipation of the journey that lay ahead. After dinner, we all hung out for awhile and Mary Ellen then left for home. She had no children and only needed to report to crew quarters for our three daily meals and for any other scheduled crew activities or training sessions. While Tom, Kevin, Nancy and I were all a little envious of her freedom, the crew quarters provided a sort of peace and quiet not possible at home, so we all took advantage of this unique opportunity. Soon we all headed to our rooms, where we made calls home to see how everything was going. I had left my house only two hours earlier, but still felt the need to check up on things.

Quarantine at JSC always reminded me of a sort of minimum security prison. I could pretty much come and go as I pleased, but I was severely restricted as to where I was permitted to go. I felt like I could see freedom just on the other side of the fence, but that I wasn't allowed to experience it on my own.

Because the astronauts frequently sleep shifted to align their day with the scheduled launch time, it sometimes meant going to bed at 3:00 p.m. and getting up at 11:00 p.m., pretty much out of sync with the rest of the world. It made it more difficult for spouses to visit.

Sometimes Simone would come over for a visit and I would ask her to drive me around a bit. "Let's just get away from JSC and NASA for a little while," I'd say, and we'd get in the car and drive around Clear Lake. We would go nowhere in particular, but it felt so good to get away for awhile. We couldn't stop anywhere because I was still in quarantine and had to stay away from people who were not cleared, but the time in the car felt good.

Working these somewhat odd hours did have one advantage: we could go to our office on the sixth floor of Building 4S in the middle of the night, because the building was empty and there was little chance of bumping into someone in the hall or on the elevator. Sometimes I would drive over just to check my mail or look through emails. It was another quiet place where I could get away.

But we always had to be present in crew quarters for meals and we could only eat the specially prepared food. There was no ordering pizza or Chinese takeout. The procedure was to eat only food prepared by the crew quarters' staff to make sure it was always fresh and to prevent any astronaut from getting sick from some form of food poisoning. It seemed a little overly cautious, but it was never a big deal for most of us. Besides, the food was excellent and the staff worked hard to accommodate any special requests to make sure everyone was happy. There always were plenty of snacks and other junk food, including the seemingly bottomless bowl of peanut M&Ms. The bowl was never empty, even though you ate as many as you wanted. The staff took excellent care of us.

Besides the kitchen and dining room, the crew quarters' main meeting place was known as the large conference room; it had large tables put together to accommodate twenty to thirty people. There we also met with various members of our training team to do refresher training on medical procedures or get updates on potential sites of interest around the world from our Earth Observation folks. They would tell us about the latest volcanoes, hurricanes, forest fires, or other things of interest. The room was also the crew's main hangout area, much like a common space in a dormitory or your family room at home. When nothing else was going on, we would watch a movie together, play cards, or simply talk.

A unique feature of this room was its bright lights that we would use for sleep shifting, the adjustment of our day/night cycle in preparation for specific launch times. The ceiling was loaded with seemingly hundreds of fluorescent lights that could provide lighting that ranged from simple ceiling lighting—like you use at home—to the full intensity of sitting in bright sunlight in the middle of summer. Our flight surgeons would work out a schedule for when, how long, and how intense these sleep shift sessions would be, and we would sit in the room under intense sunshine conditions for four to five hours each of the days. The intent was to trick our natural circadian rhythm to adjust to a new day/night cycle that we would use during our mission, which was mainly determined by our launch time. While the intent was good and many researchers told NASA that this stuff works, most astronauts found the bright lights only annoying. I sure did. It would be so bright in the room that you couldn't even watch television because you couldn't tell whether the television was on.

During our time under the bright lights, many of us would work on our small spiral-bound crew notebooks, which most astronauts use in space to provide additional information about something or to serve as a place for reminders. I would write a list of things I wanted to take pictures of during the flight, or make a few notes about schools I would be talking with during the mission. Each crew member could use the notebook any way they wanted, and I never met an astronaut who didn't use one. In the hectic last months and weeks before launch, I never seemed to have the time to organize my notebook. Any free time I had I would try to spend with my family. But during quarantine was an ideal time to work on it. It also was relaxing and a bit therapeutic. It helped provide an activity for us to focus on, much like a craft project, with glue, scissors, rulers, and colored markers. It provided a bit of reassurance that things we might forget were in the notebook; it was sort of a confidence builder.

If you wanted some quiet time, you could always leave this common area and go to your room. They were like standard dormitory rooms with not-too-fancy government-issue furnishings, with pictures hanging on the walls that were purchased at starving artist sales at local hotels. The rooms weren't exactly sterile, but they weren't very cozy either—they were something in-between. Each room had a full-size bed, a desk, and a dresser. There were no windows in any of these rooms, which made it easier when we were sleep shifting and sleeping during the middle of the day. The staff would try to close off our bedroom areas from anyone coming into the crew quarters to keep it quiet, which also helped with our sleep shifting.

On Sunday, we had a traditional dinner with our spouses, which was a late breakfast or early lunch for them. It was a relaxing time. Early Monday morning, they would be flying to KSC in one of the NASA Gulfstream jets. We were scheduled to fly in on our T-38 jets in the early afternoon, with the hope that our spouses would be there to greet us when we arrived. This was also somewhat of a NASA tradition. Then we would all ride over to the KSC Astronaut Crew Quarters together on a small NASA bus. Any opportunity to visit, however short, was always welcomed and treasured!

Woodpeckers' Revenge?

Finally, it was time for us to leave crew quarters at JSC and head to KSC for launch. But first, Tom, Kevin, Nancy, and I had a final ascent training session in the motion based simulator. After going through a series of runs where Tim Terry, the STS-70 training team lead, and the rest of our instructor team broke anything and everything, which kept us quite busy working through malfunction procedures, we ended up with the traditional final launch simulation run. It was identical to our first ascent training run ten months earlier, shortly after our crew was first assigned. It was a perfectly normal ascent; nothing broke and there were no problems, just a 'quiet' ride to orbit. The training teams traditionally ended with this run as a reminder to the crew what the actual launch should be like for us. This was what everyone hoped for, but we were well trained for any problems that might arise. I think all of us just sat back and enjoyed the simulated malfunction-free liftoff.

After a quick debrief with our training team and a round of thanks and farewells, the four of us headed to the parking lot for the drive to Ellington, where we would meet up with Mary Ellen for our T-38 flights to KSC. As we walked out of the instructor station into the hallway of Building 5, we passed a number of people also walking in the hallway. Normally, it was not an issue. But when you were in quarantine the week before launch, you were prohibited from coming into contact with people who did not possess PCC badges. As we passed all these folks in the hall, we held our breath best we could because we didn't want to inhale any floating germs. And we laughed, "Oh well, so much for quarantine!"

From Building 5, we drove in Kevin's big old white Chevy Suburban back to crew quarters, changed into our blue flight suits, picked up our bags, and headed to Ellington. That much of the trip was uneventful. But after arriving at Ellington, official quarantine was broken a few more times as people came up to us or walked right past us in the Duty Office where we were checking on the weather and prebriefing for our flight before walking out to our T-38s.

Because it was midday in July, we always had to deal with the chance of afternoon thunderstorms along the Gulf Coast and at KSC. We considered the possibility of refueling at Eglin Air Force Base, near Valparaiso in Florida's panhandle, in case we couldn't make the trip nonstop. Kevin was based there years ago and was excited that might happen.

When we got to our waiting jets, there was a surprise. It was long-standing NASA tradition that parachutes and helmet bags were carried to the T-38s for the astronaut crew that was next up for launch, the so-called prime crew. Normally we went to the parachute room and grabbed our chutes and helmet bags and walked out to the plane and attached our parachute fittings to the ejection seat before we did our walk-around preflight inspection of the jet. But when you were the prime crew, the ground crew loaded your helmet and parachute for you, normally as a courtesy and as a sign of respect. But since we were in quarantine, it was also done to minimize contact with other people in the NASA hangar that day.

But when we got to our planes, there were no chutes and no helmets. Oops! The ground crew apologized and quickly got our equipment. We loaded up, got our clearance, and began taxiing out to the runway. We were finally on our way. Florida here we come!

We taxied in formation. Tom, with Nancy in his backseat, was leading in NASA 907, followed by Kevin and Mary Ellen in NASA 914, and then NASA pilot Mark 'Forger' Stucky and I in the third T-38, NASA 955. Suddenly I heard Nancy say something about smoke in the cockpit. Tom stopped the plane and within seconds the canopies of both Tom's and Nancy's cockpits opened and both were climbing out and jumping to the ground.

Kevin alerted the ground controllers, who sent out a fire truck, while NASA sent over a few members of the ground team to check out the jet. There was no way that plane would be flying to KSC that day, especially with the shuttle commander on board. So while Nancy and Tom walked the hundred yards back to the NASA ramp to get another jet, our other two T-38s remained on the taxiway, engines running, as we waited for Tom and Nancy to return. That took more than a few minutes. They were able to get another jet checked out (NASA 902) and were taxiing up alongside us and we were once again ready to proceed. We were soon airborne and on our way to KSC.

At this point, we had another complication. Because we had been sitting on the taxiway with our engines running while Tom and Nancy got their new jet, we used up precious fuel. We no longer had enough fuel to even try to get into KSC nonstop. So after about an hour, we landed at Eglin, refueled, and used our extra NASA pilot to clear the way for us so that we could use the restrooms without breaking quarantine yet again. Everything went smoothly and soon we were on our final leg to KSC. "Whew, that went smoother than things had been going so far," I said.

The weather forecast for KSC was not too favorable as we departed Eglin, but Tom decided that we would fly over that way and check it out, knowing that we now had enough fuel to divert to another airport if needed.

As we were approaching the Orlando area, we started checking the weather at the Shuttle Landing Facility (SLF) where we were hoping to land. Unfortunately, things weren't looking good. Streaks of bright, intense, and definitely scary lightning filled the sky. The accompanying booming thunder shook the area and rattled those already at the SLF waiting to greet the crew. Rain fell in torrents as high winds swept across the tarmac. The welcoming ceremony was a washout—and more heavy thunderstorms were forecast for later in the day.

Hearing that, Tom and Kevin decided to get as close as we could to KSC. So we landed at Orlando International Airport, where we could either wait out the weather or have KSC send a van or bus to pick us up. To our surprise, an Orlando TV station filmed our unscheduled and not-so-glamorous arrival. They came right up to us looking for an interview and we had to ask them to stand back and respect our quarantine precautions. We walked into the fixed base operator's office and called KSC.

We thought it would be an easy matter for NASA to send a few cars, vans, or a bus to pick us up. We were only about an hour's drive away—sixty miles from our destination. Tom called the Vehicle Integration and Test Team office and they suggested that it probably would be quickest—easiest for them was more like it—if we got a couple of rental cars and drove to KSC. So while Forger went to rent some sort of a vehicle, we laughed about our unprecedented series of events and tried our best to stay away from people so we didn't break quarantine any more than we already had.

After a short while, Forger showed up with the rental van and we were ready to load up. It was quite a sight, watching Tom pushing the luggage cart loaded with our bags over to the van. We were prime crew, but our commander was acting as a bellhop. If I had had any cash handy, I would have tipped him a few bucks for his service. Also about this time we received a visit from a couple of airport security officers from the Orlando Police Department; they came up to us to check out what was going on. We told them the situation. I knew this was a great photo opportunity, so I asked them if we could take a picture with them to document the occasion. I pulled out the NASA Nikon F4 camera I had brought along and took a great group shot. One of the officers posed, pretending he was holding a gun on us, while the other held two pairs of handcuffs. And right next to them stood Forger, Tom, Kevin, Nancy, and Mary Ellen. Except for the big smiles on everyone's faces, you would have thought it was a real bust.

With our T-38s secure and our bags loaded into the van, our adventure continued. With Forger at the wheel and Tom riding shotgun, Kevin and Mary Ellen in the middle row, and Nancy and me in the back, we left Orlando for the final, yet most challenging leg, of our journey—our launch.

We could have done one of three things—laughed, cried, or gotten totally angry. We decided to laugh and laughed our way along the bullet-straight Beeline Expressway that connects Orlando with the Cape. We joked about driving through the tollbooths without paying (to protect our quarantine, of course) but we ended up stopping twice to pay our tolls. *Two more times where we busted crew quarantine that day*, I counted to myself. Oh, what the heck! We had been exposed to between fifty and one hundred people in the past six hours. What's another two working the tollbooths?

Finally, KSC!

We arrived at KSC Operations and Checkout Building, where crew quarters is located, close to 9:00 p.m.—about five hours later than planned. There were two or three reporters there for a brief photo opportunity as we walked into the building, but no traditional crew arrival press event. We had planned to spend a little time with our spouses that afternoon before going to bed, but the delays had changed all that. Still, we met them in crew quarters and had a quick dinner with them and hung out another thirty minutes before we had to go to bed and they had to leave.

I drank a beer after dinner and someone piled up eight empty beer bottles and cans in front of Simone and I and took a picture of us, as if I had passed out leaning on her shoulder, to illustrate what a rough day it had been. I wished we would have had more time to visit that evening, but those were the breaks. What a day it had been! One of the more interesting and unusual trips to KSC that I have taken. To this day, I haven't heard of a better story about flying to KSC for launch!

I awakened at 4:00 the next morning, ate breakfast, and had a few hours of free time before our scheduled family brunch. I went to the gym and spent about thirty-five minutes or so working out. At one point, Mary Ellen and I were walking down the hall alone when she asked me, "Are you scared at all?" I think she was a bit surprised when I replied, "Hell yes, I'm scared!" I told her it was natural to be a little scared and that I thought it was normal. She seemed a bit relieved after hearing that.

I went back to my room, showered, and changed clothes, then got one of the rental cars assigned to the crew to head to the beach house for some quiet time. I drove by the launch pad on the way and could see *Discovery* lit up in the distance. The rotating service structure was surrounding *Discovery*, but it was still magnificent to see the top of the orange external tank and the two bright white solid rocket boosters. Man, what a sight! It really got the adrenaline flowing and at the same time started the butterflies fluttering around in my stomach.

The beach house was located only two hundred feet from the Atlantic Ocean, just south of the shuttle launch pads and near Launch Pads 40 and 41. It was a two-story structure, a remnant of a former development known as the Neptune Beach subdivision. It was built in 1962, a year before NASA purchased the land as it prepared to build a launch pad for the Saturn V rockets—Pad 39A—a few miles away.

The beach house has always been a sort of quiet refuge or sanctuary for the astronauts. We were permitted to go there just about any time we wanted, except during launches when the area was closed for safety and security reasons. I always found it quiet and relaxing. It was an ideal location to get away from everything and everybody, and to enjoy some time walking the beach, swimming, or jogging along the surf. It was a great place to clear your head of the pressures of the mission and an ideal location to connect with nature.

The beach was beautiful and from time to time you could watch sea turtles crawling across the sand to make a nest or pelicans skimming over the water. I found the pounding surf and crashing waves very soothing. The water was clean and cool and there was nobody else around for miles and miles. Its isolation and tranquil setting always drew me there.

Even though it was just an old house, it was treasured and appreciated by all the astronauts. It was never that fancy, but it was very comfortable and welcoming. Some astronauts say the furniture looked like it came from the 1960s-era television show *I Dream of Jeannie*. You could feel the history there and the presence of all the astronauts who preceded you. There was a nice wooden porch off the back that overlooked the Atlantic and to the north the shuttle launch pads were visible. It was the view of the shuttle launch pads that always brought my mind back to the reality of where I was and why I was there.

I arrived at the beach house, turned on a few lights, and took a quick walk out to the beach. The night sky was filled with stars. I stood there for a few minutes, gazing up at them and thought that soon we would be up there among them. It was a beautiful, peaceful, and relaxing moment that helped calm me down a bit. After five to ten minutes on the beach, I headed back to the house and watched a couple of old *Hogan's Heroes* episodes that I had on videotape.

Fellow STS-65 crew mates Carl Walz, Leroy Chiao, and I watched these and laughed hysterically during quarantine before that mission and I thought the laughter and memories of watching them would be a good calming influence now as well. And once again I laughed as I watched them. I'm sure they didn't warrant that magnitude of laughter, but it was a great way to relieve a bit of preflight stress and jitters.

After being in quarantine for nearly five days, it felt great to be alone at the beach house that night. The peace, quiet, and absolute solitude were intoxicating. I read a bit, took another walk on the beach to check out the stars, and then it was time to head back to crew quarters for our lunch. I ate with the rest of the crew and talked with Nancy and Kevin a bit, then headed back to the beach house with the rest of the crew to meet our spouses there before our families and friends arrived.

When the spouses arrived, Simone and I hung out with everyone for a few minutes and then took off for a walk on the beach to get away for a bit and get some private time together. It's a rare moment when in quarantine, so we took advantage of it. We slowly walked up the beach for twenty minutes or so, heading north.

Simone and I slowly headed back to the house where we found Mary Ellen, always the adventurer, trying to learn to use a surf board. Nancy and her husband Dave joined us in watching the show. We all chuckled and shook our heads when she fell off once, making a big splash and striking her head on the board in the process. Thankfully she didn't get hurt, but she was a bit dazed as she came out of the water. That quickly put an end to her surf lesson!

In another ten or fifteen minutes, the buses pulled up with our families aboard. It was so exciting to see them! I had invited my mom, Irene; my fraternal twin brother, Dennis; my sister, Cindy, and my other brother, Bill. It was comforting to have Simone and the rest of my family there that afternoon. We had a few beers and the traditional NASA feast of barbeque beef, potato salad, and baked beans. After eating and visiting a bit, everyone went outside on the back deck. We had a few more beers out there, took a round of pictures with various families and the entire crew, and then most of us made our way down to the beach. Nancy had invited the priest from her church back in Houston and anyone interested was invited to join them for a few prayers. I'm not much of a religious person, but I joined the group out of respect for Nancy.

It's always great hanging out at the beach. It was a nice relaxing moment for all of us, but at the same time I was starting to feel a bit of tension and apprehension knowing that our goodbyes were coming up soon. I enjoyed these final few moments with my family, but hated what came next. At 5:30 p.m. we heard: "Time to say goodbye and load up the busses." The butterfly activity in my stomach ratcheted up a notch as I said goodbye to everyone.

I said goodbye first to my mom and gave her a big hug and a kiss and told her I would see her soon. I also took a few moments to thank her again for helping make this day possible for me. I hugged Cindy and Bill and said goodbye, and then toughest of all was saying goodbye to my twin, Den. I reached up and he leaned down to hug. I told him I loved him and he told me "Fly safe and have a blast." I was looking forward to the flight, but with all the preflight stress at this point I had forgotten about having fun on the flight. We gave each other another long hug and with my eyes and his filled with tears, we separated and he and the others walked onto the bus. I stood there holding Simone's hand with my eyes fixed on Den. I followed him as he sat down and made a few funny faces or whatever to ease the tension of the moment as the busses slowly pulled away. I gave him one final wave and a thumbs-up sign and he did the same.

I took a deep, deep breath and let out a long sigh. These goodbyes were pretty tough and never got any easier. There was some relief that these goodbyes were behind me, but in the back of my head loomed the thought of one more goodbye that had to be said to Simone. I squeezed her hand even tighter than I already had been and took another deep breath. There had been enough goodbyes for me for one day. The one with Simone wasn't until tomorrow and although time seemed to be flying during these last few days before launch, our final goodbye seemed a million miles away.

We had another hour to hang out with our spouses and then it was time for us to head back to crew quarters to start getting ready for bed. I said goodbye to Simone, knowing that I would next see her again tomorrow at the beach house. Kevin, Nancy, and I got in our car and headed back to crew quarters.

The next morning we were scheduled to wake up at 4:45 and after a quick breakfast, Tom, Kevin, and Nancy headed out to the Shuttle Landing Facility for some landing approaches in the Shuttle Training Aircraft. Mary Ellen and I stayed behind in crew quarters and caught up on email, worked out a bit, and just relaxed as best we could.

Beginning at 8:30 that morning, we had a final briefing from Steve Smith, our dedicated astronaut support person, who reviewed one last time our procedures for getting into *Discovery*. Following that, we had a visit from various NASA VIPs.

KSC Director Jay Honeycutt, KSC Director of Shuttle Operations Bob Sieck, and a few NASA managers from headquarters in Washington, D.C., stopped by to wish us a safe and successful mission. Next was a thirty-minute briefing giving us the latest updates on the status of both *Discovery* and our TDRS. The launch team was working very few problems and everything was looking good for our flight.

We then received a briefing from the KSC weather team about the forecasts for launch the next day at KSC and our primary abort landing sites overseas. Once again, there were no concerns and everything was looking as good as it could be, considering we were trying to launch from Florida in July. After a quick telecom with JSC Director Carolyn Huntoon, who wished us a safe journey, the STS-70 crew quickly changed into shorts and sandals and we all headed to the beach house for a final visit with our spouses.

On our way to the beach house, we stopped by Pad 39B, where *Discovery* was in her final stages of preparation. The normal tradition the evening before launch is for the crew to make a visit to the pad during night viewing, when a limited number of family members and friends are invited to see the shuttle up close, with it brilliantly illuminated against the pitch black sky. The crew stands on one side of the road and our guests stand on the other side, maintaining a safe quarantine distance. It is a time to say a few more goodbyes.

But since our scheduled bedtime the evening before launch was 8:45, our crew wouldn't be awake during the normal night viewing time, typically 10:00 or so in the summer months. So instead NASA scheduled a pad visit for our family members and friends at 11:30 that morning. We were each allowed twenty-eight guests that NASA bussed out to the pad for the visit. My mom was there, along with my brother Dennis and his daughter Asja.

So from across the road, just a few hundred yards from *Discovery*, we were able to wave to and chat momentarily with our designated guests. Afterward, the crew and our spouses headed to the beach house where we spent a final afternoon walking the long stretches of unspoiled beach, listening to the rhythmic noise of the waves crashing, enjoying a warm ocean breeze coming off the Atlantic, and swimming in its cool water. It was such a relaxing afternoon for everyone, but looming in the back of everyone's mind was the thought of our final goodbyes to our spouses. At 4:00 p.m. it was time for Becky, Tom's wife; David, Nancy's husband; and Jerry, Mary Ellen's husband; to depart for their individually arranged night-before-launch parties. Since Simone and Jeanne, Kevin's wife, had decided not to hold one of the traditional parties, they were allowed to stay with us at the beach house where we all had dinner. At 7:00 p.m., it was time for them to leave.

Simone and I, both with tears in our eyes, dreaded this moment for each of my flights. We stood there hugging each other tightly, almost silently, except for repeated "I love you so much" and "Have a safe flight." My last words to Simone were, "I'll see you back here in nine days," when we were scheduled to land at KSC. After hugging so tightly until we had no strength left, and with moist eyes, we parted. Simone and Jeanne got in a car to drive back to their hotel, while the crew got into our cars for the drive back to crew quarters for our scheduled sleep period.

While we had been out at the beach house that afternoon, technicians performed a prelaunch debris inspection of the Launch Pad 39B. Everything looked great and there were no anomalies noted. We were still 'Go' for launch!

Shortly after 8:45 p.m., I was asleep. The next time I would go to sleep would be in space. What a thought!! How can you possibly fall asleep the night before launch? I was exhausted and emotionally drained after the visits and goodbyes to Simone and my family and friends. And with a light sleeping pill as an aid, I got a good night's sleep. I would be ready and well-rested the next morning for launch.

As the STS-70 crew slept, some of our family members and friends were treated to the official night viewing of space shuttle *Discovery*. Simone and Kai, who was wearing a Cleveland Indians outfit, along with my mom, brothers and sister, nieces and nephews, friends from both high school and college, and our neighbors in Houston, made the trip to Pad 39B that evening. I only wish we could have been there to join them!

As I drifted off to sleep, I thought it had been one heck of a journey for us to get to this point. Woodpeckers. Quarantine infractions. Smoke in the T-38 cockpit. Thunderstorms at KSC. Searching for change at the tollbooths between Orlando and KSC. I was full of hope that everything would go a bit smoother for us from this point on. Later that night, technicians began loading the half a million gallons of liquid oxygen and liquid hydrogen into *Discovery*'s external tank. Things were almost ready for us to get out of town.

Chapter Two
In the Footsteps of Glenn and Armstrong

The trip to the Kennedy Space Center began much earlier than the woodpecker-interrupted, quarantine-challenged, weather-marred flight from Houston. That I wanted to be an astronaut wasn't too unusual for boys growing up in the early 1960s. There were many kids in my class who professed to having the avocation. But what set me apart was that the dream didn't fade. The desire kept growing stronger and stronger.

My interest was piqued when I was a soon-to-be six-year-old enrolled in the half-day morning kindergarten program at Independence Elementary School, located about ten miles south of downtown Cleveland, Ohio. It involved an assembly in the school's gymnasium, the flickering image on a small black-and-white television set, a candle-like structure topped by a black bell-shaped object, and a guy named Alan B. Shepard.

It was May 5, 1961, and it already was a special day for my twin brother, Dennis, and me because the next day was our sixth birthday. Since that was on a Saturday, we would be celebrating at school that Friday morning.

On your birthday, you were supposed to bring a treat for a small party with your classmates. Growing up in a lower-middle class family only allowed us to bring small treats. And since my brother and I shared the same birthday, one treat was good enough to cover for the both of us. My mom had packed a bag of Tootsie Roll Pops for us to pass out. Tootsie Roll Pops were only two cents each in those days, so for our class of about twenty-five students, a round of suckers was only fifty cents! I was excited about turning six and looking forward to the class party before we went home at lunchtime.

About 9:15 a.m., our teacher told us that we would be going to the gymnasium for a special program. It seems that the United States was trying to launch its first astronaut into space. So in the best elementary-school tradition, we lined up single file and headed down to the gym, where we and a few hundred students in grades K–4 sat in quiet anticipation. In front of us was the single black-and-white TV on a metal stand. The hazy picture showed a Redstone rocket and its Mercury capsule with Shepard, one of NASA's seven original astronauts, inside.

Only three weeks earlier, the Soviet Union had successfully launched and orbited Cosmonaut Yuri Gagarin. The world—me and my classmates included—were watching closely to see the US response.

As I sat on the hard floor with my legs crossed, you could see the rocket with smoke—in reality, condensation from the liquid oxygen—coming from it. Soon we reached the final seconds of the countdown: "10…9…8…7…6…5…4…3…2…1…" At the command "ignition," you could see the flames and smoke spewing from the base of the rocket and almost immediately it took off. Flames, smoke, and lots of noise—great stuff for an almost six-year-old boy! I don't remember whether we cheered, but I know my eyes were riveted to the TV.

Within seconds, the rocket got smaller and smaller, with only the long flame visible. Soon it was just a point of light that bounced around as the cameraman struggled to track the rocket during its launch. A few moments later, it was out of camera range.

Shepard's mission lasted fifteen minutes and twenty-two seconds. His highest altitude was 116 miles and his total distance traveled was 302 miles. His peak speed was an amazing 5,143 mph and he experienced eleven times the force of gravity during his reentry. It was a historic day.

By today's standards, it wasn't much as a flight. But to an impressionable boy, it was spectacular. If a single event can dramatically change the course of a life, that was mine. I knew then and there what I wanted to do later in life. I owe a debt of thanks to whomever had the idea to assemble all the students in the gym to watch that launch. That seemingly insignificant decision changed my life. I wanted to be an astronaut. I wanted to fly in space.

So as my fellow kindergartners headed back to our classroom, you could hear a few of them doing countdowns, followed by imitations of blast-off noises and other rocket sounds. Later that morning, my teacher passed out the Tootsie Roll Pops and we had a short birthday celebration. We shared the celebration with another classmate, Tommy Amato, whose actual birthday was that day. My mom picked us up from school shortly before noon and we headed home for the weekend.

Deciding what you want to do when you grow up is one of life's most difficult decisions. On the last day that I was five years old, I made that monumental decision. The easy part was done. Now the only question that remained was "How do I become an astronaut?" Luckily, I had years and years to find that answer.

The first NASA mission that featured an all-Ohio crew occurred February 20, 1962, when astronaut John Glenn, of New Concord, Ohio, traveling all by himself, became the third American to go into space. I very clearly remember watching Glenn's launch in the same school gym where we had watched Shepard's launch, on the same black-and-white TV. I was in first grade now, and really excited because it would be the first US orbital flight manned by someone from my Buckeye State. The countdown proceeded on schedule and, at 9:47 a.m., the Atlas rocket roared to life. "Godspeed, John Glenn."

He was on his way and taking a little piece of each of us along with him.

We all watched attentively. Once again, a few moments later, only a small bright dot that jumped around a lot was visible on the TV. Within minutes John Glenn was in orbit and we all headed back to our classrooms.

From the moment I got back to my desk until we were excused for lunch about two hours later, I spent nearly every minute looking out my classroom window. My desk was in the third row, right next to the windows, so it was easy to sit and daydream as I scanned the skies above the Cleveland area, straining my eyes to see John Glenn and his *Friendship 7* Mercury capsule.

Glenn only made three orbits of the Earth and never flew anywhere near Cleveland or the rest of Ohio, for that matter. But I didn't know his flight path and kept watching for him, hoping to catch a glimpse. Had his orbit taken him over Cleveland, he never would have been visible, but such is the mind of a six-year-old. I didn't know any better and just hoped I would be able to see him.

One of my classmates brought a small rocket model to school that day. Mark Pifer, with some help from our teacher, explained how the rocket worked (nothing too technical: "It blasted the astronaut into space"). Our teacher then had Mark and my brother (Mark was our best friend) take the model to the other first grade classrooms to do a show-and-tell about how John Glenn had gotten to space. I was so envious! Why couldn't I go around and explain about the rocket? After all, I was the one who was going to be an astronaut. Mark and my brother went from class to class and explained all they knew about orbital mechanics and reentry theory while I sat at my desk. My only consolation was that by sitting at my window-seat desk all morning I wouldn't miss seeing John Glenn when he passed over Cleveland.

Glenn splashed down safely in the Atlantic Ocean a few hours later and became an instant hero. That evening the newspapers were filled with stories about Glenn, his Mercury capsule, and his incredible flight. A few days later, *Life* magazine published beautiful color pictures of his launch and recovery. I saved the magazine, cut out the pictures, and glued them in a scrapbook. Glenn became a hero of mine, along with twenty million other young kids across the United States and hundreds of millions of others around the world.

In one issue of *Life*, there was an article about the Mercury capsule. At the time, my father was a salesman at National Screw and Manufacturing Company in Cleveland and he told me that his company made some of the screws and bolts that were used in the Mercury capsule. That, I thought, was pretty exciting. My dad actually helped make part of the capsule. Wow—wait till I tell all my friends!

During the early days of America's space program, astronauts got a lot of publicity. There were lots of stories in the newspapers, magazines, and on TV about their selection process and training. The things that stuck with me were that the Mercury capsule was very cramped, that space was pitch black, and that the astronauts needed to spend hour after hour in isolation as part of their training for space.

Since I was going to be an astronaut, too, I figured I might as well start training as soon as possible. So my brother and I would take turns staying inside our homemade Mercury capsule. It was nothing more than the kneehole area of a desk. We would throw some pillows underneath and one of us would crawl in.

The other would take some chairs and blankets and close off the opening. By adding more pillows and blankets, we could get it quite dark in there—just like in the real Mercury capsule, we surmised. The whole idea was to spend as much time as possible under the desk in our homemade space capsule. It was a tight space, but each of us would spend an hour or two under there until we got bored or distracted by something else going on in the neighborhood. I can't remember which of us holds the record for desk kneehole spaceflight, but it probably is something in the neighborhood of two hours or so. It was always great to open the hatch by removing the blankets and chairs to get some fresh air. Our simulator wasn't equipped with a functioning life support system.

Life in Cleveland

While I had other passing fantasies, like becoming a professional football player for my beloved Cleveland Browns, I never wavered on the idea of flying in space. After each Mercury and then Gemini flight, I couldn't wait to get the next issue of *Life* magazine, which always featured multipaged photo-packed stories of the missions. I was blown away by the issue that featured highlights from Gemini 4. On the cover was a spectacular color image of astronaut Ed White floating outside his Gemini capsule, with the blue Earth far below him. I saved that special issue and every other *Life* magazine that had any pictures or stories about the astronauts. But the issue with Ed White remains one of my all-time favorites.

I wasn't a particularly gifted student, probably average by most standards. In elementary school, I was a B student, getting an occasional A and an occasional C. I struggled more with my reading than my twin brother and found it hard to keep up with him. In middle school, I started doing better and got nearly all As and Bs. I made up for my average intelligence by working harder than the other students in my class.

One night when I was in the fourth grade, I got some shocking news from my dad. After dinner, he gathered the family in the living room and told us that he and my mom would be getting a divorce and that he would be moving out of the house. "What do you mean moving out and not living with us?" I wondered. I clearly didn't understand the concept of divorce with all its ramifications. I started to cry. My dad kept repeating that he would still be our father and that we would still be able to see him, but his words didn't ease the uncertainty and trauma of the moment. What was even more upsetting for me was watching my father start to cry. I had never seen him cry before and now he was almost sobbing.

As far back as I can remember, my parents never got along. I can't recall ever seeing them holding hands or kissing. My mom always slept on the living room couch; my dad slept in their bedroom. This was normal around our house. Also normal was yelling most evenings. After dinner, my brothers and sister would go to our rooms or watch TV in the living room. I have vivid memories of my mom and dad yelling at each other. While that was taking place, I would turn up the volume on the TV and press a few pillows close to my head to drown out the noise. It was my way to escape the yelling. A few times, I remember things escalating to the point of my father pushing and shoving my mom. On these few occasions, my mom would gather up us kids and we'd head to my grandmother's apartment where we would spend the night. I was too young at the time to appreciate what was going on. For Dennis and me, it was merely an exciting sleepover where we got to camp out in my grandmother's living room and sleep on the floor. This was my normal—a way of life I was used to.

A few months after announcing their divorce, my dad moved into an apartment in a neighboring city and on most Sundays he would pick up Dennis and me for a visit. We'd go to a park and hike or visit a museum. He had a sailboat, so some days we would spend time with him on his boat. When Lake Erie was rough, I would get seasick. But I still wanted to be out there on the lake with my dad, because it was the only opportunity I had to spend time with him.

What made all of this a bit harder to deal with was the fact that we were told to keep my parents' divorce a secret—not to tell any of our friends. Divorce was not very common in the early 1960s and there was a stigma associated with your parents' divorcing, especially while living in a heavily Catholic neighborhood. Occasionally, when a friend would spend the night and ask where our father was, we were trained to lie—"He's out of town on a business trip." Thankfully, none of my friends pushed the point and inquired any deeper. I think they probably heard the rumors or knew the truth and politely accepted our misrepresentation of the facts.

When I was in sixth grade, I was devastated to hear the news that three astronauts—Gus Grissom, the veteran Mercury astronaut who followed Shepard into space; Roger Chaffee, a rookie preparing for his first flight, and veteran Ed White, my spacewalking hero from

Gemini 4—died in a fire on the launch pad as they were training for the first Apollo mission. It brought home to millions across the country that the space business was hazardous, but that never deterred me from wanting to become an astronaut.

Less than two years later, I was at my grandmother's apartment watching the special Christmas Eve broadcast from the crew of Apollo 8. It was absolutely amazing that men were orbiting the Moon. What made it even more amazing was that they were able to transmit live images of the Moon's cratered surface passing beneath them only sixty miles away, as well as views of the distant Earth. It was exciting stuff for a young astronaut wannabe.

During the summer of 1969, in between middle school and high school, we moved from Independence to Cleveland Heights, a suburb on the east side of Cleveland. I hated the move; it was difficult leaving all my friends behind. That first summer in our new house without any friends might have been pretty boring, except that I kept myself occupied much of the time with the Apollo 11 mission—man's first landing on the Moon! I watched the crew of Neil Armstrong, a fellow Ohioan; Buzz Aldrin; and Michael Collins take off in their Saturn V rocket on our recently purchased color television.

My father remarried and was living in an apartment about a half-hour's drive from our house. My brother Dennis and I were still spending occasional Sunday afternoons with him and on July 20, 1969, a beautiful sunny midwestern summer day, we sat in his living room watching CBS's televised coverage of the Apollo 11 Moon landing. Walter Cronkite and former astronaut Wally Schirra were explaining the events as they took place with the help of some rudimentary special effects, which included a model of the lunar module descending to the surface. On the screen was superimposed a countdown clock with the notation: Time to Lunar Landing.

When the clock hit zero, it was clear that Armstrong and Aldrin had not yet touched down. It turned out that the computers were flying the lunar module to a landing site in the middle of a boulder field. Armstrong took control and was trying to find a safe level place.

We were on the edge of our seats. Then we heard it: "Houston. Tranquility Base. The Eagle has landed." It was an incredibly exciting moment.

We had dinner and my father then drove us home, where I continued to watch every minute of the television coverage, which included an interview with Armstrong's parents from Wapakoneta, Ohio.

Originally, Armstrong and Aldrin were scheduled for a five-hour sleep period before venturing onto the lunar surface. But it soon became clear there would be no sleep after the huge adrenaline rush from landing on the Moon. So just shy of seven hours after landing, Armstrong left the lunar module.

It was unbelievable that—while the TV images from the Moon were grainy and fuzzy—we could watch Armstrong climbing down the ladder and stepping on the surface of the Moon. Then, as he stepped onto the surface, Armstrong spoke his immortal words: "That's one small step for a man, one giant leap for mankind."

Sitting in front of the television along with my mom and my brother Dennis, I felt like I was right there with Armstrong. I watched for the next two hours as Armstrong and Aldrin planted the American flag, collected moon rocks, described the surface and view of the Moon, unveiled a special plaque on a leg of the lunar lander, set up various experiments, and spoke with President Richard Nixon, who was at the White House.

After the astronauts were safely inside the lunar module, I went outside and looked up at the Moon, a quarter of a million miles away. It was hard to believe there were humans up there. I went inside—it was close to 3:00 a.m.—and went to bed with visions of moonwalks dancing in my head.

Armstrong and Aldrin spent less than twenty-four hours on the Moon before firing the lunar module's ascent engine and heading back to *Columbia*, the Apollo capsule. Three days later, they splashed down in the Pacific Ocean to worldwide acclaim. It was one of the most exciting weeks up to that point in my life. I was even more committed to becoming an astronaut and I hoped to one day visit the Moon.

At the end of the summer, Dennis and I started ninth grade at Roxboro Junior High. It was a rough year because neither one of us made many friends. We were real outsiders, but thankfully we had each other. The students were much more politically active than in our previous school. Near the height of the Vietnam War, we were even excused from class one day so we could attend an antiwar rally in downtown Cleveland. I felt alone and focused much of my energy on school work and spent a lot of my free time reading about the Moon program. I mowed lawns and got a part-time job working after school and on Saturdays, which also helped keep me busy and out of trouble.

Also about that time, my dad suddenly moved to Chicago without telling us in advance. He simply wrote a letter one day announcing his new address. Dennis and I visited him there in the summer of 1970. It was the last time I saw him for many years. Two years later, he moved to Florida, a state well known at the time as being a safe haven for divorced fathers, because it did not enforce child support and alimony payment orders from other states. My father stopped paying my mom any child support the day he set foot in Florida. She was forced to raise us on her own. While living in a single parent household had its challenges, it also provided unique opportunities. I had one less parent telling me what I could or couldn't do and had much more freedom as a teenager than most.

My parents' divorce was the most traumatic event of my childhood and had an impact on my life for many years. The situation forced me to grow up quicker than I might have and to be more independent and self-sufficient. If I wanted a job, I had to find it on my own. When it was time to choose a college, my brother and I did it on our own. My mom worked hard as an underpaid secretary during those early years of the women's movement and did her best to provide for us. But what I appreciated most about her was making me feel loved—I never felt abandoned and alone. While I didn't grow up in ideal surroundings, I always felt loved, which along with my dream of becoming an astronaut, helped keep me on the right path.

After the Apollo Moon missions, NASA launched Skylab, its first space station. The program used hardware left over from the last couple of lunar missions, which had been canceled by Congress and the Nixon administration. Skylab was all about science, which was exciting for me because NASA began flying some of the science astronauts on those flights. It was clear to me that this would be my path to space. During Mercury, Gemini, and Apollo, the vast majority of the astronauts were military test pilots. Now NASA was starting to fly science astronauts. I focused even more on my studies and dedicated myself to becoming one of NASA's science astronauts.

During my three years at Cleveland Heights High School, I made a few very good friends in the tenth and eleventh grades. I ran cross-country my sophomore year, but that was about the only activity I participated in at school. I attended no other sports events, dances, or proms. It was a very liberal school with essentially no dress code. Without a strict father at home telling me to get a haircut, I stopped going to the barbershop. My hair grew straight and long, nearly down to my waist. That led many of my teachers to assume that I was a druggie. One day in tenth grade, as report cards were handed out, my homeroom teacher looked at my straight A grades and blurted out in disbelief to everyone in the room: "Hey Thomas! You're smart?!" He couldn't believe my grades. In his eyes—because of my hair length—I was a loser, a drug user, or worse. I laughed and thought, *What an idiot!* I hated being prejudged like that, but it was a part of having long hair at the time.

During my senior year, with most of my graduation requirements already met, I worked thirty-two hours a week in a warehouse stocking shelves, sweeping floors, and cleaning restrooms. I was finished with school by 11:00 a.m. each day, so I didn't enjoy much of a traditional senior year. My focus was on saving money for college. I graduated from Cleveland Heights High School in 1973 with nearly straight As. I had only one B, from a grumpy old English teacher with whom I never quite saw eye-to-eye. I can only assume that she was never a big fan of or a strong advocate for teenage boys growing their hair long. My brother Dennis had a similar experience with his eleventh grade English teacher at the school and also picked up his one and only B. Out of a graduating class of twelve hundred students, we tied in our class standing—we both ranked No. 7. At the time, I didn't see my high school graduation as marking much of an accomplishment, so I chose not to attend. I had my sights already set on greater accomplishments, such as graduating from college.

The College Years

Growing up in Cleveland, I dreamed of attending Case Western Reserve University, a private school with strong science and engineering programs. This was the only college that I applied to and I was thrilled when I got the letter of acceptance. With no financial support from my father and with my mom barely getting by on a secretary's salary, I was pretty much on my own to get through college. Thankfully, through government loans and other programs that were more plentiful at the time, along with a few scholarships and work-study programs, I was able to attend the university of my choice.

My four years at Case were by far the hardest four years of my life. The competition was far greater than it had been in high school and it made me work that much harder. I chose to major in physics because that was my favorite subject in high school. I enjoyed the idea of being able to explain the world around me that physics provided. I took a particular interest in solid state physics and later in materials science. During NASA's Skylab space station program, the astronauts conducted many crystal growth experiments. That helped cement my future academic direction. Here was a field in which I was not only interested, but also might be useful in a future astronaut selection. I never worked so hard in my life as I did in my four years at Case. I graduated with honors in May 1977 and I proudly attended that graduation. I was excited that I now possessed NASA's minimum degree requirement to become an astronaut, a bachelor of science degree in a technical field. But I knew that just meeting the bare minimum requirements probably wasn't going to be good enough for me to make it into the astronaut program, so I decided to continue my education in graduate school.

I applied to and was accepted at Cornell University, in Ithaca, New York, where I majored in materials science and engineering, with a minor in applied physics. Fortunately, through a research assistantship, I attended the Ivy League Cornell at virtually no cost. I worked on my research and got paid four hundred dollars a month to do so.

As I was finishing up my doctorate at Cornell, NASA launched the first space shuttle, *Columbia*, on April 12, 1981. I watched the launch on my small black-and-white television in my apartment. The shuttle wasn't as tall as the towering Saturn V rocket, but with the two solid rocket boosters and three main engines, it put on quite a show at liftoff. Because it had never flown before, I didn't know what to expect. I was not disappointed as *Columbia* roared into space. The dream was still alive inside me and I couldn't wait to get a chance to do it myself one day. Two days later, I went home early from school to watch *Columbia* land at Edwards Air Force Base in California. Again, the image on my TV was small, but it was oh so powerful. The shuttle program had started and I dreamed of being a part of it!

NASA had selected some science astronauts in the mid-1960s, but never intended to fly any of them. NASA was used to dealing with military test pilots and was always a bit uncertain about scientists flying in space. They viewed scientists through the old stereotype of loners who were not very good team players. But slowly the stereotypes faded. Harrison Schmidt, a lunar geologist, landed on the Moon as part of the Apollo 17 mission. Then each of the three Skylab missions had one or two science astronauts on each of their three person crews. NASA realized that the space shuttle program was all about science and they would need to include more scientists in their astronaut classes.

In 1978, NASA selected its eighth group of astronauts—thirty-five people, more than a third of them civilian scientists, including

Ohioan Judy Resnik. As I worked my way through graduate school, I started to think that getting selected as a mission specialist might be a real possibility! I pressed on with my studies and, in 1982, graduated from Cornell with my doctorate in materials science.

Accentuating the Possibility

I hoped to get a job at one of the NASA centers when I finished school, but in the early 1980s most of them had hiring freezes. I contacted the Marshall Space Flight Center and was told that I better consider going back to school and entering a co-op program. "We don't hire anybody from the outside," I was told. That was a bit hard to believe. I had my PhD and they wanted me to go back to school and enter a cooperative work program where I would work at NASA for a semester and then go to school for a semester. After being in college for nine and a half years, that seemed like a less-than-desirable path.

Instead, I opted for a job at Western Electric, which later became part of AT&T Bell Laboratories. I worked at the Engineering Research Center outside Princeton, New Jersey. Shortly after I started, NASA decided to choose a new group of astronaut candidates. I was thrilled. I had not applied to NASA earlier because I wanted to finish my PhD. While not a requirement, if you were a civilian, having a doctorate significantly increased your chances of getting selected. I sent my application to the Johnson Space Center and waited. I heard nothing. Months later, while reading the *New York Times*, I saw a small article with the headline "NASA Selects New Astronauts." I thought that was odd, because NASA had not yet contacted me. I quickly scanned the article and my worst fear was confirmed. My name was not among the seventeen listed. I had not been selected.

That was a real eye-opener. Even though I now had the necessary qualifications to be a mission specialist and even though I had gone to great schools and was working for a well-respected company, I now realized that getting into the astronaut program was going to be much harder than I imagined. I kept my application current and submitted it again the following year, 1985, when there was another call for prospective astronauts. Once again, I heard nothing until I saw the names of the thirteen astronauts candidates in the paper. Once again, my name wasn't on the list. Weeks later I received a postcard from NASA that read in part, "Dear Sir and/or Madam, Thank you for your interest in the NASA astronaut program. We had a lot of good candidates and while we did not select you, we wish you all the best in the future." I had just received a form letter rejection postcard from NASA. I was so far away from making it into the finals.

Knowing that it probably would be a few years before the next selection process, I decided I better work more on building up my background in an effort to improve my chances of being selected the next time around. By looking at those chosen, I noticed that there were some things that, while not a requirement, definitely seemed to help. Parachuting, skydiving, flying a small plane, scuba diving, and teaching a university class were all things that NASA seemed to be looking for in its new astronauts. I decided to try and tackle as many of them as I could.

I started off skydiving. One day I took off from work and drove to an airpark near Trenton, New Jersey, that offered skydiving lessons. Without telling anyone what I was doing, I signed up and after a few hours of training on the parachute landing fall and what to do if the main chute didn't open, I found myself airborne with a parachute on my back. That first jump would be a static-line jump, which meant that my ripcord would be attached to the plane, so that when I jumped the chute would open automatically. There would be no free-fall on that first attempt; usually it took ten jumps or so before you were allowed to free fall.

When it was my turn to jump, the instructor told me to crawl out of the hatch and stand on the small platform above one of the landing gear and to hold on to the wing strut until he gave me the word. So with my first experience at having 75 mph winds hitting me, I crawled out and held on. After a few seconds with my heart pounding, the instructor yelled "Jump!" You have to be a little crazy to let go at that point, but I was ready and I released my grip. After a few seconds, I looked upward and saw my canopy fully opened, which was great news. I was three thousand feet above the ground and without an open chute I would have had only thirteen seconds to live.

What a moment of exhilaration it was. The view was incredible; I was no longer looking out the small window of a plane, but was floating free with nothing but the Earth below. It was a gentle ride all the way down. And with the sun beginning to set, I enjoyed the beautiful, colorful sky. During the final hundred feet, the ground really seemed to rush up at me, so I fixed my gaze on the horizon as I had been taught and prepared for impact. BOOM! As my feet hit, I immediately initiated my landing fall.

The whole experience was an incredible adrenaline rush, and I wanted to immediately get back in line and do it again. But the sun was setting and that would be the last jump for the day. I drove home exhilarated and talked to anyone who would listen at work the next day about how I had spent my day off.

I next wanted to learn how to fly, so I signed up for lessons at Trenton Airport, which was less than five miles from my apartment. That was another incredible rush. As the plane lifted from the runway, the instructor pilot gave me the controls. I was flying a plane! We were bouncing all over the place and I asked if that was turbulence we were experiencing. The instructor told me to let go of the controls and see what happened. The plane flew smoothly. He then had me retake the controls. The plane again bounced through the air. I was told I was overcorrecting and overflying the plane, something all new pilots have to get used to. Bouncy ride or not, I was hooked. I signed up for more lessons.

As we were taxiing in after one of my lessons a few weeks later, the instructor asked me to stop next to the control tower. He opened the door and hopped out, telling me to take it up myself for a few touch-and-go landings. It was time for me to solo. I knew this day was coming, but I had no clue it would be that day. So with a mixture of excitement, fear, and trepidation, I taxied to the runway and took off. It was strange and a bit scary not to see the instructor sitting next to me. I was all alone and it became painfully clear that it would be up to me to get the plane back on the ground safely. I went around the traffic pattern, made the final approach, and brought the plane down to a picture perfect landing. What a great confidence builder that was. I applied full power and immediately lifted off for my second go-around.

The second landing was not as good as my first. I had a bit of a bounce and most pilots would jokingly count that as two landings, but I made it. I once again applied full power to go around for my third and final attempt. My third landing wasn't quite as good as my first, but definitely was better than my second. After landing, I taxied to the control tower where my instructor joined me and we taxied to a parking area. After signing off in my pilot logbook that I was now cleared for solo flight, he congratulated me. But there was no formal ceremony as was sometimes the tradition, where they cut off the tail of your shirt. Regardless, I was elated. I drove home and called my brother Dennis with the good news. The next day he came out to visit and we celebrated.

I continued flying as much as I could afford. I had a car payment and was paying off my student loans, but almost all the rest of my money went into flying. I soon had my private pilot's license and continued taking lessons to get an instrument rating. I really enjoyed the challenge of flying 'under the hood'—wearing goggles that only allowed you to see the instruments. While it took even more of my money, I figured it was making me a better pilot and I was hoping that one day all this experience might pay off when NASA decided to select another group of potential astronauts. As I continued flying, I updated my astronaut selection application package to make sure NASA knew of my flying and skydiving experience. I wanted to be ready for the next astronaut selection which I figured would be coming soon.

The Possibility

In late March 1987, I was sitting at my desk at Bell Labs when I received a call from the Astronaut Selection Office. They invited me to participate in a week of medical examinations and an interview. I was told I needed to report to the Johnson Space Center in six days. There would not be much time to prepare; that was the way NASA wanted it. I quickly agreed that I would be there and after talking over a few issues regarding airfare, hotel, etc., I hung up the phone. I was at work so I couldn't show my emotions, but inside I was screaming like crazy. I called Dennis and told him, and then told a few of my coworkers, who had long known of my dream to fly in space.

My focus immediately shifted toward preparing for the week in Houston. I had a good friend from Cornell, Roman Herschitz, who worked for RCA in Princeton and knew Bob Cenkar, a payload specialist who had flown on STS-61C in January 1986, just a few weeks before the *Challenger* accident. He arranged for me to meet with Bob so I could pick his brain about what they were looking for and what would happen during the week in Houston. Bob was incredibly reassuring and told me as we parted: "I know you are going to be selected."

I also contacted Terry Hart, a former NASA astronaut who worked for Bell Laboratories at a different location in New Jersey, so I could learn more about the astronaut selection process, the medical exams, the interview, and anything else that might be useful. Terry graciously agreed to meet with me the next day and during the hour or two that we had, he tried to tell me all he could and answer all my questions about what I should expect. I was pretty intimidated about meeting a real astronaut that day, but it began to sink in that I was going to meet him more as a potential colleague than as an average astronaut-wannabe space fan that I had been since I was six years old. Terry was a tremendous help.

So after preparing myself as best I could, and after getting as much info as I could from Bob Cenkar and Terry Hart, I boarded a plane for Houston. I was so nervous and excited—I was a wreck! With the huge butterflies flying around in my stomach and with my thoughts high in the clouds, I barely needed the jet to get to Houston.

I checked in at the Kings Inn Sheraton, located less than a mile down NASA Road 1 from the Johnson Space Center. I was instructed to meet in one of their meeting rooms at seven that evening, where I joined nineteen other potential candidates. Immediately, we were all checking each other out, evaluating the competition. One guy had PhD and MD degrees, plus a couple of master's degrees. That was a bit intimidating to say the least. What was I doing competing with people like that? What chance would I ever have of getting selected? Many of the other candidates were from the military, some test pilots and some flight engineers. I was a bit more relieved when I met a few fellow scientists like myself, most of whom only had one PhD. It was definitely an intimidating group of highly accomplished individuals. One of the shuttle commanders, Dan Brandenstein, who at the time was head of the Astronaut Office, was there, along with some of the other shuttle veterans. The biggest thrill for most of us that evening was meeting John Young, the ninth man to walk on the Moon. Young had flown on two Gemini, two Apollo, and two space shuttle missions. I couldn't believe what I was doing!

At the introduction, we were given our individual schedules for the week, learning when our all-important interview would be, along with details about some of the more unpleasant examinations that they would be conducting, with the proctosigmoidoscopy (don't ask!) at the top of the list.

We also were told to write an essay on "Why I want to be an astronaut" that we would need to submit immediately before our interview. Many of the other candidates, especially those from the military, already knew they would ask for that and had prepared theirs, beautifully typed, well in advance. I would have to write mine on the spot in my hotel room. Without access to a typewriter or computer, I would need to handwrite my essay. Fortunately, my handwriting is fairly neat, but I felt at a bit of a disadvantage not having that already prepared. I went back to my room that night and wrote a one-page essay about how I had wanted to do this my whole life and how much I wanted to not simply watch NASA, but to participate in its endeavors. It wasn't funny or cute, but it was sincere and straight from my heart.

The medical exams were endless and exhaustive, but they all seemed to go very well. To test whether I was claustrophobic, they wired me up to monitor my heartbeat and breathing, then took me into a room that contained a small canvas ball, about three feet in diameter. I recognized this from books I read. It was called the Crew Rescue Sphere or the Personal Rescue System. It was designed to transfer astronauts from one space shuttle to another without needing to use a spacesuit. An astronaut would go inside the sphere with a small oxygen tank, and after zipping it closed, that astronaut could be safely transferred from shuttle to shuttle. The sphere was only thirty-four inches in diameter—its size limited by the shuttles' forty-inch diameter hatches. It had to be smaller than that to fit someone through the hatch during a transfer. The sphere was constructed of gas-tight, thermal protective fabric to keep the crew member inside alive during the process.

So after wiring me up, I was put inside the sphere and they zipped it shut. I was told they would be monitoring my reactions and that they would get me out "in a while." Ten minutes? Thirty minutes? An hour? I had no clue. They then shut the lights off in the room and left me in the dark. Inside the sphere things were quite cramped. I sat with my legs crossed and my head tucked forward. There wasn't much room at all. One of the most important features of the sphere was a small half-inch hole through which they passed fresh air. So with my nose straining toward the hole, I sat and waited. I found the experience almost relaxing. But remember, this was familiar territory for me. After all, I spent countless hours crammed underneath a desk at home as a boy, simulating the cramped interior of a Mercury spacecraft.

While not the least bit comfortable crammed into the rescue sphere, I found it a lot more pleasant than some of the other tests they had been doing on me. Since nobody was poking or probing me, I decided I had it pretty good. The test was designed to evaluate claustrophobia. If you did not like being in small spaces in the total dark, you definitely would not like this experience! Fortunately, I am not claustrophobic. After about thirty minutes, the lights came on and someone entered the room to let me out. No rapid heart rate. No heavy breathing. No screams. I passed another test.

I was given a questionnaire by a psychologist, asking me fifty different ways if I ever had the urge to throw a brick through a window and whether I preferred sitting in a room all by myself or being at a big party. "How stupid would you have to be to not see through these questions?" I wondered. No, I had never felt like throwing a brick through a window and yes, I did prefer being with people over sitting in a room by myself.

They checked my eyes, my hearing, my breathing, my blood, my saliva, my urine, my brainwaves, and much more. There was some series of exams or tests each of the five days. I was scheduled for my interview at 1:30 p.m. Thursday. I was to report to Building 259, which housed both the Astronaut Selection Office and the Astronaut Crew Quarters. Before the interview, I went back to my hotel room, put on my suit and tie, brushed my teeth, combed my hair, and then headed back to NASA. I got there about twenty minutes before my interview and I couldn't have been more nervous. As I waited, I alternated between getting drinks of water and going to the restroom. One then the other, over and over. As the start time for my interview approached, someone took my handwritten essay and gave it to the selection board. It was quiet for a few minutes and then I heard laughter. *I guess they found my essay amusing*, I thought. I had poured out my heart and soul and all I heard was laughter. It definitely didn't help my nerves. Finally, the conference room door opened and I was told: "Don, they're ready for you. Come on in." Seconds earlier, I had been totally nervous. I didn't see how I was going to get through this. There was so much at stake, so much riding on the outcome. But as I heard them call my name, I experienced a sudden peace. I took a deep breath, exhaled, and said to myself, *Don, this is what you have always dreamed of. You have worked hard to get here. Go tell them why you would make a great astronaut!* I was ready. *Let's get on with this.*

The next hour went by fairly rapidly and the interview was a relatively friendly one. "Tell me about what you have done since high school," was the leadoff question. Why did you go to Cornell? Why did you choose to work at AT&T? Why would you want to become an astronaut after the *Challenger* blew up just a little more than a year ago? Some of the questions were off the wall like, "What shoe do you put on first in the morning?" followed by, "Do you prefer oil-based paint over latex?" Some of the questions seemed ludicrous, like they were just wasting time, but I soon realized that they were intentionally messing with me. By asking me a series of stupid questions, they were watching and assessing how I would react and respond. Inside I wanted to tell them that we only had an hour for this most important interview and to quit wasting the valuable time asking these stupid questions, but I just played along and answered each one.

At the end, George Abbey, who was director of Flight Crew Operations and headed the selection committee, asked if I knew who won the NCAA basketball championship the night before. They kept us pretty busy the entire week and each evening most of the candidates went out to dinner as a group, so there was very little time to watch TV. But by chance, when I returned to my room the night before, I turned on the TV and saw that the championship game had just ended. I hadn't channel-surfed for it, it just came up that way. I saw players from Indiana jumping up and down, so I assumed they won. I didn't know the score, but I could tell them that I thought the Hoosiers won the national championship. Evidently that was my current events question to see if I knew what was going on in the rest of the world. I passed.

Overall, I thought the interview went very well. On Friday, our final day, I met with the medical personnel who said they were still going over the results, but told me that they expected I would have no problem passing the physical. I headed home to New Jersey feeling pretty good. I felt like no matter what happened, I did my best. Then the waiting began. I had a good feeling that I'd be among the chosen ones.

Things started looking better when some of my friends let me know that the FBI had called them and were asking questions about me. NASA was doing a thorough background investigation, interviewing friends, employers, neighbors, and landlords. I figured there was no way NASA would be doing that for all hundred of the candidates they interviewed. I figured I must be on a short list. I expected to get selected and I anxiously awaited their call.

Rejection, Then Limited Acceptance

It was well known that if Abbey called, you were selected. If anyone else was on the telephone, you were not. After a few weeks, my home phone rang. It was Pinky Nelson from the Astronaut Office. He was a member of the selection board and was one of the astronauts who interviewed me. It didn't sink in. My heart raced as I waited for the good news. Then, in an instant, reality intervened and my world came crashing down. Pinky thanked me for applying, but told me I had not been selected. He asked if I had any questions for him. I was speechless. I had none. He wished me well and said goodbye. I was numb.

For a few brief seconds, I stared off into space (no pun intended), then I started crying. The news broke my heart. I was thirty-two years old and had dreamed of becoming an astronaut for the past twenty-six years. I worked hard to attain my goal. I passed all the medical exams. They chose me for an interview. They spent time looking carefully at me and getting to know me. And then they said no. I had been rejected. It was devastating news—NASA had selected fifteen new astronauts that day, and I wasn't one of them.

I cried—sobbed was more like it—for about ten minutes. I wasn't mad at myself—I wasn't mad at anybody—I was just very, very sad.

A little boy's dream had just been crushed. I decided that night that I needed a new plan for the rest of my life. I concluded that I needed to forget about my childish dream of flying in space and I needed to face reality. I decided that I'd begin figuring out my future the next morning—after a good night's sleep. So I went to bed one thoroughly inspected, recently rejected, and very much dejected astronaut wannabe.

When I woke up the next morning, my first thought was: *I still want to be an astronaut.* Nothing had changed. It was like a curse—always present, always driving me. In reviewing the list of selectees, it was clear that nearly all the civilians chosen already worked for NASA—most of them at the Johnson Space Center.

I was working in New Jersey for AT&T; that would have to change. I decided to go to Houston to try and get an engineering job with NASA. I met first with George Abbey and asked if he could help find me a position within the space agency, but he was pretty noncommittal. I also scheduled a few minutes with Pinky Nelson to see if I might get any clues as to why I wasn't selected. He couldn't tell me outright, but he strongly hinted that working for Bell Laboratories was not helping my case. Terry Hart worked for Bell Labs and took a leave of absence from the company when he became an astronaut. After he flew on STS-41C, he was offered another flight but decided instead to return to Bell Labs. I was another Bell Labs guy and NASA wondered whether I would do the same—leave NASA after one flight. I told Pinky that I was looking to get a job at JSC and he strongly concurred with my plan. That was it—I needed to leave Bell Labs and find a job in Houston.

I never heard back from NASA about a job, so I decided to try on my own. I put an ad in the *Houston Chronicle* newspaper under "Positions Wanted," seeking a job in the aerospace field and hoping one of the NASA contractors might pick me up. The ad ran for about a week and I finally got a call from Al Copeland, a chemist at JSC. He said he had been reading the newspaper's classified ads when he saw mine under "Lost Pets." He said he felt so sorry for me and asked me to send him my resume and he would see what he could do. He passed it along to the materials group in the engineering directorate and soon Bud Castner, who headed up the metals section, called. He liked my background and needed somebody exactly like me. When I told him I interviewed with the last group of astronauts, things changed dramatically—and not for the best. Castner explained that if he did hire me and I was selected for the next astronaut class, he would have to go through the hiring process all over again. He wasn't sure he wanted to do that. I told him that the chances of me being selected were pretty slim and that I simply wanted to work for NASA. He politely told me he would think about it over the weekend. I kept my fingers crossed.

He called me back the following Monday and told me he decided against hiring me because it would be too much trouble for him if I was selected for a future astronaut class. I was annoyed that he didn't give me a shot, but he did offer to pass my resume along to Lockheed Engineering and Science in Houston. They were a huge contractor for NASA and did a lot of work for Castner and the rest of the engineering groups at NASA. A month or so later, I got a call inviting me to Houston for an interview. A few weeks later, I was offered a job—which I quickly accepted. I had just gotten one foot in the door.

When I told my coworkers at Bell Labs that I was quitting and moving to Houston, most thought I was crazy. I was just a few months away from receiving my much coveted third week of vacation and getting vested in the company's retirement program. I tried to explain that I needed to chase this dream of mine. While a few people understood, most did not. I worked through the end of 1987 and in January 1988, drove to Texas.

When I went to the Johnson Space Center to get my identification badge on my first day of work, I drove past a huge Saturn V rocket lying on its side. Needless to say, there was an equally huge smile on my face. I was working at NASA; I was only a contractor, but I was still working at NASA. I became part of the NASA team that day, and I will never forget that special moment. I would spend the next twenty years of my life at JSC.

It was quite a transition, going from research and development at Bell Labs to working at Lockheed. My first surprise was the five thousand dollar a year pay cut. Then I was informed that I would not get any vacation time my first six months. I think I was most surprised when I saw my new office. At Bell Labs, I shared a nice office with another engineer. My new office was one of those tiny cubicles you see all the time in comic strips. I also had to fill out a time card every day. I wasn't punching a time clock, but it was the next closest thing. I guess everybody has to make some sacrifices and compromises to work for NASA and I was willing to take those hits for the cause.

To put it mildly, my job with Lockheed wasn't the most glamorous. Each experiment that was scheduled to fly on the space shuttle needed to submit a list of all the materials used in its construction. Every nut, every bolt, every washer, every battery, every everything needed to be listed. My job was to take that list, which was sometimes hundreds and hundreds of pages long, and make sure each material being used was on the list of NASA-approved materials.

Unfortunately, there were no computer databases at the time, which would have made the job relatively simple. I did it all by hand. I would take the first material listed, which might be a special alloy of aluminum such as 2019, and I would then see if it was on the list of NASA-approved materials. If it was not, I needed to make a note of that. If it was, I moved on to the next item on the list. The work was tedious, boring, and required almost no original thought. I reminded myself many times that I was lucky to be working for NASA. And while the work was not very interesting or challenging, I have to admit I was still thrilled to be working as part of 'the team.' Everyone associated with NASA was working hard to get things ready for the STS-26 mission of space shuttle *Discovery*, the first mission after the *Challenger* accident. So it was an exciting time, and I really felt like I was contributing to the nation's space program.

After a few weeks working in the cubicle, Lockheed moved me across the street to Building 13 at the Johnson Space Center. This was my first step up the ladder. I was issued an old gray Steelcase surplus desk and shared an interior office with six other Lockheed employees. There were no windows, but I was out of my cubicle and working on-site!

The NASA civil servant engineers and technicians had nice offices with newer furniture. The phrase used to describe the difference between being a contractor for NASA and a civil servant working for the government was 'contractor scum.' Hey, I might have been considered scum, but I felt pretty lucky to be space program-related scum.

On to NASA

I had been working for Lockheed about six months when my NASA boss, who was in charge of monitoring the contractor's work, told me he had a civil servant position he needed to fill within a week or he would lose it. Since he didn't have time to advertise and interview applicants, he asked me if I'd be interested. It didn't take more than a second to say yes. The next day, he submitted the paperwork to hire me as an official NASA civil servant, effective July 16, 1988. The funny part was that my new boss was Bud Castner, the same Bud Castner who didn't hire me six months earlier because he knew I wanted to be an astronaut. So after six months in Houston, I was now working for the guy who didn't want to hire me. I knew he still didn't want me hired, but he didn't have much choice. If he didn't hire me, he was going to lose that position anyway. It's a great feeling knowing that you're better than nothing, even if it's just a little better than nothing.

I turned in my Lockheed contractor identification badge, and got my NASA civil servant ID badge. It felt pretty good having that official NASA badge. The one drawback of my new job, I soon learned, was that I would have to repay Lockheed eleven thousand dollars for my moving expenses. Even though I had been working at NASA all along and only changed badges from contractor to civil servant, there was nothing NASA could do about it. I was required to pay back Lockheed, which I did, a few hundred dollars a month. Just another of those cause-related hits.

A few weeks later, they moved me from the contractor office to an office with other civil servant materials engineers. I was the only one in the group with a PhD and a few of the others resented me, especially so when they learned I had interviewed for the astronaut position the year earlier. They all kept me at arm's length.

Even though I was now working directly for Bud Castner, who headed up the metallic materials branch at NASA, he didn't want to give me any work responsibility. "I don't want you to have too much responsibility around here in case you eventually get selected to be an astronaut one day," he told me. He advised me to go and find my own projects to work on. "See what you can come up with." It was an unbelievable opportunity, but also a tough situation. I so desperately wanted to be part of the NASA team and work on some of the space shuttles' materials issues. There were constant problems with cracks and corrosion that needed to be solved, but I would not be allowed to work on any of those.

Eventually, they asked me to look into lifetime predictions for the graphite epoxy materials they wanted to use in making new lightweight pressure tanks for the space station. We had years and years of data and experience with metal pressure vessels, but only a few years of testing with these new graphite epoxy materials. Since the space station was being designed for a twenty-year life, NASA needed to know if pressure vessels made with those new materials would survive that long without rupturing or bursting. Castner asked me to look into that because he had nobody else on his staff qualified to do so. I spent a few weeks researching what already had been done in that field and was able to put together some results that indicated the materials would survive twenty years. It was the only fun assignment I had from Castner. And now, with the International Space Station already in space for fourteen years, I only have six more years to go to prove whether my results were correct. So far, the tanks are performing well, as I predicted.

Castner also asked me to look into a corrosion problem with the turbine blades of the space shuttles' auxiliary power units (APUs). Those blades were exposed to highly corrosive hydrazine, which was causing cracks in the blades. NASA had been working on the problem for almost a year and the only solution they could derive was to double the thickness of the blades. By making them thicker, the cracks would take longer to propagate before the blades would fail. I spent a few days in the NASA library looking through technical journals and found the solution. Years earlier, someone discovered that coating the blades with a thin layer of gold halted the corrosion. It was a simple, lightweight solution to the problem. When I pointed this out to Castner, he thanked me and said they already decided on the solution—to double the thickness of the turbine blades.

When they give you lemons, make lemonade. Since I was given no real assignments and responsibility within the materials group and was told to find my own projects to work on, I decided to seize the opportunity and see if I could do something fun. I went over to the Astronaut Office and talked with Bonnie Dunbar, one of the only astronauts at the time with a background in materials science. I told her my situation and asked if she knew of any flight-related materials processing work that I could get involved with. She immediately had some ideas for an experiment that was flying on her next mission, STS-32. Bonnie and I agreed to be coinvestigators on a project we named the Microgravity Disturbances Experiment, which looked at the affects of disturbances aboard the space shuttle, such as engine burns and crew exercise equipment, on the quality of crystals grown in space. With Bonnie working on getting things approved by NASA management and me doing much of the engineering and paperwork, we made a pretty good team.

I was the envy of the materials branch because nearly everyone wanted to get more involved with the astronauts. I had the chance to work closely with the STS-32 crew, which was commanded by Dan Brandenstein, who was also the chief of the Astronaut Office. Working closely with Bonnie and having the opportunity for Brandenstein to get to know me wouldn't hurt me a bit in the next astronaut selection process—which was coming up soon.

In August 1989, as I was busy working on the Microgravity Disturbances Experiment, I received a phone call from the Astronaut Selection Office, inviting me to join a group of a hundred people for medical tests and interviews for the next astronaut class. I was thrilled, but didn't overreact because I was in the office. I went down the hall and told Castner that I would not be at work next week because I was invited to be a part of the next astronaut selection group. He wasn't exactly thrilled, but wished me well.

So I spent the following week at JSC interviewing to be an astronaut. Since I had been through the process once before, I pretty much knew what to expect. Because of that, it was not as nerve-racking as the last interview two years earlier. At our introductory briefing on Sunday evening, we were told about writing the "Why I Want to be an Astronaut" essay. I knew this might be coming, so I had found a copy of the one I had written for the last selection. I read it and it still sounded pretty good; I changed a few words and that was it. My reason for becoming an astronaut hadn't changed at all!

It was a solid week of medical tests, which all went well. When it came time for the interview, I was much more relaxed than the last time. But overall, I felt I hadn't done as well as I had the last time. I just didn't have a good feeling about it. It just seemed to me like I had an off day.

But there was one bright spot during the interview. The final question was: "Is there anything else about yourself that you would like us to know?" My first instinct was to just say no and get the heck out of there. Instead, I explained that I was a volunteer at a Big Brother-type program, and told them a bit about it.

Once I had settled into my new job with Lockheed and life in Houston, I found time to pursue a long-held interest in volunteering and mentoring. I always felt a bit lucky to have survived my parents' divorce and growing up in a single-parent household, and felt a sense of responsibility to try to help some of the other kids out there who found themselves in a similar challenging situation. I soon became connected to a volunteer organization in Houston called Friends of Students, part of a broader organization known as Juvenile Court Volunteers. The organization was very similar in its mission to the much better known Big Brothers and Big Sisters programs, but with a slight difference. Friends of Students paired adult volunteers with young teens who were not yet juvenile offenders, but were considered troubled youths who struggled in school or at home and were believed to be heading in the wrong direction—potentially toward becoming juvenile offenders.

I filled out all their forms and went through their training program before getting assigned to my first student. I was full of optimism that I could influence and change anyone and could even help them to become president of the United States one day. I was green, naïve, and more than a bit overly optimistic.

The student and I would meet once a week and go to a movie, get something to eat, shoot some pool, or just hang out at a park. He lived with his mom in my neighborhood—only a few blocks from where I lived. His parents were divorced and he no longer had any contact with his father. Slowly over time, meeting with him week by week, the walls came down and we connected. After about six months, he told me he and his mom were moving to Arkansas. They left the following week. I was sad to see him go and his sudden departure emphasized the challenges he was facing.

I was assigned to another student, who was extremely bright—a nearly straight-A student. He lived with his mom and two younger siblings in a house filled with bugs—there were no screens on the open windows. A social worker told me his grades dropped suddenly, probably because the electricity to his house had been shut off, and he couldn't do his homework at night. I reflected on my background and better appreciated how easy I had it growing up. After only a few months, his mother packed up their belongings and left the state. I never saw or heard from him again.

I soon was assigned to my third student, fourteen-year-old John Pickering—a twin like me whose parents also had divorced. He had very little contact with his father and grandmother, who were living in Florida. He was a bit rebellious and frequently in and out of the principal's office at school. John was pretty unsure of me the first time we met in 1990. His first question: "You're not a KIKKer, are you?"

He was asking if I liked listening to country-western music on a popular Houston radio station, KIKK-FM, apparently a bit concerned about my possible musical preferences. I assured him I was not a KIKKer—that I liked rock and roll. It took years for the walls John had built around himself to slowly come down. I never gave up on him and continued scheduling times with him to get together, even though early on he'd frequently not show up. Eventually, John and I became great friends and remain so today—nearly twenty-five years later.

John worked at a marina near NASA for a few years after high school before joining the Army. He was stationed at Guantanamo Bay, Cuba, long before the facility became famous following the 9/11 attacks. Years later, while aboard space shuttle *Discovery*, I would look out the windows searching for him down below as we passed overhead. Today, he is a college graduate and a successful financial officer. I played only a miniscule role in his life, but know I helped him along the way and I am so proud of all he has accomplished. It was important for me to try to help others as I continued to work toward my dream.

As I finished telling the story about John and my volunteer work, I could tell by the looks on the interviewers' faces and their reactions that they clearly liked what I told them. While I felt like I hadn't done my best on the entire interview, I thought I might have hit a home run with that answer.

Now there was nothing more I could do other than wait. We were told it would be months before NASA announced the new class. Thankfully, I was so busy preparing the experiment for STS-32 that it kept my mind occupied. I didn't have much time to sit and wonder whether I would get selected.

The Dream Becomes Reality

STS-32 launched January 9, 1990, and I was working long hours in Mission Control during that flight. My mom was in town for a few days, so I went home early January 16, after my shift in Mission Control. The red light was blinking on my telephone answering machine; I had a message. My mom was sitting a few feet away having a cup of coffee as I listened: "Don. This is Don Puddy. I'd like to talk with you when you get a chance. Give me a call." Don Puddy was the director of Flight Crew Operations and the boss of all the astronauts. I knew that if Don Puddy was calling, it meant that I had been selected for the next astronaut class. I looked over at my mom in disbelief and told her: "Mom, I think I made it into the astronaut program!" She, like me, was just as shocked and happy. Soon I was dialing the phone and speaking with Don Puddy. He asked me if I was still interested in taking a position as an astronaut, because he was offering me the job. I don't know what my exact words were, but I stuttered a bit before stammering out "Yes!" And that was that. I hung up the phone and yelled over at my mom, "I made it! I got selected!" It was an unbelievable moment, made even more special by having my mom there to share it. She raised my brothers and sister and me by herself, and now she was one proud mother of a future astronaut. What a moment it was to share with her.

I soon learned that I would be reporting for astronaut training on July 16, 1990, so I had a few months to finish up my work on the Microgravity Disturbances Experiment. I also knew that my life was about to change dramatically and that there wouldn't be much time for vacations or visiting because of training. So I took some time off that spring to visit my mom in Rangoon, Burma, where she was working for the US State Department.

After my graduation from Case and move to Cornell in Ithaca, my mom, who was nearly fifty at the time and a secretary for a Cleveland manufacturing company, realized that her house was now empty. With three of her children living in New York State and one in San Diego, she could pursue one of her lifelong passions—traveling. She applied to the State Department for a secretarial position at one of the US Embassies and soon found herself accepted and assigned to her first post—the US Consulate in Istanbul, Turkey. She packed up and soon was on her way to the first of six countries where she worked before retiring. In addition to Turkey, they included Syria, Tunisia, Indonesia, Burma, and India. It amazes me how much of the world my mom saw and experienced before I ever got to space. I'd like to think that part of my adventurous spirit and desire to explore and experience new things came from her.

Soon I was cleaning out my desk in Building 13 and reporting for duty with the other twenty-two new astronauts. We were called the Astronaut Class of 1990, the thirteenth group of astronauts selected. None of us knew when we might fly, but we were all excited. It was a pretty distinguished group—seven pilots and sixteen mission specialists. There were five women, including Eileen Collins, who would become the first female shuttle commander; Ellen Ochoa, the first Hispanic woman to fly in space; and Nancy Currie, a US Army helicopter pilot who would be my crewmate on STS-70. Also, three of the class members were born in Cleveland.

On our first day, NASA held a news conference to introduce us to the world. It was far from the fanfare generated when the early astronaut selections were announced. By the time NASA got to Group 13, it was all a bit ho-hum. The only real press interest focused on Eileen, who was in line to become the first female shuttle pilot. The rest of us were just fine with blending into the background and not being the center of attention.

We were given our training schedules for the next six months. Noticeably absent was any vacation time. But, once again, that was more than fine with everyone. We were glad to have the opportunity to train to be astronauts and, quite frankly, I'll bet all preferred training over vacationing. Most of us worked hard our entire lives to get where we were and the last thing we wanted or needed was some time off.

After reveling in our first day as astronauts, NASA quickly put us in our place. We were, after all, not yet real astronauts, merely astronaut candidates. In typical NASA fashion, that title was shortened to Ascans. Yep, there was no mistaking our social order in the Astronaut Office. We were Ascans and nobody let us forget that.

Our training began in earnest. We were issued workbooks and other training materials, and started numerous classroom sessions to learn about NASA and how the space shuttle operated. In preparation for flying in NASA's T-38 jets, we were sent to Vance Air Force Base in Enid, Oklahoma, for parachute and ejection seat training. We went to Pensacola Naval Air Station in Florida for water survival training and then to Fairchild Air Force Base in Washington for land survival training.

While challenging, exciting, and fun, it was not always as rigorous as you might expect. As we were heading out into the woods for our two-day land survival experience, our bus stopped at a convenience store where all of us loaded our pockets with candy bars, chips, and other necessities. One of our biggest complaints was that the wine we purchased and smuggled in was not quite the right temperature for that evening's dinner. We learned to build shelters using parachutes and made sleeping bags from parachute material stuffed with dry grass. We shot off signal flares and learned to use the other survival gear we would eject with in case of an emergency in the T-38. But mainly the training allowed us to bond with one another. Fellow Clevelander Ron Sega and I were teamed up at Fairchild and we had a great time together and became close friends in the Astronaut Office.

Whether in the classroom, flying in the T-38s or zero-gravity aircraft, or learning the details of the switches and reviewing procedures in the space shuttle simulators, training was an absolute blast. Everything was interesting and everything was exciting. Some of the most fun was spent underwater, during the numerous training sessions at JSC's Weightless Environmental Test Facility (WETF), a big, deep swimming pool where we trained for spacewalks. I'll never forget the first time I was suited up and lowered by a crane into the water. Looking out my visor and seeing a mockup of the space shuttle payload bay in front of me made me feel like a real spaceman. It was the moment for me when all the training started to seem like the real thing. I was an astronaut now and would be heading to space one day. I had a big smile as I practiced maneuvering in simulated weightlessness and learning how to operate some of the equipment we might use for real someday.

Not all of the training was that much fun, however. Some of the most dreaded—but highly useful—training occurred during a week set aside to learn how to deal with the media. During that week we were taught how to act in front of the cameras and how to answer tough questions from less-than-friendly reporters. There were many embarrassing moments for each of us that week. As we practiced various interview styles, we were always videotaped. After each session, we would review the interview with the help of all of our classmates. They were a rough group—there was no slack for anybody giving a stupid answer to a question or unable to think on his/her feet. A low point for me was during training for what was called the ambush interview. It takes place unexpectedly—suddenly, you find yourself in front of the lights and cameras, with a reporter sticking a microphone in your face asking for your comments on some issue. We spent nearly an hour talking about the strategies for how to best handle that type of situation and the most appropriate response to a surprise interview. After sitting in the conference room for an hour, the instructor suggested we take a ten-minute break to stretch our legs and then we were to report to another room for our next training session. I spent my ten minutes getting a soda and laughing in the hall with some of my classmates.

As I opened the door to the room for our last session before lunch, bright lights hit my face, a microphone was shoved in front of me, and a mock-reporter started asking me a series of highly politically-charged and sensitive questions. After sitting through the previous session, where we had just learned the strategies for dealing with this type of scenario, one would expect that my training would automatically kick in and that I would coolly and calmly deal with the situation. Unfortunately, no—I adopted a classic deer-in-the-headlights look and stuttered and stammered some less-than-ideal answers to the questions. I hadn't learned a darned thing, but I was about to. Minutes later, I found myself reviewing my embarrassing performance in front of my classmates. They were predictably merciless—bordering on brutal—in their comments, which were sprinkled with laughter, as we watched the replay. It was truly one of those embarrassing moments that you vow will never happen again. That day—the second time around—I learned how to handle the ambush interview.

Cancer

The Ascan training was all going pretty well and I was having the time of my life, until one day in October 1990, when I went in for a routine NASA physical. During that examination, one of the NASA doctors discovered a small bump on my thyroid. Just the size of a small pea, it was decided that we needed to have it checked out and soon I was scheduled to see a specialist and have a biopsy. The results came back a week or so later and I will never forget that phone call as long as I live. The doctor told me the small bump was cancer. The diagnosis sent chills through my body. It was devastating news in more than one way—my immediate response was: "Is this going to kill me? Will I die from this? Is this treatable?" It was some of the worst news I had ever gotten in my life.

And almost immediately after pondering my fate came another question: "Is this going to eliminate me from the astronaut program?" I was pretty scared upon hearing the news. I had worked so hard to get to where I was and now there was a possibility that I was going to wash out before ever getting to fly in space. I was totally numb.

Fortunately, my NASA flight surgeon assured me that the cancer was treatable and told me that if I had to have any type of cancer, this one was the one to have. It seemed a bit reassuring, but I didn't exactly consider that good news at the time. The plan was for me to undergo surgery, during which they would remove my entire thyroid, small bump and all. As with everything else at NASA, the surgery was scheduled around my training schedule to minimize the time lost at work, so that I would not fall behind my classmates. We were scheduled to take a trip out to the West Coast in mid-November to visit NASA's Jet Propulsion Laboratory in Pasadena, California, and the Rockwell facility in Downey, just outside Los Angeles, where the space shuttles were built. It was decided that I would miss that trip and stay in Houston and have the surgery during that time.

The surgery went extremely well and, with my thyroid removed, I now sported a big scar on my neck. Recovery went smoothly and the following week I was able to rejoin my class as we resumed training in Houston. It was a life-threatening setback for me, to be sure, but fortunately it never impacted my astronaut spaceflight status. After a few weeks, I was cleared to fly in the T-38 jets once again and everything else was back to normal.

I have to give a lot of credit to the NASA flight surgeons who looked after my health and that of all the other astronauts at the Johnson Space Center. They easily could have declared me unfit for flight and that would have been the end of my astronaut career. However, after carefully examining the case, they concluded that I was fine, with an excellent prognosis for a full recovery. Having had cancer and having my thyroid removed was not considered grounds for elimination from the program. With a great feeling of relief, as if I had literally dodged a bullet, I put it all behind me and pressed on with my training. While many other astronauts have been diagnosed with cancer over the years, I believe that I was the first cancer survivor to ever fly in space, and I was proud to represent my fellow cancer survivors in doing so.

After a year of training, we graduated from Ascan to official astronaut status, which meant we were now eligible for space shuttle flight assignments. For most of us, that didn't come anytime soon. But a few lucky members of my class were assigned a few weeks later to their first flight. As part of the Ascan graduation, we were given a silver astronaut lapel pin. I was so proud to receive it. I had big hopes that one day I would receive the gold pin given to astronauts once they flew in space. But for the time being, I was mighty proud just to have made it to flight assignment eligibility.

I took my silver pin and had it made into a necklace, which I presented to my mom soon afterward. I told her how much the pin meant to me and that I wanted her to have it for always being there for me when I was growing up. I laughed a few weeks later when I received a bill from NASA for thirty-five dollars to cover the cost of the silver astronaut pin. I guess they just don't just give those things to the new astronauts.

Even though we now had our silver pins, it was made clear to us that unofficially we were still considered Ascans by the rest of the Astronaut Office. And we would remain Ascans until we flew in space. Ascan or not, silver or gold, I was just happy to be where I was and patiently waited for my chance to be assigned to a space shuttle mission.

Even though we had graduated from the Ascan ranks, there was still a lot to learn. We continued our training, spending more and more time in the shuttle simulators. In addition, we were each given technical assignments. These involved various jobs that supported the space shuttle program or crews in training. Some got assigned to be Capcoms, the astronauts who speak to the crews in space from Mission Control. Some were assigned as astronaut support personnel, to strap the crews into the shuttle on launch morning and help out with other duties at the Kennedy Space Center. These were both the most sought after job assignments. And then there were others, sometimes less glamorous, but absolutely no less important to the space shuttle program.

I was assigned to the Shuttle Safety Review Panel (SSRP), which evaluated issues or problems that arose and ultimately cleared each mission as safe and ready to launch. I had a few classmates console me on this less-than-most-desired assignment, but I didn't want and didn't need any of their sympathy. I knew I was playing a role as an important member of the space shuttle team. Watching the SSRP at work and participating in many of its decisions gave me great confidence in the safety of the system. I worked very closely with many engineers and managers on various safety issues. I knew that when my moment came to strap into the space shuttle for my launch that I would do so with great confidence, knowing that everyone behind the scenes had done their job to the best of their ability to help guarantee a safe and successful flight.

I had other assignments over time as well. I became the space shuttle propulsion system representative and followed some of the developments in making the space shuttle main engines (SSME) safer and the efforts to develop the lightweight external tank, which allowed us to launch greater payload weight. I frequently traveled to the Rockwell International facility in Downey, California, where I watched them assemble the crew compartment for the space shuttle *Endeavour*, which was being built to replace the lost *Challenger*. I visited one of the NASA contractor manufacturing sites outside Canton, Ohio. Here in a big barn more than a hundred years old, turbine blades for the SSMEs were being manufactured. Parts for the space shuttle made in a barn! It was an impressive operation and, again, meeting the dedicated employees who were actually designing, manufacturing, and assembling these things was a great confidence builder for me.

Meeting and Marrying Simone

One day in 1992, I found myself with a free afternoon. Usually, when I found myself with a free block of time, I would look for an opportunity to fly in a T-38. Unfortunately, there was no availability, so I decided to go for a ride on NASA's KC-135 zero-gravity aircraft. The four-engine KC-135 was an older stripped-down cargo aircraft, similar to a Boeing 707, with only a few rows of seats. Otherwise, it was a pretty large cavernous space where researchers could test the operations of the experiments they were planning to fly aboard a space shuttle. It also was a great place for new astronauts to learn how to move around in zero-gravity, which is why I signed up for the flight.

Flying parabolic arcs over the Gulf of Mexico, it was possible to create about twenty seconds of zero-gravity each time the plane pitched over the top of an arc and began its fall. In principal, it's similar to a roller coaster going over the top of the hill and creating a brief moment of weightlessness. The KC-135 was just an upscale version of a roller coaster, flying up and then pitching down, until finally pitching back up again to pull out of the dive and begin the climb upward for the next parabola and another twenty seconds of weightlessness.

As the KC-135 pulled out of its dive at the bottom of each parabola, everyone inside the plane was subjected to twice the normal pull of gravity. A typical session aboard the KC-135 would involve fifty consecutive parabolas. Up and down, up and down, first brief twenty second periods of zero-gravity followed by thirty seconds of twice the force of gravity. Inside you found yourself either floating or smashed to the floor—there was little else in between. The flights would last about two hours and could be pretty rough on the stomach from all the motion. It was unusual to fly on the KC-135 and not see at least a few of the participants get airsick; hence the nickname the Vomit Comet.

And in this most romantic of settings is where Simone Lehmann and I met. She was working as a training coordinator for the German astronaut office and had been living in Houston for a few months already, helping to get the German astronauts ready for their STS-55 D-2 Spacelab flight. Tom Henricks was the pilot on that flight, so Simone already knew him. Tom, of course, would become the commander of the All-Ohio STS-70 mission.

With the STS-55 launch a year away, some of the German researchers were scheduled to test their experiments and hardware on this particular KC-135 flight and Simone was with them to help out.

As I boarded the KC-135, I stopped by the cockpit and said hello to the flight crew, some of the same pilots that I regularly flew with in the T-38s. I then headed back to the cabin, where everyone was busy installing and preparing their hardware. As I was walking through the cabin, headed to the seats in the far back, I first saw Simone. Even in her baggy drab olive-green flight suit, she was beautiful and had an eye-catching smile. We shook hands and introduced ourselves, speaking only for a moment before the final preparations were made for takeoff.

During the next two hours, we flew out over the Gulf and completed our fifty parabolas before heading back to Ellington Field. Simone and the other German researchers were busy the entire flight working on their experiments, while I just practiced floating around. Given that the Vomit Comet is not exactly a romantic location, with a good number of the passengers, well, vomiting, I think it's safe to say that Simone and I probably are the only couple who met on the KC-135 and later married.

The next time I saw Simone was at the Outpost Tavern, an astronaut hangout dive bar a mile or so from JSC's main gate. It was as run-down a bar as they get, but the place where astronauts hosted their shuttle flight assignment parties. Simone accompanied Tom to the Outpost one evening for one of the crew assignment parties and I soon caught a glimpse of her from across the bar. She was even more beautiful all dressed up than when I first saw her on the KC-135. I went up to her and introduced myself once again.

She was born in the small town of Göppingen in Southern Germany, about twenty miles east of Stuttgart. She had a degree in animal science and was now working for the German space agency, Deutschen Zentrums für Luft-und Raumfahrt (DLR), at its astronaut training center in Cologne. She was scheduled to be in Houston for only six months as the STS-55 crew completed their final months of training. After that, she would be heading back to Germany.

We became friends and would regularly meet for a beer after work at Molly's Pub, another popular NASA hangout near JSC. Our friendship soon got pretty serious. Fortunately for me, the STS-55 mission experienced numerous launch delays, which meant we had the opportunity to spend more time together in Houston before she was scheduled to return to Germany. Finally launching on April 26, 1993,

Tom and the rest of the STS-55 crew successfully completed their D-2 Spacelab mission before landing on May 6. Soon Simone was en route to Germany.

After our whirlwind romance, which we somehow were able to conduct despite my hectic training schedule and Simone's work with the STS-55 mission, we finally got engaged in the summer of 1993. Then we looked carefully at my schedule to find a few days when we might be able to squeeze in a wedding. A window opened for us in October, following one of my training sessions in Germany, and we decided to get married then. We were married October 7, 1993, in a civil ceremony at Cologne city hall, but unfortunately didn't have enough time for a honeymoon. After nearly twenty years of marriage, we're still looking for that window of opportunity for a honeymoon. For now, it remains on our bucket list.

Also in the summer of 1993, I was assigned to work in Mission Control as a Capcom. The experience of working with the flight directors, NASA management, and the rest of the flight control team was an asset. I was familiar with the Apollo 13 mission in April 1970, during which an onboard explosion threatened the safety of the crew. The heroic efforts of the Mission Control team to safely get the crew back to Earth have been well-documented in numerous books and in Ron Howard's movie *Apollo 13*. That the Mission Control team was extremely talented was not a surprise to me and having the opportunity to work with them was an amazing experience.

My first Capcom assignment was STS-47, the eight-day Spacelab J mission, which involved the crew working in two shifts around-the-clock in the European-built Spacelab science module, situated in space shuttle *Endeavour*'s payload bay. One of the more unusual events I ever experienced in Mission Control took place on one of the first days of that mission. I arrived for my shift about two hours earlier and was just settling in when the entire Mission Control Room went dark. All power was lost, all communications were down, and all the lights were out except for a few dim emergency lights. We had trained for just about every scenario, but never for anything like that. Because NASA had countless backup systems, losing all power wasn't considered a credible scenario. But here we were sitting in the dark. Each of the controllers stood up and looked at the flight director seated at the console next to me. Nobody knew quite what to do. Quickly, I had an idea—I remembered that I had written down the phone number for the payload operations control center at the Marshall Space Flight Center in Huntsville. Those folks were in charge of the payload operations during many of the Spacelab flights and they had their own set of nonastronaut payload Capcoms, which they called Crew Interface Coordinators (CICs). Those people were mainly responsible for talking with the astronauts performing the various scientific investigations in the Spacelab module during the flight. I picked up the phone and dialed my counterpart in Huntsville, a warm and friendly Southerner named Homer Hickam. I told Homer about our problem and asked him to keep the line open so we could use him to relay messages to the crew. Power was restored in Mission Control about five minutes later. After that small burst of excitement, I settled in for what turned out to be routine operations for the rest of the shift and the rest of the mission. Homer went on to fame himself as the author of *October Sky*, which was eventually turned into an award-winning film.

My next Capcom flight was STS-52, which featured a set of payload bay-based materials processing experiments. One day during that mission, astronaut Bill 'Shep' Shepherd notified Mission Control of a small problem. He was trying to take some electrical measurements with a handheld multimeter device but the unit wasn't working properly. It was nothing critical to the mission in any way. But once he reported the problem, Mission Control went into action. A small team was formed to investigate the problem and after a day or two they had it solved. In their research, they discovered that the multimeter manufacturer had known of this potential issue and had tracked the problem down to a bad batch of resistor components. They came up with an easy fix which involved folding a small piece of paper inside the back cover of the multimeter over the area of the suspect resistor and then closing the back cover. The little added pressure on the resistor from the folded paper fixed it. We sent Shep the repair procedure, and it worked perfectly after that.

I learned two important lessons from that experience. The first was that there is no problem too big or too small for Mission Control. The second was the realization that the same caliber of people who saved the Apollo 13 crew were still working in Mission Control.

Working as a Capcom involved long hours day after day, but was incredibly fun. I couldn't imagine being any more involved with a mission or being as close to the crew and knowing exactly how they were doing than I felt as a Capcom, short of being up there myself.

STS-65, My First Flight

One evening, as I was settling in for a ten-hour shift during the STS-52 mission, the Capcom phone rang. I answered, and it was Dave Leestma, who at the time was the acting chief of the Astronaut Office. He quickly informed me that he was assigning me as a science mission specialist aboard the STS-65 mission. The scheduled launch date was nearly two years away. Rick Hieb had been assigned months earlier as the payload commander; so had Payload Specialist Chiaki Mukai. Leestma told me that Leroy Chiao was also being assigned as a mission specialist. The commander, pilot, and our mission specialist 2, he said, would be named later—sometime in the next few months. He told me NASA would issue a press release the next day announcing the new assignments, and asked that I keep it quiet until then. I thanked him and hung up.

Boy, was it hard to get back to work. There I was, just assigned to my own space mission, my very first assignment, and I had to coolly and calmly sit in Mission Control for the next eight to nine hours. I wanted to yell, scream, jump up and down. I wanted to call my mom and my brother Dennis. But I couldn't do any of that until my shift was over.

The STS-65 mission would involve flying the Spacelab science module in the payload bay of *Columbia*, where we would perform more than a hundred experiments from around the world. It was the second flight of the International Microgravity Laboratory, or IML-2 for short. My commander would be Bob Cabana and the pilot was Jim Halsell. Other crew members included me, fellow Clevelander Carl Walz, Hieb, Chiao, and Mukai, the first Japanese woman to fly in space. Jim, Carl, Leroy, and I were all members of the Class of 1990.

Before I could begin training for STS-65, I needed to finish up one last Capcom assignment. I was the designated lead Capcom for the STS-53 mission. That Department of Defense mission would be deploying a secret satellite from space shuttle *Discovery* six hours into the mission.

I started working my shift about an hour after the launch on December 2, 1992, after handing over from the ascent Capcom. Immediately, my shift's Mission Control team started working with the crew on the deployment, which after a few-hour checkout, occurred without a hitch.

While I had been working the STS-53 mission, my STS-65 payload crew members were already training in the Netherlands and in Germany. I was anxious to join them. So after the successful satellite deployment, I was given permission to join my crew in Germany. That was my final shift as a Capcom, and I wished the STS-53 crew well with the rest of their mission and then headed home to sleep, shower, and pack. I left the next day and met up with my crew in Cologne, where my nearly nineteen months of training for STS-65 began.

Because this was an international mission, we would be flying experiments from many different countries. That meant I had the opportunity to travel around the world training for this flight. First Germany, then Canada, Japan, France, and Italy. Then back to Canada and Japan, followed by more trips to Germany, France, and Italy. We met with the scientists whose experiments we would be conducting to learn more about the science behind their work and to gain some expertise in their labs. The first four months of training were hectic because we shifted from Europe to Japan and then back to the United States. It was a lot of hard work and involved a lot of time away from home, but there is nothing quite like the opportunity to be training for a spaceflight.

We also spent week after week training at NASA's Marshall Space Flight Center in Huntsville, Alabama, where a Spacelab training simulator was located. There, we conducted a multitude of experiments replicating what we would be doing during our mission. We practiced working through the experiment procedures and learned how to operate each piece of scientific equipment we would be flying. There was so much to learn that we compared it to drinking out of a fire hose. Much of what we would learn leaked out of our ears, so we repeated the training over and over until all of it became familiar and routine. That was a well-proven method of training used by NASA and designed to minimize mistakes we might make while on orbit.

STS-65 would be the seventeenth flight of space shuttle *Columbia*. It was the first shuttle to fly, was one of the workhorses of the shuttle program, and had earned a reputation during its previous sixteen missions for having a significant number of launch delays. But as we marched through the months and months of training, the targeted launch date held pretty firm.

Soon it was July 8, 1994, STS-65 launch day. And there I was, strapped into *Columbia*'s middeck, only a few minutes from liftoff. I waited my whole life for this moment—my dream was close to becoming true. What an amazing feeling. The clock continued to count down and time raced like a truck going downhill without brakes. There was no stopping it. Each second seemed shorter than the one before it. It felt to me like time was accelerating.

The final two minutes were by far the fastest of my life. At T-30 seconds, I heard the call: "Go for auto sequence start." At that point, the shuttle switched to internal control and was operated by its five redundant computers. The seconds flashed by on the countdown clock I had positioned in front of me. How can time be passing so quickly?

"Twenty (seconds remaining)," Cabana reported. About six seconds before liftoff, I could feel and hear the three main engines coming to life. "Here we go," Cabana said. The vibrations were awesome and caught me by surprise. We never trained with vibrations like those. The entire shuttle began to shake—wildly, but not violently. A definite firm shaking, rocking and rolling. I don't believe I heard the final few seconds of the countdown and don't remember watching the clock either, but I do clearly remember the enormous burst of vibrations that occurred at T-0, the moment the solid rocket boosters lit and we lifted off the launch pad. The vibrations were jarring. I was glad I was firmly strapped into my seat. Then I felt a strong force. It felt like someone was pushing me with their hand in the middle of my back. We definitely were on our way.

I screamed inside my helmet, "YaaaaaaaaHoooooooooooooooo!" and "Yes!" and "GO! GO! GO!" No one could hear me but myself. Rick Hieb was only a foot away from me, but with the visors on our helmets down and locked, and with the noise of the engines, it was impossible for anyone but me to hear me. I realized I was on my way to space. My countdown clock mounted in front of me was now counting up. A few seconds after liftoff, I could sense the roll of *Columbia* as Cabana called to the ground: "Houston, *Columbia*. Roll program."

Down in the middeck, there was nothing for me, Rick, and Chiaki to do but carefully listen to how things were going and monitor our progress to orbit. After two minutes, the solid rocket boosters separated and dropped away, leaving us with our three main engines still roaring. The ride immediately got smoother at SRB separation and remained so for the rest of our climb to orbit.

A personal major milestone for me was reached three minutes and forty seconds into the flight as *Columbia* passed through the fifty-mile mark above the Earth. It was great of Cabana to call out to all of us over our intercom circuit: "We've got several new astronauts on board. Fifty miles." The tradition since the early days of rocket planes and rocket ships is that fifty miles constituted the boundary between Earth and space. Once you crossed that artificial boundary, you were considered an astronaut. At that moment, I became the 312th human in space. And while it was nice to officially qualify as a real astronaut, my attention was still on the remaining five minutes of powered flight that would get us safely to orbit.

Things proceeded smoothly for the rest of the launch. Right on schedule, eight and a half minutes after liftoff, I entered the world of weightlessness. I could see things starting to float. I had tethers attached to my kneeboard and I could see that they were floating. "Welcome to space, you guys," Cabana said.

For the next two weeks, we conducted our science operations in the Spacelab module. Each day, we performed ten to fifteen experiments. Some were fun to perform, others a bit tedious and boring. The schedulers packed our timelines, allowing almost no free time. We were given an hour off for lunch, during which Carl, Leroy, and I would get together on the flight deck, look out the windows at the Earth passing below us, and marvel at the sights.

The STS-65/IML-2 mission was an extremely busy one, and each day we felt more and more fatigued. The pace of work was unrelenting—there seldom were moments to kick back, relax, and savor the experience. Science operations continued twenty-four hours a day—twelve hours for the Red Shift and twelve hours for my Blue Shift. As we neared the end of the mission, almost everyone was exhausted. We were looking forward to returning home.

We did our deorbit burn and for the next fifty-five minutes, fell back to Earth with the gravity level increasing ever so slowly from zero to its maximum value of 1.5 *gs*. I felt incredibly heavy, as if I weighed a couple of thousand pounds. After floating weightless for the past two weeks, it was going to take some adjustment to living with gravity once again.

There was nothing for us to see or do in the middeck, so we just listened to the calls between Cabana or Halsell and our Capcom in Mission Control. It was welcome news when Cabana reported: "Houston, *Columbia*. I have the field in sight," meaning that he could see the runway at the Kennedy Space Center's Shuttle Landing Facility. Everything was looking perfect. As Halsell called out our ever-decreasing altitude, I suddenly felt a huge thump as *Columbia*'s main landing gear touched down. Halsell deployed the drag parachute to help us slow down. When we rolled to a stop, Cabana reported: "Wheels stop, Houston."

Our mission was officially over. I was totally elated. I was glad to be home and amazed at how good I felt. I was dizzy and a little weak, but otherwise felt pretty good.

After a short ride in the Astrovan, we were greeted at crew quarters by our families. It was so great to see Simone and we hugged in a tight embrace. I was still a bit unstable and wobbly, so it was a good thing Simone was there to hold me upright. We were helped out of our launch and entry suits, examined by some doctors, took hot showers for the first time in two weeks, and enjoyed our first non-freeze-dried meal. While I missed being in space, I liked being back on terra firma!

An hour later, we held a postlanding press conference and then met with some of our other family members. Only my mom was there for me, but it was great to see her and share the excitement of my journey. I thanked her for all her support while I was growing up and for taking care of me and my brothers and sister. Without her, becoming an astronaut and flying in space would not have been possible.

Soon afterward, we boarded a plane for the flight to Ellington Field. After a brief 'Welcome Home' ceremony, NASA loaded us up on buses and drove us to our homes. It felt great as I walked in the front door of our house and it closed behind me. I made it home!

It was a pretty good feeling being a 'flown' astronaut. I had the same thought the day STS-65 landed as I had the first time I went on a roller coaster at Cedar Point in Sandusky, Ohio when I was ten years old: *Man, I want to do that again!* So it was back in line to await my next flight assignment. It wouldn't be that long.

Chapter Three
The All-Ohio Shuttle Crew

It started off as just another day. Exactly one month earlier—July 23, 1994—my STS-65 crew mates and I landed at the Kennedy Space Center aboard *Columbia* after our International Microgravity Laboratory mission that set a shuttle program flight duration record—14 days, 17 hours, and 55 minutes. Now we were in the middle of our postflight debriefings, which included visits to other NASA centers. But on that day, I was at my home base, the Johnson Space Center.

I was in my office doing paperwork when I got a call that Hoot Gibson, chief of the Astronaut Office, wanted to see me. My initial reaction was, "What have I done now?"

Since Hoot's office was only three or four from mine, it took me only about a minute to get there. He told me I was being assigned to the STS-70 mission, tentatively scheduled for sometime the following summer. The primary objective, he said, was to deploy a Tracking and Data Relay Satellite (TDRS). It would be a five-day mission with five crew members. He wondered whether I had any questions, so I asked the obvious: "Who are the other four?"

"I can't tell you now," he replied. "You'll find out tomorrow at the all-hands meeting when we officially announce it to the rest of the [Astronaut] Office."

I thought it was a bit strange and didn't understand the whole secrecy thing, but I didn't much care. All I knew was that I was thrilled to be assigned to another space shuttle mission, and that I would be flying in space again very soon.

I left Hoot's office with the telltale signs of having just been assigned to a flight: a big smile and an effort to act like nothing new was happening in my life. I went back to the office where my STS-65 crew mates—Rick Hieb, Leroy Chiao, and Chiaki Mukai—were working and didn't say a word about my meeting. I finished up the day and went home to tell Simone the good news.

I walked through the front door and saw her standing in the kitchen. "I have something to tell you," I said, a smile on my face. She had no idea what I was about to say, but recalled later that she was a bit apprehensive. "I've just been assigned to the STS-70 mission."

Without any hesitation, she asked: "When is the launch date?"

"Sometime next summer, maybe June or July," I replied.

After a deep breath and a big sigh, she said, "At least the baby will be born by then." She was happy for me that I was getting another chance to fly in space, but wondered to herself why NASA couldn't have given us a bit of a break. After all, it had only been thirty days since I landed from my first flight and here I was immediately being thrown into training for my second. No time to relax with the family. No time to decompress.

The next morning, I saw Kevin Kregel, a member of the Astronaut Class of 1992. I knew Kevin before he became an astronaut when he was one of the T-38 instructor pilots at Ellington Field. Kevin was a nice guy, always friendly. But that morning something was different. He was staring at me with a huge smile on his face—a much, much bigger smile than the one with which he normally greeted me. The only thing I could think was that he, too, must have been assigned to STS-70.

"You on seventy, Kevin?" I asked.

"Yeah, I sure am," he replied. "Do you know who else is on the crew?" I asked.

"Yeah, sure," he continued. "Tom Henricks is commander, and Nancy Currie and Mary Ellen Weber are mission specialists."

I was impressed that he knew all this. I'll never understand why I wasn't told the day before like Kevin apparently was, but at eleven fifteen that morning the whole office would find out who was on the crew.

The all-hands meeting took place in the same large conference room used for our Monday morning astronaut office tag-up sessions and debriefings. With about a third of the astronauts present, Hoot walked in and announced the names for the STS-70 crew: Tom, Kevin, Nancy, Mary Ellen, and myself. After a round of applause and congratulations, the meeting was over. A few people came up to shake my hand and congratulate me; others did the same for my new crew mates. It was especially great news for Kevin and Mary Ellen, the crew's rookies. We slowly left the conference room and got back to our jobs. The next day, I headed to KSC for our STS-65 postmission visit, and later that day NASA issued a press release announcing the STS-70 crew.

The STS-70 Crew

Tom Henricks was the most experienced member of the STS-70 crew. While it would be his first mission as a space shuttle commander, he previously flew as a pilot, or second in command, on STS-44 and STS-55. His was the typical rotation in the astronaut office: fly twice as a pilot and then get assigned as a commander, which would give him the opportunity to accomplish something he'd wanted to do since becoming an astronaut, something that is the sole domain of a commander—land the shuttle.

But being named a commander meant more to him than that. "It's a progression through a career that's been very exciting. It's an honor to be selected to make any spaceflight and, for me, to make it as a commander is the pinnacle of my career," Henricks said.

He said Buckeye State astronauts John Glenn and Neil Armstrong had some direct influence on his career choice. "While growing up in Ohio, we had excellent astronaut role models. I remember while in grade school watching John Glenn and the Mercury program, then watching Neil Armstrong step on the Moon while I was in high school. Those were very exciting times and, at that point in my life, I thought, 'What a dream. That can't happen to some boy from Ohio.' And as it turns out, this country allows you to strive for such goals and achieve them."

He said, "Major credit needs to be given to the school systems that produce people who are interested in following careers in math and science. Other than that, it is the Midwestern work ethic that allows people to excel in whatever career field they choose."

I found Tom to be one of NASA's best commanders. Yes, he definitely was the commander—very business-like when he needed to be, make no mistake about that—but he was also a very good friend to everyone on the crew and treated us with great respect. Tom worked as hard as anybody I ever met—he referred to himself as a recovering workaholic—but he also liked to have fun. Whatever you did with Tom, you knew you'd be working hard, but you also were pretty much guaranteed a good time.

Tom was born July 5, 1952, in Bryan, a small city about sixty miles west of Toledo, but spent most of his childhood growing up in Woodville, which he considers his hometown. All are located in Northwest Ohio. He graduated from Woodmore High School in 1970 and was appointed to the US Air Force Academy, from which he received a bachelor of science degree in civil engineering in 1974. He also holds a master's degree in public administration from Golden Gate University, which he received in 1982.

After pilot training, he flew F-4s in fighter squadrons in England and Iceland. He then was assigned to the Air Force Test Pilot School from which he graduated in 1983 and where he remained as an F-16C test pilot and chief of the 57th Fighter Weapons Wing Operating Location until he was selected to become an astronaut. Tom flew more than thirty different types of aircraft and logged more than six thousand hours of flying time. He was a master parachutist with 749 jumps and held an FAA commercial pilot rating. "My Air Force career allowed me to go step-by-step to meet the prerequisites to apply to be an astronaut. NASA felt I was qualified." Tom was selected as a pilot in NASA's eleventh group of astronauts in June 1985.

His first mission was aboard space shuttle *Atlantis* in November 1991. During that Department of Defense mission, the crew deployed a Defense Support Program (DSP) satellite, completed one hundred and ten orbits of the Earth, and traveled nearly three million miles.

Tom also flew as the pilot of space shuttle *Columbia* on the Spacelab D-2 mission in April 1993. That ten-day German science mission featured eighty-nine experiments in materials and life sciences, robotics, astronomy, and Earth mapping. On that one, Tom completed 160 orbits of the Earth traveling 4.2 million miles.

Tom is married to the former Rebecca Grantham. He has three children from an earlier marriage. He said his first job was helping at his grandfather's farm near Montpelier, Ohio, where he drove a tractor before he was old enough to drive a car. He said he could live without mail, considered the Bible a great reading experience, and greatly enjoyed his wife's singing. He also noted that his mother's maiden name was Thomas, setting off speculation that he and I might be related. A definitive answer was never determined.

Kevin Kregel, the only non-Ohioan on the STS-70 crew, was born September 16, 1956, in the small town of Amityville on New York's Long Island, and graduated from Amityville Memorial High School in 1974. His first job as a teenager was as a caddy at a golf course at Bethpage State Park. He was selected as a member of NASA's fourteenth astronaut class in 1992, and STS-70 would be his first spaceflight.

He decided to become an astronaut while growing up on Long Island after being inspired by the early astronauts—Mercury, Gemini, and those going to the Moon as part of the Apollo missions. His interest also was heightened by his close proximity to Grumman Aerospace, of Bethpage on Long Island, which received significant media coverage for its work building the Apollo Lunar Module that carried Americans to the surface of the Moon.

"So it's something I've always wanted to do since I was a little child. I aspired to go the same route as the original astronauts and that is why I became an Air Force test pilot, got an engineering degree, and applied to the [space] program," Kevin said.

Kevin graduated from the US Air Force Academy in 1978, with a bachelor of science degree in astronautical engineering. He received his pilot wings the following year at Williams Air Force Base in Arizona and, from 1980 to 1983, flew F-111 aircraft at RAF Lakenheath in England. As part of an Air Force-Navy exchange program, he flew A-6E aircraft at Naval Air Station Whidbey Island near Seattle and also served aboard the USS *Kitty Hawk* where he made sixty-six carrier landings during a cruise in the Western Pacific.

"The Air Force selected me to go to test pilot school, but they sent me to the Navy test pilot school at Patuxent River [in Maryland]. I did that for a year, and then I went down to Eglin Air Force Base [in Florida] and did flight tests on F-15, F-15E, and F-111 airplanes."

He received a master's degree in public administration from Troy State University in 1988.

"Then in 1990 I left the service and got hired by NASA as an instructor pilot and test pilot at the Johnson Space Center flying in aircraft ops, flying T-38s and the Shuttle Training Aircraft. Then I was fortunate enough in 1992 to be selected as an astronaut," he said. Kevin logged over five thousand flight hours in thirty different types of aircraft during his military career.

The first time I met Kevin was in the fall of 1990 at Ellington Field, when he was my instructor pilot for one of my early T-38 flights. I went on to fly with him on many occasions. From the first time I met him, he made a strong impression on me as being an outstanding pilot. He was so comfortable in the cockpit and never got nervous when he let me take the controls. He was very confident in his flying abilities and was an excellent instructor, always trying to make me a better pilot in the T-38.

Kevin was one of the taller astronauts and always seemed to have a smile on his face. He was a very friendly and an easy to get to know guy. Maybe it was because of his small town, middle-class family upbringing, but he always treated everyone with respect. He was hardworking and loved to fly, and I always enjoyed flying with him.

Kevin and his wife, Jeanne, had four children. He said one of his most memorable learning experiences was being taught how to cook by his grandfather. He also said he could live without lima beans, thought *Shogun* by James Clavell was a great reading experience, enjoyed the music of Creedence Clearwater Revival and Eric Clapton, and admitted that his favorite movie was *The Sound of Music*.

Nancy Currie and I became friends when we were both selected for NASA's thirteenth group of astronauts in 1990. Like Tom and Kevin, she is an intensely hard worker and a proud aviator. She was born December 29, 1958, in Wilmington, Delaware, but considers Troy, Ohio, near Dayton, her hometown. She graduated from Troy High School in 1977 and received a bachelor's degree with honors in biological science from Ohio State University in 1980.

Nancy joined the US Army in 1981, completed helicopter pilot training, and was an instructor pilot at the Army's Aviation Center, becoming a master Army aviator with more than four thousand hours flying rotary and fixed wing aircraft. She received a master of science degree in safety engineering from the University of Southern California in 1985. Nancy was assigned to the Johnson Space Center in 1987 as a flight simulation engineer before being selected as an astronaut.

She said becoming an astronaut was not a concrete goal until much later in her life. "I think it's kind of interesting, especially for a woman my age, because when we were kids growing up, women weren't military pilots. Women weren't astronauts," Nancy said, noting that many of her male counterparts probably would say something like, "From the time I was four years old, I wanted to be an astronaut."

"I'd say from a very early age, I knew I wanted to fly. I mean, I just dreamed about flying probably from the time I could walk." So her career became sort of a natural progression. She joined the military, became a military aviator, got an advanced education, and then became an astronaut. "It seems like my lifetime has always kind of been in the right place at the right time, because it was later on in high school and the beginning of college when they started allowing women to be military aviators. And then, I was already in college before the first females became astronauts," she said. "So by the time I got to that point in my life, the doors were wide open and it was a very easy process at that point."

"When I was in high school, I used to dream about and read about Army aviators flying helicopters and extracting the wounded in Vietnam," she said. "And so, that was really a goal of mine—to become an aviator and then maybe later on a doctor in the military. About the time I was graduating from college, the military opened up the opportunity for women to be appointed in the combat arms. So I actually became an air defense artillery officer and went directly to flight school.

"I had never flown anything in my entire life, but I knew I wanted to do it. And the first day in a helicopter I said, 'This is for me! This is what I wanted to do.'"

Nancy enjoyed flying so much that after her training she ended up staying at Fort Rucker, Alabama, as an instructor pilot. "To this day, other than the job I'm in presently [astronaut], it was the most fun, and the most enjoyable job I've ever had in my entire life! I just loved training flight students and sharing with them the same enjoyment of flying and the discipline of flying that I had."

Nancy's first spaceflight was aboard the shuttle *Endeavour* on the STS-57 mission in June 1993. She was the primary robotic arm operator and successfully retrieved the European Retrievable Carrier satellite known as EURECA. That mission also featured the first flight of Spacehab, a commercially provided middeck augmentation module used to conducting microgravity experiments. Spacehab modules flew in the forward area of the shuttles' payload bay. During that mission, she completed 155 orbits of the Earth and traveled 4.1 million miles. STS-70 would be her second spaceflight.

Nancy was always a kind, caring, and an extremely compassionate person. Each Christmas she led an effort in our astronaut class to collect food and presents to donate to needy families in the Clear Lake area near the Johnson Space Center. I have the utmost respect for her as a human being. Like Tom, she is also a fun-loving person. She loved working out; whether running, swimming, or weight training, not a day would pass for Nancy without some form of exercise. She could be tough as nails, but was also as kind as anyone I have ever met.

When asked to describe herself in three words before our mission, she wrote humorous, serious, and short. She was the smallest member of our astronaut class and was known affectionately as Smurf. But everyone in our class had the utmost respect for her abilities. She was one of the best at everything she did.

Nancy said her most cherished accomplishment was her daughter, Stephanie, and that she could live without bad weather on flying days. She liked reading Tom Clancy novels, said her musical preferences tended toward Kenny G. and David Sanborn, and that her favorite movie was *Sleepless in Seattle*. I was thrilled to have the opportunity to fly with her on STS-70.

Mary Ellen, like Kevin, was a member of NASA's fourteenth group of astronauts. She was born August 24, 1962 in Cleveland, but considers Bedford Heights, a Cleveland suburb, her hometown. She graduated from Bedford High School in 1980 and received a bachelor of science degree with honors in chemical engineering from Purdue University. In 1988, she received her PhD in physical chemistry from the University of California at Berkley.

Mary Ellen was an avid skydiver, her real passion in life. She logged over four thousand jumps since she started in 1983 while at Purdue University and was a nine-time silver/bronze medalist at the US National Skydiving Championships.

I met Mary Ellen when she joined the astronaut office in 1992. She was one of the seven who at the time were from the Cleveland area—Ken Cameron, David Low, Greg Harbaugh, Ron Sega, Carl Walz, Mary

Ellen, and myself. It was unheard of to have so many active astronauts from one city. Not even New York City could match Cleveland's total.

As a child, Mary Ellen never considered being an astronaut. "That was not something women did. It didn't even enter my mind as a possibility. It wasn't until I was in graduate school in chemistry and I was very much into science and research. I'd also gotten involved in a lot of aspects of aviation, and space just seemed like the perfect adventure and the perfect mix of all the things that I loved."

As a student, Mary Ellen liked problem solving. "I liked trying to figure out how things worked, why things worked, trying to understand the details, the physics, behind things. I really enjoyed that in school, and so I went on to college and into graduate school just pursuing those same things."

She also got involved in aviation. "Actually, skydiving was the first [aspect] of aviation that I got involved with, and it was exhilarating. It was exciting. It was challenging—very challenging. It's a sport that you can do for many years and there's always room for improvement and always another challenge around the corner. That's what I liked about both fields, science and aviation."

Mary Ellen got her pilot's license after graduate school.

"I'm the kind of person that just likes to try new things and likes to experiment and likes to be bold. And I think all of those experiences led me to try the biggest, the grandest adventure that I could imagine. I applied to NASA and went through their selection process, which is a somewhat involved process, and I was very fortunate, very lucky, to be selected in 1992."

Mary Ellen, who was married to Jerry Elkind, said her first job was as a hostess at the Brown Derby restaurant. She also said her most memorable learning experience was doing her graduate student research, that she could live without having to pay bills, thought *Interview with a Vampire* and *The Making of the Atomic Bomb* were great reading experiences, enjoyed the music of Steely Dan and R.E.M., thought *A Christmas Story* was a great movie, and that her domesticated fauna consisted of silk plants.

STS-70 would be her first spaceflight, and she was thrilled to be a member of the crew.

Primary Payload; Secondary Experiments

Before the age of communication satellites, which are very common today, NASA used a series of ground stations located around the Earth to communicate with its spacecraft and astronauts in orbit through what was called the Global Tracking Network. There were communication dish antennas in Florida, Bermuda, Grand Canary Island, Kano in Nigeria, Zanzibar in southern Africa, Muchea and Woomera in Australia, Canton Island and Hawaii in the Pacific, and numerous sites across the United States. That network of tracking sites allowed Mission Control to communicate with orbiting spacecraft for only about fifteen minutes of every ninety minute orbit of the Earth—less than 20 percent of the time.

Only when a spacecraft flew over one of the tracking stations could Mission Control talk with the astronaut and receive data about the spacecraft's onboard systems. The ground tracking network was used during Mercury, Gemini, Apollo, Skylab, and the early years of the space shuttle. It was only when the TDRS satellites were deployed in geosynchronous orbits that the beginning of near-continuous communications in low-Earth orbit was made possible.

The first step in building a series of geosynchronous communications satellites was accomplished during the STS-6 mission—the first flight of space shuttle *Challenger*, which was launched April 4, 1983. It was memorable for me because it was the first time I witnessed firsthand the launch of a rocket. Before that, I had only seen launches on television. For STS-6, with the help of Senator Glenn's office, I obtained a special pass to view the launch from the NASA Causeway viewing area, located about ten miles from the launch pad. As I watched *Challenger* climb skyward on its first trip to space, I was mesmerized. The roar of the engines—their sound waves pounding on my chest—and the bright flames from the solid rocket boosters bombarded my senses. My knees were shaking as was the ground on which I was standing. It was absolutely amazing and mind-boggling. That huge machine was leaving Earth at an amazing speed. After fifteen seconds, with my eyes firmly fixed on the shuttle, the thought finally hit me that there were people inside that vehicle. Seven astronauts riding the river of flame that flowed from *Challenger*, pushing it toward space. It was an amazing, awe-inspiring experience that only heightened my resolve to follow in their footsteps.

A few hours into the flight, the crew successfully deployed the first Tracking and Data Relay Satellite, known as TDRS-1 or TDRS-A. The second was to be delivered to space aboard the space shuttle *Challenger* on January 28, 1986, but an explosion seventy-three seconds after launch destroyed the shuttle and the satellite and killed the seven-member crew, including Ohioan Judy Resnik.

NASA placed such a high importance on developing the communications network that two of the first three flights following the *Challenger* disaster were devoted to deploying TDRS satellites—STS-26, the Return to Flight mission two and a half years later on September 29, 1988, and STS-29 on March 13, 1989. Two other missions—STS-43 (August 1991) and STS-54 (January 1993) delivered additional TDRSs.

Theoretically, by placing three communications satellites equally spaced around the Earth in geosynchronous orbit at an altitude of about 22,230 miles, continuous communications could be achieved. At that altitude, the rate of orbit exactly matches the Earth's rotation so the satellite essentially remains stationary over a specific point on the Earth's surface.

Instead of the full three satellite configuration, NASA decided to implement a two satellite system, known as TDRS-East and TDRS-West. With only two satellites, there would be a narrow timeframe during each orbit when there would be no communications. That was known as the Zone of Exclusion (ZOE), and it lasted between two to eleven minutes for each orbit.

The satellites at those two locations transmitted their signals to one of two ground stations—Las Cruces, New Mexico, or Guam.

Signals from those two sites were then sent to NASA's Goddard Space Flight Center in Greenbelt, Maryland, before their ultimate destination—Mission Control in Houston.

Each satellite weighed about five thousand pounds and measured fifty-seven feet across the fully deployed solar panels. It had a six-foot dish antenna and two steerable sixteen-foot parabolic antennas, thirty electronic beam array antennas for S-band and K-band communications, and a pair of large solar panels to power the spacecraft. The dish antenna was used for all communications between the TDRS and the ground sites. The parabolic and beam array antennas were used for communicating with orbiting spacecraft. At its highest transmission rate, a TDRS can transfer the entire contents of a twenty-volume encyclopedia—an average of forty-eight million words or nearly two million tweets—in one second.

The TDRS system represented a major advancement for communications with orbiting spacecraft and scientific data transfer. Many of NASA's large scientific satellite missions rely on TDRS satellites to relay their data to Earth. All of the spectacular images taken with the Hubble Space Telescope have been sent to Earth via one of the TDRS satellites. Other satellite users have included the Compton Gamma Ray Observatory, Upper Atmospheric Research Satellite, Topographical Ocean Explorer, and the Extreme Ultraviolet Explorer spacecrafts among many others.

STS-70 would be deploying TDRS-7 (aka TDRS-G), the replacement for the TDRS destroyed when *Challenger* exploded. Our mission was one of the last steps of recovery from that tragedy, and our crew felt a deep sense of pride and responsibility that we were finishing the work started by the *Challenger* crew.

The satellite was mounted atop an Inertial Upper Stage rocket which would boost TDRS from the space shuttle about 175 miles above the Earth to its operational geosynchronous orbit of 22,300 miles. The IUS, a two-stage rocket that used solid fuel as its propellant, was nearly seventeen feet in length and over nine feet in diameter with an overall weight of about 32,500 pounds.

The first stage contained a single large solid rocket motor with 21,400 pounds of propellant that developed 42,000 pounds of thrust. The second stage contained 6,000 pounds of propellant and provided 18,000 pounds of thrust. One unique feature of the IUS was that its first stage was the longest firing solid rocket motor ever developed for spaceflight with a maximum burn time of 150 seconds.

TDRS-7 was planned to be the last satellite in the series that would be deployed by the space shuttle. All future TDRS satellites would be launched aboard unmanned Atlas rockets.

While NASA continues to use the satellites to communicate with the International Space Station and to receive images from the Hubble Space Telescope and other scientific satellites, most of the satellite bandwidth and use is devoted to the Pentagon which pays for most of its operating costs and was the main driving force in establishing the requirements for the satellites. Besides use for scientific purposes, the Tracking and Data Relay Satellite System (TRDSS) is a vital part of our country's national security.

We also would be carrying fourteen secondary payloads—small experiments that covered a wide range of topics—including a study of the development of unborn baby rats in microgravity and the Shuttle Amateur Radio Experiment which allowed us to communicate with amateur operators around the world. The payloads, which also included study of the growth of colon cancer cells as well as protein crystals in space, measurement of different characteristics of the human eye in the microgravity, and operation of a Department of Defense experiment that used a special camera to take pictures of ground targets as it recorded accurate coordinates for later analysis, would keep us busy during the flight. Most of them were relatively simple and would be performed on the middeck or the flight deck. They were much less complicated than many of the experiments we performed during my STS-65 Spacelab mission and didn't require a lot of training. But we still needed to be trained to conduct all of them.

Block I Engine

An important engineering test during the STS-70 mission was flying a new turbopump on one of the three main engines at the aft end of space shuttle *Discovery*. NASA thought about flying the new turbopump assemblies on all three of the shuttle's main engines, but decided it would be safer and more prudent to fly only one of them to test it before committing to flying them on all three engines in the future. The new pumps underwent ground testing that was the equivalent of forty shuttle flights. For each flight, the pumps ran approximately nine minutes. So with six hours of actual runtime, NASA committed to flying one of the pumps as part of our mission.

The high-pressure liquid oxygen pump used in the older main engines needed to be removed for inspection after each flight. The new pump would not require inspections until after ten shuttle missions. The new pumps also were expected to increase the safety margins and reliability of the main engines which each provide half a million pounds of thrust during a shuttle launch. Another key element of the new pumps was that the pump housing was produced through a casting process, which eliminated nearly three hundred welds in each pump—a great and significant safety enhancement.

Besides the new turbopump design, the new Block I engine included a two-duct powerhead (sounds mighty impressive, doesn't it?) which increased the flow of liquid oxygen to the engine while decreasing the operating pressure and loads. The new engine design also featured a single-coil heat exchanger in the powerhead which eliminated all seven weld joints inside the engine which were extremely difficult to inspect.

Did I fully understand all the changes and modifications made to this new Block I engine that we would be flying? Not at all. But I did understand fewer welds in an engine meant potentially fewer weld failures and leaks, both critical factors when dealing with engines that have extremely high speed turbopumps feeding extremely low temperature cryogenic fuels to the combustion chamber of the engine. All this attached to a vehicle that would be shaking and vibrating like crazy from the two solid rocket boosters. Fewer welds. Safer engines. I was all for that!

First Crew Meeting

Our mission preparation got underway almost immediately; our training shortly thereafter.

On the Monday after we were assigned, we received our first briefing regarding the mission's proposed medical experiments. And at 8:00 a.m. that Friday, Tom, Kevin, Nancy, Mary Ellen, and I held our first crew meeting in Tom's office to talk about the mission and decide our roles and responsibilities. During the course of the discussion, Tom began assigning various tasks.

With her impressive flying background, Nancy had a strong interest in being mission specialist 2. That person sits behind the commander and pilot during launch and entry and acts as a flight engineer, helping oversee the shuttle's performance and systems. Without hesitation, she was given the assignment.

I was interested in being the lead mission specialist responsible for the checkout and deployment of TDRS. After all, that was the mission's main objective. Without much discussion, I was given that task. Mary Ellen would be my backup and would assist me during the deploy.

Determining the two crew members who would be assigned to train for contingency extra-vehicular activity (EVA), or spacewalk, was not very complicated. Typically, mission specialists are chosen. Unfortunately, there wasn't a spacewalk spacesuit small enough to properly fit Nancy, so the assignment fell to Mary Ellen and me. I had previously trained for contingency EVA during STS-65, so Tom made me the EVA lead, known as EV1. Mary Ellen was assigned as EV2.

While I was very experienced in space shuttle experiments, having just completed the STS-65 International Microgravity Laboratory Spacelab flight, Tom thought it best and fair to assign Mary Ellen as the payloads lead for STS-70. I would assist her with the experiments.

That sounded good to me. I was thrilled to be the lead in deploying TDRS and happy to be assigned as EV1.

One last thing that needed to be decided was where Mary Ellen and I would sit during launch and landing. Tom, Kevin, and Nancy would be on the flight deck for both. Tom wanted to know if I had been on the flight deck for either launch or landing of STS-65. I told him I hadn't and said I would like to sit there for launch. Tom thought it most fair to let one of us sit there for launch and the other for landing. Once again my limited seniority helped me get my first choice to be on the flight deck for launch. Mary Ellen would ride uphill by herself in the middeck and we would swap seats for the return to Earth.

So in very short order, without much discussion and with no argument or dissent, we got our primary assignments. Overall, I was very satisfied with how things worked out. I was really thrilled to have a chance to experience a launch from the shuttle's flight deck. After sitting in the middeck for launch and landing of STS-65, I was looking forward to the new experience of launching on the flight deck where there were windows.

Tom noted the flight's duration was scheduled to be only five days, but that consideration was being given to extend it to an eight-day mission. The first mention of such an extension was in a letter from NASA Headquarters to JSC proposing that the flight be lengthened "to enable NASA to manifest some space shuttle middeck secondary payloads that require, or whose research objectives would be enhanced by, a flight extension to eight days."

Five days would have been a quick mission, to say the least. Tom was in favor of the added days, as was everyone else, so he went to work advocating for the mission extension. He had his eye on trying to get an experiment with rats sponsored by the National Institute of Health. That one required a minimum of eight days, and Tom was checking to see if they could use even more time on orbit to extend us further. So at that point, even though we were officially a five-day mission, it was pretty likely that it would become at least an eight-day flight.

Also at that point, NASA listed our launch date as June 26, 1995, and had us flying aboard space shuttle *Atlantis*. Very soon afterward, however, NASA changed things. *Atlantis* would become the main space shuttle used in the planned series of dockings with the Russian *Mir* space station, the first of which would be STS-71 set for launch on June 29, 1995. STS-70 was moved to space shuttle *Discovery* and our launch date was moved forward to June 8.

A few hours after our initial meeting, we hosted the STS-70 crew announcement party at the Outpost, the bar less than a mile from JSC's main gate and the traditional place for the astronaut office to hold crew announcement parties. The place wasn't fancy, but there was lots of atmosphere and space program-related memorabilia. The crew sprang for a keg of beer and most of the astronaut office showed up for the celebration and socializing that lasted about two hours.

The Making of the All-Ohio Shuttle Crew

Almost as soon as we were assigned to STS-70, we realized that four of us were Ohioans.

Only Kevin was the odd man out.

I had an idea. Since 80 percent of the crew was from Ohio, why not write the governor of the Buckeye State and ask him to declare Kevin an 'Honorary Ohioan.' Then, maybe, we could claim to be 'The All-Ohio Crew.' I talked it over with my colleagues and everyone thought it had potential. Kevin was an extremely good sport about it and agreed to go along. He figured he could handle being an Ohioan for a while.

I called Governor George Voinovich's office in Columbus and explained that I was a NASA astronaut and was going to fly on space shuttle *Discovery* with three other Ohioans. I then explained that the remaining crew member was from New York and we wondered whether it would be possible for the governor to make Kevin an honorary Ohioan so we could say we had an 'All-Ohio' crew.

At that point, I was transferred to another person and began my explanation once again. Then I was transferred again and presented my case a third time after which the person asked "So who are you?" followed by "And what do you want?" followed by "So who are you?" and then we once again discussed what I wanted.

Finally, it became clear to them who I was and what I wanted. "That should be no problem," I was told. Just send them a letter with the details, such as Kevin's name and the shuttle mission number, and they would make sure we got a proclamation. "I'm sure the governor would be happy to do this for you."

I thanked the aide for her help and promised to send her the letter. I ended the conversation with an invitation for the governor to attend our launch. I was politely told that while they were sure the governor would appreciate such an invitation, he was much too busy to attend such a function. "Thanks, but no thanks," was the courteous reply. I was a bit surprised because I thought it would be something the governor would want to do. But on that day, that was not the assessment of his aide in charge of proclamations.

It was with great pride and a big smile that I told the rest of the crew later that day that Governor Voinovich would be making Kevin an 'Honorary Ohioan.' Kevin, needless to say, was quite honored.

The next day, I sent the request to the governor's office. A month went by, no proclamation. Then a second month, still nothing. I called the governor's office to check on the status of our request, and was told that it was too early to issue the proclamation. They assured me that it would be coming. All we could do was wait.

In the meantime, Governor Voinovich started to get more and more interested in our mission. He even requested that the state flag that flew in front of the Ohio Capital on March 1, 1995, the 192nd anniversary of Ohio's statehood, be sent to NASA and flown on the mission. We planned to present the flag to the governor after we returned.

Mike Dawson, the governor's press secretary, said, "We're extremely proud that four of the five crew members are Ohioans, and the governor's going to make the fifth crew member an honorary Ohioan, so we consider it to be an All-Ohio crew." It was now

The All-Ohio Shuttle Crew

clear that we finally got the governor's attention. The idea of an Ohio shuttle flight was starting to take hold.

More months passed and we were getting closer to our launch date, maybe three months remaining. Still no proclamation or word from the governor's office. I was a bit concerned, so I decided to call them again to make sure they hadn't forgotten.

They promised me that it would be arriving soon. By that time, our mission was picking up a bit of notoriety because of the four Ohioans on the crew. In mid-May, we finally received the proclamation.

We kept it a secret from Kevin until our next crew meeting a few days later. There, with some fanfare, we presented it to him. With that, he became an Ohioan of sorts, and we officially became the 'All-Ohio Space Shuttle Crew.'

But after reading the part about now having "all rights and privileges held dear by all citizens of our great state," Kevin asked facetiously, "Do I have to pay taxes in Ohio from now on?"

Included with the proclamation was a resolution from Governor Voinovich acknowledging our crew as "exemplary role models for the youth of Ohio and throughout the nation" and whose "backgrounds and experiences make them an inspiration through their qualities of hard work and determination." The resolution added that the governor joined "with all Ohioans in expressing our sincere pride in the achievements and contributions of STS-70, The Ohio Shuttle Flight." It was signed by the governor on May 5, exactly thirty-four years after America's first spaceflight, Alan Shepard's fifteen minute suborbital Mercury mission.

Not everyone was thrilled about STS-70 being declared 'The All-Ohio Space Shuttle Mission.' In a friendly but competitive gesture, Don Campbell, the director of NASA's Lewis Research Center in Cleveland, wrote a letter to Amityville Mayor Emil Pavlik inviting him to the launch noting that the mayor's presence "would be significant for the five person Ohio crew." Mr. Campbell added, "I look forward to your participation in supporting the 'All-Ohio Space Crew' and witnessing the launch of this special space mission."

The Village of Amityville shot right back with a good-natured letter of its own. The response offered a slightly different perspective: "The fact is that Kevin Kregel is a New Yorker through and through, and WE lay claim to him! Be assured that no matter how strong a PR campaign you run, when STS-70 lifts off, ten thousand Amityville residents will be praying for ONE New Yorker and only FOUR Ohioans."

The STS-70 Crew Patch

One of the first things crews needed to deal with after being assigned to a mission was to come up with a crew patch. Kevin took the lead on this and told us he would work with an artist friend to come up with a few ideas. A week or so later, he showed us a preliminary design we ended up using as the foundation for our patch. It featured a space shuttle above a grouping of three stars and a looping ribbon in a triangular arrangement. We all liked it. We added a blue and green Earth, placing it behind the space shuttle. There was some discussion that the ribbon was a little too similar to those used to designate various causes, so we went with red, white, and blue in our design. The three stars and the triangular ribbon didn't have anything to do with our mission, but were nice design elements. So we creatively developed some deep symbolism about the triangle and three stars representing the TDRS triad (our satellite would complete the three-satellite network planned for the TDRS system, each equally spaced one hundred and twenty degrees apart around the Earth, each one capable of transmitting via line of sight to the other two). Viola! A patch design that was rich with mission specific symbolism. We were just about done.

As we were going over the final details of our patch, we were becoming more aware of the significance of having four of the five crew members with Ohio hometowns. (Kevin, unfortunately, was not yet an Honorary Ohioan.) Nancy had the great idea to make the patch in the shape of a large block letter 'O' for Ohio—very similar to the large block letter 'O' used by Ohio State University, of which

Nancy was an alumna. Nobody else had gone to Ohio State. But since that university was probably the most recognized school in the entire state of Ohio, it seemed like a good choice. Kevin said he didn't mind shaping the patch like a block 'O,' so our design was nearly complete. Nancy had one more suggestion—add red trim to outline the block 'O' (again a color heavily associated with Ohio State) and that was that.

The design was completed. Well, almost. We would need to gain approval for the patch from Chief of the Astronaut Office, Bob Cabana; Head of the Flight Operations Directorate, David Leestma; and Director of the Johnson Space Center, George Abbey. Then it would be sent to NASA Headquarters for approval by NASA Administrator Dan Goldin. Usually the approval was a rubber stamp process. Only for the most important missions would public affairs or others want to get involved in the design of the crew patch. Otherwise, they left it up to each crew and let them do pretty much what they wanted.

We decided not to mention anything about the 'O' shape representing Ohio since we figured NASA would not be interested in snubbing the other forty-nine states which also were footing the bill for the space shuttle program. After a few weeks, we got word that our patch design was approved by headquarters. It was with great pride that all of us, even Kevin, had the new patch sewn onto our blue flight suits. For the STS-70 flight and the fun we would have with the 'All-Ohio Space Shuttle Mission,' this patch captured our pride in our state and celebrated our 'Ohioness.'

The first public mention that our block 'O' shaped patch design honored Ohio was in an article published in *The Blade* newspaper of Toledo, Ohio, on March 6, 1995, below the headline "Shuttle's 'O' insignia lifts Buckeye pride on space mission." In an interview with the newspaper, Tom told them that "We designed it in the shape of an 'O' to signify that four of the five crew members are from Ohio." He also added that "Kevin was sport enough to go along with it." Tom went on to explain that Nancy originally suggested that the border of the patch that contained the crew's names be colored scarlet and gray, the official colors of Ohio State. Tom pointed out that the crew thought doing that would be too supportive of a single school, but the article went on to discuss further subtleties and speculated that those may have represented compromise: "The crew members' names that surround the outer edge of the insignia are in gray on a black background bordered in red, which some might describe as cardinal while others might prefer to call it scarlet," the article stated. "The cardinal is the state bird, and the state flower is the scarlet carnation. But there are no buckeyes anywhere in the insignia. That, Colonel Henricks said, was a concession to non-Buckeye Mr. Kregel."

Another newspaper reported that the red border and gray names also represented the school colors of Amityville Memorial High School, Kevin's alma mater. We were proud to have a patch that was diverse enough to honor both the great state of Ohio and Amityville High!

A Time to Train; A Time to (Super Bowl) Party

To get things started, I met with the four people responsible for our TDRS/IUS training. The group was headed by Bill Preston, a veteran trainer, and supported by Ginny Young, Robert Graubard, and Allen Burge: all young, bright, and enthusiastic instructors. They gave me some manuals and other material to review. A few days later, I had my first course—Inertial Upper Stage Operations—that reviewed the timeline and procedures we would be using to deploy the TDRS satellite. It was a two-hour class in what we called the single system trainer (SST), a small room that contained a mockup of the shuttle's instrument panels, switches, and dials that roughly represented the layout of the flight deck. We used the SST to learn one system at a time. Later, we'd move to more sophisticated simulators to undergo multisystem integrated training.

Initially, Mary Ellen and I took the lessons separately. By focusing on one-on-one training, I was able to learn everything inside and out. Later, Mary Ellen and I would take lessons together in order to develop our crew coordination and refine the roles and responsibilities of our tasks during the critical deployment operation.

After some much deserved time off over the Christmas and New Year's holidays, the tempo of training picked up quite a bit in January. Besides learning about TDRS deployment, there were countless other courses on various shuttle systems and refreshers of all sorts. Some of these activities began to involve the entire crew, while others would involve only two or three of us. There was a lot to learn and we were all rolling up our sleeves.

Tom and Kevin were scheduled to fly the Shuttle Training Aircraft (STA) at White Sands Missile Range in New Mexico in mid-January. NASA had a facility at White Sands and part of its responsibility was being the prime ground station for the TDRS network. The White Sands TDRS team was a small but important group, and I wanted to visit them to see what they did and to talk to them about our mission. Kevin offered to go with me, so we flew to El Paso aboard a T-38 and then drove to Las Cruces where the facility was located. We met many of the people involved and toured numerous rooms filled with computers. While it was not the most exciting tour I had ever taken, it was important to visit with the TDRS personnel and thank them for their support.

After spending about two hours with the TDRS team, Kevin and I headed back to the El Paso airport for Kevin's STA flight. I flew with him that afternoon as he completed ten practice landing approaches in one of the NASA Gulfstream jets that simulated the flight characteristics of a shuttle. Nancy flew out with Tom and accompanied him on his STA flight that same afternoon. After the flights were completed, we flew back to Ellington, arriving about 10:00 p.m. Training was to continue the next morning at 8:00.

We knew the months leading up to launch would be hectic, so we looked for any opportunity to unwind. The head of our training team, Tim Terry, suggested we have a Super Bowl party and offered to host it at his house in the city of El Lago, just two miles from the front gate of the Johnson Space Center. So the crew and training team along with our spouses or significant others met on January 29 to watch Super Bowl XXIX. As is Super Bowl Sunday tradition, most people wore some sort of attire signifying their favorite team. Naturally, Simone and I attended representing the Cleveland Browns. The San Francisco 49ers beat the San Diego Chargers, 49–26, and while most of us didn't care much about who won the game we still had a great time—even Simone who was less than two weeks away from our baby's delivery date.

It was a good team building experience and time for us to meet with our training team in a more relaxed setting. The training teams were some of the most important and underrated parts of NASA's shuttle team.

Let's Have a Baby!

Simone's due date of February 11 came and went with no baby. The doctor decided that if the baby wasn't born by February 13 or 14, he would induce labor February 16. Being able to schedule the birth of your child fits in well with NASA's rigid training schedules. When February 14 came and went without the baby's arrival, we went ahead and scheduled the induced labor for February 16.

I had a full day of training scheduled for that day, including six hours with the entire crew. In the morning, we had a water survival class to talk about bailing out of the shuttle in the event that it was unable to make the runway on landing. That was followed by actual bailout training in the Weightless Environment Training Facility (WETF) swimming pool. The pool was twenty-five feet deep and twenty-five yards long and normally used for EVA training. It was also used for bailout training where we practiced splashing into the water in our survival gear and getting out from underneath a parachute and getting into life rafts and using our survival communications equipment. In the afternoon, I had a meeting to discuss our first full simulation of the TRDS deploy with the entire team participating followed by two-and-a-half hours of training with Mary Ellen on the Bioreactor experiment.

When Kai's birth was scheduled for February 16, I informed Tom. Without any hesitation he told me to take the entire day off and to make sure I was at the hospital with Simone. He had the training team reschedule our water survival and bailout training for another day and told me the rest of the crew would cover for me at the simulation discussion meeting. He cleared my schedule for that day so I could be with my wife when our first child was born. I can assure you that not every shuttle commander would have done that, and I was really impressed with Tom for doing it for me. He earned my deepest respect for that. A good commander takes care of his crew, and Tom was one of the greatest.

I drove Simone to the hospital early that day and they began the process to induce labor. Much later that day, a little after 6:00 p.m., Kai Daniel Thomas was born. I was able to be with Simone the entire day thanks to Tom. The next day (a Friday) was a light one for me. Tom, Kevin, Nancy, and I had a two-hour contingency abort class and then I had a two-hour review class on the deployment of the TDRS/IUS. That afternoon, I only had 'admin' time scheduled and spent it back at the hospital with Simone and Kai.

The next morning, after a great night's sleep, I drove to the hospital and picked up Simone and Kai to take them home. Less than four months from launch and I was a new father. My life was already busy preparing for the mission and it was about to get even busier.

Just a few days earlier, on February 11, space shuttle *Discovery* arrived at the Kennedy Space Center's Orbiter Processing Facility to begin its preparations for the STS-70 mission. It had just completed an eight-day mission during which it rendezvoused with the Russian Space Station *Mir*.

Launch Invitations, Some Touring, and EVA Training

As we approached T minus three months and counting, it was time to send our official launch invitations to family, friends, and others.

Since our mission would be honoring Ohio, I decided to invite the first All-Ohio crew, one of the Buckeye State's most famous astronauts, and a hero of mine since childhood, John Glenn. Because he was a US Senator, he was invited by NASA to every shuttle launch. But I decided to write a personal letter to him as well, hoping that it would be more effective than a NASA VIP form letter that he would have received for all previous shuttle launches.

So I composed a short letter telling him that I was an Ohio astronaut, that he had inspired me to become an astronaut, that four of the five crew members were from Ohio, and that it would mean a lot to all of us if he would attend the launch as my guest. As I sealed the envelope I thought to myself, *He probably gets a bunch of these letters.* But I always believed things like my invitation were worth a try.

Three weeks after Kai was born, things had settled down enough for me to travel a bit. So on March 6, the crew flew to the West Coast to tour of the facilities where TDRS and the IUS upper stage that we would be deploying were built. After arriving in Los Angeles, we drove the short distance over to Space Park, the TRW facility that was building our TDRS.

There we met TDRS Flight 7 Manager, George Ruppucci, who gave us an overview of the satellite, describing how the spacecraft was constructed, how it deployed, and how it operates. Ruppucci, who described the satellite as a "mechanical and electronics wonder," was so proud of 'his' satellite, as was the rest of his team present that morning.

We also met one of the workers, Ralph Ibarra, a TDRS mechanical test conductor who worked on the program since its inception. He pointed out that our visit signaled the start of the prelaunch preparations. Every crew deploying a TDRS from the shuttle visited the facility before the satellites were shipped to KSC. Our spacecraft, TDRS 7, was scheduled to make the trip the following month, in April.

That tour marked the first time I laid eyes on the TDRS. It was huge—an impressive sight. The two large umbrella communications antennas were folded up for launch. They were gold and shined under the bright lights of the clean room. The large solar panels were similarly folded in a compact manner for launch. I could only imagine how large it would be once it was deployed and fully opened in space.

Before we departed, we had one more important task to complete. Each of us signed a piece of thermal insulation foil that would get affixed to the spacecraft. The TDRS integration and test team members, headed by Joseph Gross, also would sign the same piece of foil. "This tradition creates a permanent record of the men and women who were instrumental in the integration, test, and launch of the satellite," Gross said.

It was an awesome morning, and soon we were aboard our T-38s headed for Seattle. Tom had a brilliant idea to fly at low level—about nineteen thousand feet—to the Emerald City instead of the usual thirty-nine thousand to forty-one thousand-foot altitude so we could enjoy some of the pretty spectacular country between Los Angeles and Seattle. The bonus was the crystal clear day with nearly unlimited visibility.

I have spent nearly a thousand hours in T-38s flying all over the country, but this was by far the most spectacular scenery I had ever seen. Crater Lake in Oregon was absolutely breathtaking from the air. As we flew over other parts of Oregon and southern Washington State, I was amazed at all the forest that had been cleared. The ground had a checkerboard appearance from the air with every other square totally devoid of trees. The lumber companies weren't thinning the trees in the forest, they were leveling them. I just couldn't believe how much of the land had been totally cleared.

On our way to Seattle, we flew along the Cascade Mountain Range and almost directly over Mount St. Helens. Kevin, with whom I was flying, dipped the wing and did a sharp bank maneuver so we could look straight down at it as we passed. Half the mountain was no longer there because of a volcanic eruption in 1980. But what amazed me the most was how the trees had fallen—all lined up in a radial pattern emanating from St. Helens due to the force of the tremendous explosion that occurred. Even after fifteen years, I was taken by how little recovery had yet occurred. There was a lake at the base of St.

Helens called Spirit Lake, and it was totally choked with dead tree trunks floating on the surface. Everywhere you looked you saw fallen trees, and it appeared as if it had just happened a few weeks or months earlier—not almost a generation ago!

Next, just to the east of us, was Mount Rainier, its summit still covered with snow. It was an impressive sight against the dark blue, crystal clear sky. As we approached Seattle, I could see the famous Space Needle from the 1962 World's Fair. Then one by one, our jets landed just as the sun was setting.

I had little interest in going out to enjoy Seattle that night. All I wanted to do was take advantage of not sleeping with a newborn in the room waking me up a few times a night. Kai had been home only two weeks by that point, and he definitely was *not* sleeping through the night.

The next morning we headed to the Boeing Space Center in Kent, Washington, where they were building the Inertial Upper Stage (IUS) that would send our TDRS from low-Earth orbit—165 miles—to its final destination—a geosynchronous orbit 22,300 miles above the Earth.

After a quick introduction with the Boeing upper management, we were given a briefing on the IUS before we headed out to the labs where the hardware was built and tested. Our visit included tours of their IUS Systems Integration Lab for training with IUS hardware and then had some close-up looks at IUS stages in production in the IUS Clean Room.

The IUS that we would be flying on STS-70 was already at KSC, so while we never saw 'our' IUS that day, we were able to see an identical unit. That was just as valuable.

Once again the most important aspect of the visit was simply walking through the IUS engineering bays and chatting with the employees. Workers wanted to see real astronauts, shake our hands, get an autograph, have their picture taken with us, or simply ask a question about what it was like launching on the shuttle or what it was like in space. It was a memorable visit and I'll never forget the warmth and hospitality the Boeing IUS Team shared with us that day.

It was late in the afternoon when we began heading back to Houston, arriving home about 11:00 p.m. I visited with Simone for a few minutes, then headed to bed. Training resumed at eight the next morning. It had been a quick trip but an incredibly valuable one that allowed us to see the real flight hardware and to meet with people who built it and checked it out. The trip was amazingly well-coordinated and efficient. Nancy, Kevin, and I had children at home. None of us liked to spend more nights away from home than absolutely necessary. Tom did a great job setting up the trip and squeezing in some flying fun along the way.

Right after getting back from the West Coast, Mary Ellen and I had our first EVA training session in the Weightless Environment Training Facility (WETF) swimming pool, at the bottom of which was a mock-up of the space shuttle airlock and the payload bay, where crews practiced performing various EVA tasks. Some of these were contingency tasks, including what to do in case the shuttle payload bay doors couldn't be properly closed, and others were tasks for planned EVAs. Since STS-70 had no planned EVAs, Mary Ellen and I would train only for the contingency tasks which included a few associated with problems that might occur with the TDRS/IUS deployment mechanism.

Nancy was our designated EV3, the crew member who assists EV1 and EV2 into their suits and helps them get out of them after a spacewalk. She was at all our training sessions to practice what she needed to do on orbit.

After that training, Nancy, Mary Ellen, and I spent a day at the University of Alabama in Birmingham visiting the labs of the investigators whose protein crystal growth experiments would be on our flight. We were briefed on their research and the objectives of their experiments. It was a useful trip for learning more about the experiments. Then it was on to the Kennedy Space Center where the entire crew looked over some potential contingency EVA tasks associated with TDRS/IUS. We also spent some time doing tool fit checks to make sure the tools were indeed the ones we would need in a contingency. Then we headed over to the Air Force side of Cape Canaveral to meet with the people working on the IUS. While there, we got to see the IUS that would be flying with us on STS-70.

It had been quite a week. Leaving home Monday, we flew to Los Angeles then to Seattle. Tuesday night, it was back to Houston, where the next day we participated in EVA WETF training. On Thursday, I flew to Birmingham and then to KSC, where I was joined by the rest of the crew. Friday afternoon, we all flew back to Houston to be home for the weekend. We covered three of the four corners of the continental United States that week and traveled some six thousand miles getting nearly thirteen hours of flying time in the T-38. Such was the life of an astronaut training for a space shuttle mission. Now I had the upcoming weekend to recover.

A Wedding and Family Time

Having a baby born only four months before our scheduled launch made life a bit hectic around the Thomas household. But I wasn't the only crew member experiencing a life change. Nancy was engaged to a fellow Army officer and aviator, Dave Currie, who she met a year earlier at Fort Rucker, Alabama. Dave had been Nancy's instructor pilot for a course on the UH-60 Blackhawk helicopter. Nancy and Dave decided to get married before our flight.

So on March 18, 1995, a Saturday, the STS-70 crew took some time off from our busy training schedule to join Nancy and Dave for their wedding. They wanted their reception at the South Shore Harbor Resort on Clear Lake, only a few miles from the Johnson Space Center, to be family-friendly. So Nancy gave Simone and me a tuxedo onesie, complete with black tails. Even though Kai was only a month old, Nancy provided the baby tux knowing we would *have* to bring him along with us that night. Without a doubt, he was the best dressed newborn in attendance. It was the first time that I got to meet Dave and we all welcomed him into the STS-70 extended family. I'm not sure he knew what he was getting himself into as the spouse of an astronaut, but he quickly got up to speed as our June launch date rapidly approached.

But the wedding was not our only social event. Even though we were only a few months from launch and things were getting pretty busy for us, Tom thought it would be a good idea for our families—minus the children—to spend some time together to get to know one another a little better. So after checking with all the spouses to make sure we had a date when everybody could attend, he scheduled the get-together at the Oasis Restaurant located on the shore of Clear Lake, less than a mile from our offices. Mary Ellen's husband, Jerry, who was living in Austin, came to town so we had all the families there at one time.

It was a beautiful sunny spring day and after having lunch which consisted of some Po-Boys (the Gulf Coast version of a sub, grinder, or hoagie) we drank a few beers and played a few games of volleyball. It wasn't the most competitive game I have ever played, but one that was very polite. The idea was to have fun and relax as a group, and fortunately we were all more interested in drinking a beer than winning the games. It was a fun afternoon and we all got a bit sunburned that day. It was also a great opportunity for our spouses to be together. The better they got to know one another now, the better they could support each other on launch morning when they would all be together as a group atop the Launch Control Center at the Kennedy Space Center.

Discovery Moves to the Pad

In early May, space shuttle *Discovery* was rolled from KSC's Orbiter Processing Facility to the Vehicle Assembly Building (VAB) so it could be mated with its external tank and a pair of solid rocket boosters. The actual operation began a few weeks earlier with the stacking of the four segments of the left solid rocket booster (SRB) onto a mobile launch platform, an extremely robust eight million-pound piece of equipment designed to support the weight of the shuttle (three million pounds) and withstand the exhaust plumes and high temperatures of the two solid rocket boosters and the shuttle's three main engines when they ignited at launch.

Each SRB was 149 feet long and 12 feet in diameter and weighed 1.3 million pounds loaded with fuel. Each SRB consisted of four individual segments that got stacked on top of one another. Once the left SRB was stacked, the process was repeated to stack the right SRB. SRB stacking typically took about three weeks.

Then the external tank was lifted vertically inside the cavernous VAB and attached to the two SRBs. With the SRBs and ET in place, *Discovery*—122 feet long and also in a vertical configuration—was hoisted sixteen stories using an overhead crane capable of lifting 250 tons. After clearing a sixteenth-floor walkway, it was lowered into position for mating. After some tests to confirm that all attachments and connections were made properly, the entire shuttle stack—*Discovery*, the external tank, and the two solid rocket boosters—was ready for its final three and a half-mile journey to Launch Pad 39B aboard a massive crawler transporter. The trip began at 2:30 a.m. May 11 as the transporter literally inched its way out of the VAB at a top speed of nearly one mile per hour.

In what might have been an omen, *Discovery* successfully made the trek, but not without some difficulty. *Florida Today* newspaper writer Todd Halvorson compared the giant shuttle crawler transporter to the classic children's story character in *The Little Engine That Could*. He wrote:

> Straining under the weight of its 11 million-pound load, the 30-year-old crawler had to try, try, and try again before it was able to haul its multi-billion-dollar cargo up a steep slope to the top of launch complex 39B, an elevated concrete pad 55 feet above sea level. "They got part way up the hill twice and had to go back down off the slope" said KSC spokesperson Bruce Buckingham. "They weren't sure they had enough 'oomph' to get up the hill." (1995)

Evidently, as the crawler began its slow climb, engineers driving it noticed that a gauge indicated they weren't going fast enough to make it up the hill. They backed down, tried again, and got the same result. They decided to replace the speedometer and, on the third attempt, the crawler's twin 2,750 horse power engines had no trouble getting *Discovery* atop the launch pad. The trouble was the faulty speed gauge, not the transporter.

A second problem later that day forced NASA to evacuate dozens of workers from the pad. A three-quarter-inch line used to pressurize liquid oxygen and liquid hydrogen tanks on *Discovery* ruptured and began spewing nitrogen gas. Workers in protective spacesuit-like garments were sent to the pad to identify the source of the leak and to cap the ruptured line. Nobody was hurt, and soon preparations resumed to get *Discovery* ready to launch.

Meanwhile, over at Pad 39A, just a few miles away, space shuttle *Atlantis* was being readied for its STS-71 mission in late June. It was the first time in over three and a half years that two shuttles stood on adjacent pads simultaneously being readied for launch. And waiting in the wings was space shuttle *Endeavour* for its STS-69 mission in July. NASA had a busy launch schedule with three launches planned within a six-week timeframe. It was the shuttle program's 'golden era.'

Suddenly, it seemed, STS-70 was at the front of the line. We were less than a month from launch. And we had a bit of notoriety going for us—we were slated to be the United States' one hundredth manned spaceflight. To mark the occasion, NASA developed a special logo that showed the Earth, Moon, a big "100," "Freedom 7—STS-70," and "1961–1995." NASA even added the phrase "America's 100th Human Spaceflight Mission" in bold letters at the bottom of the STS-70 press kit that was released in early June.

It was fun for me to be associated with the one hundredth manned spaceflight for the US. I had been in kindergarten when *Freedom 7* launched in 1961, and we—NASA and me—had come a long way since then.

Preflight Press Conference

The STS-70 crew held the traditional preflight press conference at JSC on May 16. NASA typically schedules them for one of the last weeks before launch. If they are done too far from launch, nobody cares about the mission. Holding them too close to launch gets difficult because the crew schedule gets extremely busy. A good compromise was to hold them three to four weeks before a launch, which for us was still holding at June 8.

With the high visibility STS-71-*Mir* docking mission right after ours, there wasn't going to be tremendous interest in our flight. After all, it was only another TDRS satellite deploy mission. If the public got bored after the second Moon landing mission in 1969, think about the public interest in the seventh TDRS deploy mission. Yawn!

The press conference, held in the press briefing room at JSC, was a business attire event, so Tom, Kevin, and I all wore suits and ties. Nancy and Mary Ellen also dressed in business attire. We sat up on a small raised stage at a long row of tables, with our STS-70 mission patch and the NASA meatball logo on either side of the large logo designed for the nation's one hundredth human spaceflight mission. There were only a handful of reporters present in the briefing room itself, but numerous others were being tied in by telephone from the other NASA locations—Kennedy Space Center in Florida; Lewis Research Center in Cleveland, Ohio; Goddard Space Flight Center in Greenbelt, Maryland; Marshall Space Flight Center in Huntsville, Alabama; Ames Research Center in Sunnyvale, California; and NASA Headquarters in Washington, D.C.

Tom made some initial comments about our mission and introduced the crew and then we opened it up for questions. There was some talk about NASA shortening our flight from eight days back to five to help support the schedules for some of the missions that would follow ours, so Tom was asked about the potential impact on our mission. There were a few questions about TDRS and some of the experiments we would be doing. But most of the questions came from reporters from Ohio who wanted us to talk about 'The All-Ohio Space Shuttle Flight.'

Tom, asked why there were so many astronauts from Ohio, said that role models such as Ohio astronauts John Glenn and Neil Armstrong probably had a strong influence. He also credited good schools and teachers "like the ones we had growing up in Woodville, in Northwestern Ohio." And finally he gave credit to "the Midwest work ethic that allows people to excel in whatever they do."

Kevin, the honorary Ohioan, was asked about his connections to the Buckeye State. "I wasn't born in Ohio, but I went to school in Colorado. So I had to drive through Ohio back and forth. I guess that's my connection to Ohio, but it is an honor to be a member of the Ohio family."

Tom explained that Kevin's crew mates had been giving him a crash course in Ohio lore. "We explained to Kevin what a buckeye is, though we have yet to produce one—and I don't think one is manifested for the flight yet."

Some Ohio residents heard about our plight and sent us buckeyes—lots and lots of buckeyes—some of which we carried with us on the flight.

We were asked to comment on NASA's position that it was just coincidence—pure luck—that 80 percent of the crew was from Ohio. I explained that while an 80 percent Ohio crew was quite impressive, it was no record. "The first All-Ohio crew was John Glenn in 1962, and we're just proud to follow in his footsteps." He, of course, flew alone aboard *Friendship 7* in 1962, to make his flight the 100 percent All-Ohio crew!

With all the talk about Ohio, many of the reporters got into the swing of things. Before asking a question, the reporters usually stated their names and press affiliations. Soon they were including their connections to Ohio as well, such as their hometowns or schools they might have attended in Ohio.

Todd Halverson, of *Florida Today*, who let everyone know that he "was born in St. Louis, but considered Cincinnati as my hometown," wanted to know how we were planning to perform "Script Ohio" while in space and who specifically was going to take along the tuba and dot the 'i.'

"Script Ohio" is the signature formation of the Ohio State University marching band performed at all home football games. The tradition is to have a different sousaphone player dot the 'i' of Ohio.

Occasionally, celebrities were given the honor. As the only Ohio State University alumna on the crew, Nancy jumped at answering the question. She told Todd that one of her goals in life was to dot the 'i,' "preferably during a Michigan game." She added that "maybe I can do it while we're on orbit."

When asked what the flight meant to her, Mary Ellen's enthusiasm and excitement came through as she responded, "I've been waiting a long time for this and working a long time for this. To have a chance to fly aboard the shuttle is just tremendous."

Nancy talked about the value of inspiring students when she explained, "Whatever inspiration we can be to kids is worth it. It's just extremely important to show a kid that no matter what your background, no matter where you're from, you can achieve your dream."

Overall, the press conference was uneventful and we even had a bit of fun doing it.

Emergency Training and a Practice Countdown

The next day, the entire crew flew to the Kennedy Space Center to participate in some launch-related training and a practice countdown, known as the Terminal Countdown Demonstration Test, or more colloquially TCDT. Those had been conducted since the early days of the space program and allowed the crew and the Launch Control Center and the Mission Control teams to practice launch-day procedures. We went through the countdown from crew ingress right up to the moment of liftoff. The test was some of the best practice for new astronauts because everything was done just as it would be on the real launch day.

First on the agenda during our two-day stay at KSC was a stop at the high bay area of the Operations and Checkout Building to see our TDRS and the IUS Upper Stage which were now mated together in a vertical configuration. It made me appreciate the size of the satellite we would be deploying from the payload bay. We wore white clean-room 'bunny suits' and special gloves to minimize possible contamination of the satellite. We adhered to strict protocols regarding what we could touch and what we couldn't as we examined the satellite. Most of our TDRS training took place with a mock-up of the satellite, which didn't have all the detail of the real thing. We studied the TDRS/IUS for an hour or so, looking closely at some of the connectors and cables Mary Ellen and I might have to deal with during a spacewalk in the event of a malfunction during deploy.

We also spent time at Launch Pad 39B where *Discovery* was awaiting our flight. We reviewed the safety equipment around the pad and did a walkthrough of some of the emergency evacuation procedures. In the event of a problem at the pad on launch day, we needed to know where to go and what to do. For example, if there was a hydrogen fire due to a leak, a water deluge system might be activated and we would have to evacuate in a heavy downpour of water. We were trained to utilize a buddy system. Two or more astronauts would evacuate the area together with the following astronaut placing their right hand on the right shoulder of the lead astronaut so they could feel each other. Then, from the hatch of the space shuttle, they would walk to the far side of the fixed service structure at the 195-foot level where several escape baskets attached to slide wires were located. The path from the hatch to the baskets was marked clearly with a wide yellow stripe with black chevrons, or arrows, pointing the way to the baskets. "Follow the yellow brick road," we were instructed.

Once there, Nancy, Mary Ellen, and I got in one basket; Tom and Kevin in another. Once loaded, we would slap a paddle to release the basket so it could slide down the wire reaching top speeds of nearly fifty-five miles per hour. It only took a few seconds to get the few hundred yards from the launch pad to a protective bunker. Once the basket came to a stop, the astronauts would jump out and run to the nearby shelter where additional emergency equipment was located, including a special telephone to contact the rescue convoy and inform them of our condition.

It looked like it would be quite a thrilling ride, but we never got to experience it. NASA deemed the ride down as an unnecessary risk during our training. So instead, we got in the baskets, hit the paddle to release it, and the basket slid only about two feet before it stopped. We climbed out, took the elevator down, and drove to the bottom of the slide wire, climbed in one of the baskets there, and continued our training exercise. Then we jumped out of the basket and ran to the bunker. We spent a few minutes in the bunker reviewing the emergency equipment and additional evacuation procedures.

Then it was on to our next method of escape from the launch pad area, the M-113 armored personnel carrier. NASA was always quick to point out that the tracked vehicle was an armored personnel carrier, not a tank. It was used to protect us if we needed to leave the bunker and escape farther from the launch pad.

The M-113 originally was used during combat operations in Vietnam to break through heavy thickets and jungle vegetation, which made it ideal for escaping the thick brush and swampy land that surrounds Launch Pad 39B. The Army nicknamed it the 'Green Dragon,' but astronauts simply called it 'The Tank' even though we were well aware—NASA told us so—that it wasn't a tank. All I know is that it was the closest I would ever come to driving a tank.

During launch, the M-113 would be parked next to the bunker. We were given instruction on how to start the vehicle—press the start button—and shown how to drive it, steer it, and brake. It was big, noisy, and solidly built, and most astronauts looked forward to taking it out for a spin. So each of us was given the chance to drive it to get the feel for the throttle and steering. The M-113's top speed on land was about forty miles per hour, but dropped to four miles per hour in the water.

We also held a brief press conference and had a few crew photos taken with Pad 39B in the background. It was fun and exciting training that led up to the big event the next day—our practice countdown test.

We went back to crew quarters where most everyone either hit the gym, went running, or checked emails before getting ready for that night's barbeque dinner at the beach house prepared by the

Astronaut Crew Quarters staff. KSC Director Jay Honeycutt and other managers, including our launch director, Jim Harrington, attended along with members of the Vehicle Integration and Test Team (VITT), the people who helped get the shuttles ready to fly.

We got up early the next morning and followed a schedule exactly like it would be on launch day. We had about twenty minutes to shower, get dressed, and report for breakfast in the crew dining area. Once we were seated, the staff began serving us what we had requested for breakfast on launch morning. The traditional breakfast from the early days of the space program was steak and eggs. I never felt like launching with such a full stomach and preferred something a bit lighter and more in line with what I would normally eat at home. I wanted to treat it like a normal meal and not as if it were going to be my last meal. So I ate a bowl of low-fat granola cereal with skim milk and had a big glass of freshly squeezed orange juice. That was it. And as much as I wanted to drink all the juice, I refrained from doing so. I figured the less I drank at breakfast, the less likely I would need to use my diaper when we were out at the launch pad.

After breakfast, Tom, Kevin, and Nancy were given a quick weather briefing and a status update on *Discovery*. We returned to our rooms where they placed a plastic bag containing a diaper, a pair of dark blue long underwear, a pair of heavy gray wool socks, and another pair of long underwear that was part of our launch and entry suit's cooling system. Woven throughout the long underwear were a series of thin tubes through which cool water was circulated once we were strapped inside *Discovery*. After one last trip to the bathroom, I got dressed and reported to the suit-up room down the hall where Bill Todd, a good friend and an experienced suit technician, was waiting for me.

The technicians had reported for work hours earlier and had prechecked everything and tested all the equipment. I sat down in my designated comfortable leather upholstered overstuffed chair, and Bill started helping me don my suit. First the feet, then the arms, and finally the head got forced through a small rubber neck dam and the hard work was done. After connecting the water lines for the liquid cooling long underwear, Bill zipped up the suit from back to front. I sat back down and he loaded my pockets with various personal items, such as sunglasses and survival equipment we might need in the event we had to bailout during a launch emergency. He also added a few airsickness bags in case I experienced any space motion sickness once we got to orbit. Better to have them and not need them than to need them and not have them.

Then he helped me put on my communications cap, better known as the Snoopy cap, connected my communications cable inside my suit, and installed my helmet and gloves. After connecting the oxygen and water-cooling lines to my suit, I closed and locked my visor and turned on my oxygen supply. All sealed up in my suit, Bill slowly inflated it to make sure it was sealed properly and verified that there were no leaks. With the suit fully inflated, it ballooned outward and I became the equivalent of the Pillsbury Dough Boy or the Michelin Man. It became quite rigid, making it difficult to bend my arms and legs.

Once the suit checked out, the pressure was reduced, and I could open my visor. Then I would take off my helmet and gloves and just sit back and relax while they finished suiting up the rest of the crew. The whole procedure took only ten to fifteen minutes if everything was working properly, which was normally the case.

Then it was time to head to the launch pad. Tom led us down the hall to the elevator. On the way, we passed the door that led to our crew quarters. Some of the staff members were standing there to say goodbye and wish us well, just as they would on the real launch day. The elevator was waiting for us and we walked right on. After a quick trip down to the first floor, the doors opened and we walked out—Tom leading the way. Outside the Operations and Checkout Building, a few workers watched us walk to the Astrovan. On launch morning, there would be about a hundred people there, including members of the news media, to record our steps. But on that day there were only a handful. Nevertheless, we practiced our waves and smiled for the imaginary cameras.

Once aboard the Astrovan, I connected the cooling unit to my suit. It got pretty hot and uncomfortable inside the suits, especially in the summer heat of Florida.

The trip to the pad was about ten miles and took about twenty minutes. When we arrived, we pulled right up to the base of space shuttle *Discovery* and got out. We loaded our helmets, some other equipment, and ourselves on the elevator and headed to the launch tower's 195-foot level where an access arm stretched from the fixed service structure to the White Room, a small area about ten feet long and eight feet wide, which acted as the final preparation area for the astronauts entering the shuttle. We entered the White Room one at a time where our closeout crew, affectionately known as Pad Rats, helped us put on our parachute harness and removed the protective rubber booties from our boots. One by one we then climbed inside *Discovery* on our hands and knees and got into our seats.

Fellow astronaut Steve Smith was our astronaut support person (ASP). His job was to help strap us into our seats; help put on our helmets; and connect the communications, oxygen, and cooling lines. He was assisted by veteran NASA suit technician Jean Alexander on the flight deck and Lockheed suit technician Max Kandler, who strapped Mary Ellen into her seat on the middeck. First Tom, then Kevin, then me, and finally Nancy were strapped into our seats on the flight deck. Once we were all settled in, everyone else left and *Discovery*'s hatch was closed and locked. After a few communications checks, we pressurized the cabin to verify the hatch was sealed properly and that there were no leaks. We proceeded through our checklists exactly as we would on the actual launch day.

Both the Launch Control Center at KSC and Mission Control in Houston participated in the practice countdown. Besides being good training and familiarization for the astronauts, it also provided controllers with an excellent launch training opportunity. We counted down to six seconds when the three main engines would normally begin to fire and then terminated the exercise.

With the TCDT exercise finished, we continued our training by practicing an emergency egress. We unstrapped and disconnected from the hoses and communications lines, left our parachutes behind, and made our way to the hatch. Practicing our buddy system, Nancy, Mary Ellen, and I made our way to one of the escape baskets and climbed aboard. Tom and Kevin followed and did the same in the next basket. We hit the paddle and dropped two feet. With that, we completed the training.

We removed our helmets and gloves and took an elevator up to the access arm near the top of the external tank (ET). There the five of us had our picture taken standing in front of the top of the ET and the top of one of our SRBs. We spent a few minutes enjoying the view and gazing at *Discovery*. It was a unique opportunity to check out the vehicle up close. On the real launch day, we'd have no time for anything like that. Then we headed back to the Operations and Checkout Building, got out of our sweaty suits, and took much-needed showers.

After changing into our blue flight suits, we met in the crew conference room to discuss any issues and problems that arose, but there weren't any. We had a quick lunch in the dining room across the hall, grabbed our bags, departed for the Shuttle Landing Facility, checked the weather, filed our flight plans, and soon were airborne headed home to Houston.

Everything had gone smoothly. We were about three weeks from launch and, while we had some additional training to complete in Houston, we were pretty much set. TDRS was nestled in the payload bay, and *Discovery* was ready to fly. Everything was looking good for our launch on June 8. It was an exciting time.

I was flying high as I drove home. When I got there, Simone and Kai were waiting for me. He was crying his eyes out. It was a quick transition from spaceman back to the real world. I paced around the house gently rocking Kai, trying to get him asleep. A short time later, he did. And so did I.

Our Spouses Get to Launch

It's a fun NASA tradition—Spouse Day—and Simone; Becky, Tom's wife; Jeanne, Kevin's wife; David, Nancy's husband; and Jerry, Mary Ellen's husband, were all looking forward to it.

A month or so before launch, the spouses were invited to the Johnson Space Center and given the opportunity to experience shuttle launches and landings in the motion-based simulator, used solely for those purposes.

The motion-based simulator, as its name implies, moved. Through a complex series of hydraulic actuators, it could be pitched up to a ninety-degree angle so the astronauts inside were on their backs just as they would be during a real launch. And to increase the realism a bit further, the entire simulator shook during the liftoff to prepare the astronauts for the rocking and rolling they would experience. Graphics projected on the simulator's windows showed the fixed service structure at KSC. And with the loud rumbling noise of the engines piped into the simulator, it was a fairly realistic launch experience making it an excellent training aid.

To simulate landing, it could be tilted from left to right similar to banking an airplane. The nose could be pitched up or down, also like in a plane. The motions were all tied to the control stick so the crew could get used to the motions they would experience in the shuttle.

Next to the motion-based simulator was the fixed-base simulator. As its name implies, that one did not move and was used to simulate being on orbit. It consisted of a high-fidelity shuttle flight deck with overhead windows and windows overlooking the payload bay together with all the switches and instruments of the aft flight deck. There also was an interdeck access hatch through which astronauts could climb down a ladder to a much lower fidelity model of the shuttle's middeck. There were middeck lockers where equipment was stored, a galley for preparing meals, and a training unit for the waste containment system (WCS), the shuttle's toilet.

On Spouse Day, each wife or husband got to take the place of one of the astronauts to experience what it would be like inside the cockpit during launch. Besides being one of the coolest rides in the world, the sessions were great confidence builders, making the spouses more familiar with where their husband or wife would be seated and what they would be experiencing during launch.

After helping each spouse strap into their seat with full seat belts and shoulder harnesses just like the actual shuttle, the motion-based simulator was slowly tilted upright into a launch orientation, with everyone now on their backs. Normally the instructors started the simulation at about two minutes before launch. Six seconds before liftoff, the simulator began shaking to mimic the firing of the shuttle's three main engines. Upon the ignition of the solid rocket boosters, the simulator really bounced up and down to mark liftoff. The spouses, wearing headsets, listened in to the simulated conversations between the astronauts and mission control. Two minutes into the launch, the simulator slowly pitched downward slightly to imitate the tapering off of thrust before the SRBs were jettisoned. A little more than six minutes later, the three main engines shut down and simulated orbit was achieved.

Depending on how much time was available, spouses sometimes were strapped into the commander's seat and allowed to practice landing the shuttle. That was Simone's favorite activity. With Tom and Kevin talking her through it, she skillfully landed *Discovery* safely on the Kennedy Space Center runway on numerous occasions. She had a huge smile on her face as she climbed out of the simulator at the end of her session.

Spouse Day was a good time for the crew because it allowed us to share the excitement of the mission with our spouses. And it was not only fun, but was also a bit of a stress reliever for everyone.

John Glenn Visits Johnson Space Center

Shortly before we were scheduled to go into quarantine, we got word that Senator John Glenn would be visiting JSC to meet us. We learned he was planning to spend an entire day with us, following us to our training sessions. Seldom in life do you have an opportunity to meet one of your childhood heroes. It is even rarer when your hero comes to meet you.

As was typical for such events, the JSC public affairs office coordinated the visit, and scheduled time for Glenn to meet the JSC director, George Abbey. Since his visit was being scheduled for the Tuesday following the Memorial Day weekend, they also scheduled Senator Glenn to speak at the weekly astronaut meeting, normally held every Monday rescheduled that week to Tuesday because of the holiday. He was going to talk about his Mercury flight and answer questions.

I was a little disappointed that my scheduled training would prevent me from hearing him speak that morning. We had our 'launch minus ten days' physicals and some life science experiments preflight data collections from 7:30 to 11:00 a.m. that day. I finished early but was only able to catch the last few minutes of his talk.

Afterward, the STS-70 crew got to meet Glenn. A NASA photographer took a picture of the first and second 'All-Ohio' crews together. After the photo session, Glenn was hustled off to meet the upper management at JSC. Meanwhile, we met with our public affairs specialists to discuss and review our planned TV broadcasts and interview schedules. Despite the fact that ours was considered a routine not that exciting of a mission in their eyes—Tom jokingly called STS-70 NASA's stealth mission, the Ohio media inundated the space agency with interview requests.

We met up again with Senator Glenn about 1:00 p.m. in the building housing the motion-based simulator. As our commander, Tom took the lead in showing him around. We had a one-hour session listed on our training schedule as "Ascent Demo with John Glenn." We would run through a few launches and landings with him to give him a feel for what it was like.

The motion-based simulator had only five seats. We were a five-member crew plus Glenn. So Kevin, the New Yorker and honorary Ohio, graciously agreed to sit out the run. The Senator was strapped into the commander's seat, and Tom was strapped into the pilot's seat. Mary Ellen, Nancy, and I sat behind them in the mission specialist and instructor seats. I have to say, it was pretty special to be strapped inside the motion-based simulator with the iconic former astronaut.

With Glenn at the stick, we started a Return to Launch Site (RTLS) Abort run because that would allow him to experience both a launch and then a landing. With Tom talking Glenn through the landing and explaining the instruments and procedures, we watched as he brought the shuttle down for our simulated landing at KSC. Just slightly off in airspeed and slightly off centerline, we touched down with a slight bump as Glenn brought the vehicle to a stop on our simulated KSC runway.

While the training team reset the simulator, Tom taught Glenn a few more techniques and gave him some additional pointers. Within a few minutes, we were back at fifty thousand feet with Glenn at the helm heading for another KSC landing. It was clear he was having the time of his life and, on that second attempt, he greased the landing. His skills as a test pilot were evident. "If we can get rid of Kevin, I know how we can really make this the All-Ohio crew," Glenn joked.

After another run or two, we ended our session. Our training team lead, Tim Terry, presented Glenn with a printout of his landing numbers as a souvenir of his visit. As we left the simulator, veteran astronaut and Apollo 16 Moonwalker John Young was there to escort Glenn around JSC a bit more while we continued with our busy schedule that included a video-telecon with NASA Administrator Dan Goldin.

That was another of those NASA traditions during which we got to meet with the administrator to discuss our flight and reassure him we were well trained and ready to fly. He seemed particularly focused on one of our experiments, but otherwise didn't seem to know too much about our mission and didn't seem too interested in it either. No matter. Lasting less than ten minutes, the telecom was soon over and the crew had the rest of the hour free before our last training of the day, a medical refresher and mission overview meeting with Glenn. He got to watch us go through a few of the more common medical procedures that might be necessary during our flight—giving a shot or starting an IV. We finished up a little after 5:00 p.m.

NASA scheduled a dinner for Glenn that evening at Frenchie's Italian Villa Capri along the shore of Clear Lake on NASA Road 1 in El Lago. The STS-70 crew was invited to the dinner hosted by Abbey. Also among the about thirty people in attendance were other JSC upper management types and some senior astronauts.

After dinner as the event was winding down, I said a final goodbye to Senator Glenn and asked him if he would autograph a copy of an old book of his that I had had for many years called *P.S. I Listened to Your Heartbeat*. The book, published in 1964, was a collection of letters people wrote him after his flight. He gladly signed it, writing, "Best Regards, John Glenn 5-30-95."

A few weeks later, I received a note card from Glenn apologizing for not writing more in the book. He told me to keep the card in the book to make up for it. I was amazed that he remembered what he wrote in the book, and I admired him for following through with the note afterward. I regretted not having much time to really talk with him during his visit. I did ask him when he last visited JSC and he couldn't remember. "It's been years ago," he said. It was at that moment that I appreciated even more the power of the letter I had written. Glenn hadn't visited JSC in many years, but my letter brought him back.

As we talked, he told me he wasn't sure whether he would be able to attend our launch, but that he would try, depending on what was going on in the Senate that day. I thought it would be pretty cool to launch knowing that John Glenn was sitting in the crowd only three miles away cheering us on, so I was hoping his schedule would be clear on launch day.

When it was time to leave, Simone and I said goodbye; I shook his hand and thanked him for the visit. I was thrilled that he had taken the time out of his busy schedule to come to Houston to meet us, and I thoroughly enjoyed the time I got to spend with him. It was one of my most memorable moments as an astronaut—while on the ground, at least.

References

Halvorson, Todd. 1995. "Crawler glitch slows shuttle's trip to pad 39B." *Florida Today*, May 12. Pg. A-3.

Chapter Four
Discovery to Orbit

Launch Day for the STS-70 crew dawned, well, dark. That's because our scheduled wake-up time—driven by our launch time of 9:41 a.m.—was at 4:46 a.m. I set my alarm for a few minutes earlier than that and already was up when they knocked on my bedroom door to make sure I didn't oversleep—like that was even a possibility.

As with my first launch a year earlier, the butterflies in my stomach were quite active. The sensation was similar to the feeling I got before any big exam or job interview. My symptoms were the same whatever the stressful situation—stomach a bit knotted, frequent use of the bathroom (every fifteen minutes or so), and gagging while brushing my teeth.

In the back of my head was an awareness of the possibility that this could be my last day of life. Images of the *Challenger* accident were still in the back of my mind, as they probably were for every family member. I knew there was a small probability that something like that could happen on any shuttle mission. Maybe on my mission.

For years I thought long and hard about the risks. I fully accepted them. I was ready and anxious to get our flight underway.

The strict NASA schedule allotted us thirty minutes to shower and get dressed for breakfast. So after enjoying my last hot shower for what I hoped would be the next eight days or so, I shaved, brushed my teeth accompanied by the aforementioned gag reflex, and got dressed in the standard astronaut launch-day attire—a pair of casual khaki pants and our gray STS-70 crew polo shirt. I checked my crew notebook as well as some items I would carry with me aboard *Discovery* and then headed down the hall past the suit-up room and the Astronaut Crew Quarters' staff offices to the dining room. It was 5:16 a.m.—and still dark outside.

My crew mates and I took our seats at the table where the traditional launch day cake, bearing the STS-70 mission patch, was on display. We wouldn't eat any of it that morning. Instead, it would be frozen so we could enjoy it after we landed. We posed for a NASA still photographer as well as some live video that was broadcast on NASA TV. While a minor inconvenience, I thought it was good for our families to see us sitting down for breakfast before launch. Then the cameras left and we had twenty-seven minutes to enjoy our breakfast.

Using my STS-65 experience as a guide, I ate only a bit of the cereal—the kind I normally eat at home—along with some fresh strawberries. I had a few small sips of fresh-squeezed orange juice. That was it. Those butterflies weren't helping my appetite much. Besides, I didn't want to eat too much in the unlikely event that I would see the food again on my way to, or after, reaching space.

I also learned from STS-65 that it was pretty difficult to go to the bathroom in the launch pad's 195-foot level toilet before climbing into the shuttle, so I tried to restrict my fluid intake as much as possible. Less in, less out, I figured.

There was some joking around as we ate—another way to vent nervous energy. Soon it was time to move on. Precisely at 5:46 a.m., Tom, Kevin, and Nancy received a final weather briefing.

We found ourselves with the normal weather forecast for Florida in mid-July—a chance of thunderstorms sometime during the day. We had a two hour, forty minute launch window and hoped the weather would not impact our ability to get off the ground. Despite the crew's strong desire to launch, some of our nine children—Kevin had four, Tom three, and Nancy and I one each—were hoping for delays so they could enjoy more time in Florida and possibly get in a trip to Disney World. Kai, being only five months old, didn't seem to care much one way or the other.

I went back to my room to begin the suiting-up process. I brushed my teeth once more (still nervous—gagging a bit), went to the bathroom yet again (par for the launch-day course), and then started dressing: First putting on my diaper (yes, under our launch and entry suits we wore adult diapers. They were evil, but necessary because it could be seven or eight hours before we got out of our suits and activated the shuttle's toilet).

Next I donned a pair of dark blue long underwear and some heavy gray wool socks. I went to the bathroom one last time (I hoped), grabbed the things I would be carrying with me to the pad, and walked about thirty to forty feet down the hall to the suit-up room.

As I walked in, our suit technicians were busy getting our equipment ready. They greeted me with a big, "Good morning. Are you ready to launch?"

I smiled and assured them that I was more than ready. A few minutes later, Mary Ellen entered, followed by Tom, Kevin, and Nancy a few minutes after that. Then just like for launch on STS-65 and the Terminal Countdown Demonstration Test (TCDT) practice countdown we had just completed a few weeks earlier, we went through the suiting up process, doing a few pressure and leak checks before finishing and having a few minutes to sit back and relax. I loaded my pockets with my personal items and a few photos that I was carrying along with my crew notebook and the standard emergency rescue items.

Former astronaut Steve Hawley, deputy director of the Flight Crew Operations Directorate at JSC, stopped in as we suited up as did a few other NASA folks. Visitors were kept to a minimum so as to not interfere.

We were right on schedule when we got word at 6:26 a.m. that it was time to go. I swallowed hard and, with a surge of adrenaline, stood up and shook hands with my suit technician, Bill Todd. "Thanks for all your help, Bill. See you at landing," I told him.

With a pat on my back, he wished me a safe flight. Then I, and the rest of the crew, headed for the elevator. We were on our way to the pad. The crew quarters' staff was in the hall and clapped and waved as we walked by. The elevator was waiting for us, and we all loaded on. Seconds later, we were walking out of the Operations and Checkout Building and headed to the Astrovan.

The news media was there along with some NASA personnel. We smiled and waved as they applauded, cheered, and called out our names. Tom, who flashed a big smile as he hoisted aloft a buckeye held in his right hand, led the way. Mary Ellen was next, followed by Kevin and Nancy. I brought up the rear. Normal astronaut protocol has the commander and pilot walking out first, followed by the mission specialists. But Kevin didn't mind letting Mary Ellen take his place during the walkout, although I knew there was no way he would allow her to strap into the pilot's seat aboard *Discovery*.

One by one we loaded into the Astrovan. I climbed on board last, took a seat near the front with a great view out the windshield, and plugged my suit into one of the cooling units. The launch and entry suits were hot, and even after a short walk you could start overheating. It was always a good idea to stay ahead of the heating and cool down whenever you could. Among those joining us in the Astrovan was Lockheed suit technician Max Kandler, who was wearing a red and yellow Woody Woodpecker baseball hat, which brought a smile to my face.

As the Astrovan left the area, our helicopter escort, which had been hovering overhead, took the lead as we made our way to the launch pad. All traffic was stopped for us as we headed for our first stop, the Launch Control Center (LCC). People got out of their vehicles and waved as we drove by. KSC employees took that stuff pretty seriously, even though NASA had been launching nearly eight shuttles a year. Launches were a special event. After all, it was KSC's primary business.

Hawley earlier gave me my official 'launch pass,' which granted me permission to leave Planet Earth that morning. They were given only to veteran crew members, never to the rookies. When we got to the LCC, Hawley stood up, wished us well, and asked for our passes. Tom, Nancy, and I pulled ours out, while Kevin and Mary Ellen wore puzzled looks on their faces. We all laughed as they realized it was just a joke. Hawley wished us a safe flight one more time and then headed to the LCC. The Astrovan's door closed and we continued on our way.

At the Pad

We were only three miles from our next stop—Launch Complex 39, Pad B, where space shuttle *Discovery* was waiting for us. The drive only took a few minutes. And the sight out the windshield as we approached was magnificent—*Discovery* poised majestically on the pad with its protective service structure rolled back and white liquid oxygen vapors coming out the top of the external tank and venting from the main engines. It was absolutely breathtaking—made all the more significant because that shuttle was going to take me and my crew mates to space.

And it hopefully would be the last time I saw her vertical on the pad fully fueled and ready to go. If all went as planned, the next time I would see the outside of *Discovery* would be on the runway at the Shuttle Landing Facility in the horizontal position after landing.

The Astrovan pulled up to the base of *Discovery* at 6:56 a.m., and we all got out. They weren't quite ready for us to take the elevator up to the 195-foot level, so we had a few minutes to walk around the base of the launch pad standing directly below the mobile launch platform.

There was a very noticeable difference between launch morning and all the other times I had been to the pad while shuttles were being readied for launch. There were a lot fewer people around. Usually the pad was crawling with workers, busily getting the vehicle ready—much like ants working around the main entrance of an anthill. There was a flurry of activity on the days leading up to a launch. But on launch day itself, it was like a ghost town. Except for a few suit technicians and the orbiter closeout crew, nobody was within three miles of the fully fueled vehicle.

I had the feeling that *Discovery* was alive. The dominant sound was the whirring noise of the pumps of the ground support equipment. While not overly loud, it was very noticeable, very unique, and very distinct. Ba-ba-bum. Ba-ba-bum. Ba-ba-bum. It was a uniquely rhythmic sound—one that I will forever associate with being at the pad on launch morning.

I took a few pictures of *Discovery* and the rest of the crew with a small camera I was carrying before we climbed aboard the elevator for a trip that ultimately would lead us to *Discovery*'s hatch.

The elevator ride took less than a minute. It seemed old and made a bit of a disconcerting noise as it rose. We often joked that we were about to head to space aboard one of most complex machines in the world, and that the elevator ride quite possibly posed our greatest hazard to getting there. Tom got off first, followed by Nancy, me, Mary Ellen, and finally Kevin. Jean Alexander, our lead suit technician, was waiting for us and greeted each of us with a big welcoming hug.

I was scheduled to be the fourth person to board since I would be sitting on the flight deck. First Tom climbed in to get to his flight deck commander's seat, while Mary Ellen was strapped into her middeck seat. Then Kevin, me, and finally Nancy all on the flight deck.

So while standing on the 195-foot level waiting for them to strap me in, I had a few minutes to walk around a bit and look west toward the VAB as well as the adjacent LCC three and a half miles away, where Simone and Kai would watch the launch. I also looked over toward the Banana Creek viewing site, a little to the north of the VAB and LCC, where the rest of my family and friends were soon to gather.

I turned back toward *Discovery*, gazing at her momentarily, and then looked beyond to the east and watched a beautiful sunrise over the Atlantic Ocean.

I continued my walk on the platform's mesh floor, which made it possible to see the lower levels. I took a few deep breaths, slowly exhaling in an attempt to relieve a little of the tension that was building up. To the left of the crew compartment access arm, the so-called gangplank, was an excellent spot to see the shuttle. Looking up, I could see the top of the external tank. Looking straight down, I had a great view of the left solid rocket booster and the external tank and all the way down to the bottom of the flame pit. I thought about spitting over the railing and watching it plunge into the pit. I didn't, but smiled at the thought of it.

I stood there appreciating the sights, sounds, and fresh air, wondering if I would have the opportunity to enjoy that—or anything else—another day. I fully appreciated the risk I was about to take. There was no kidding myself about that.

I walked back over to the west side near the escape baskets and again looked toward the LCC and then over at the Banana Creek site. A million faces, memories, and thoughts flew through my head. I was naturally a little scared, but mainly very excited. I felt somewhat alone standing at the railing and trying to mentally connect with my family and friends. I got great comfort knowing they were there to support me. Of course, I was hopeful that I would be seeing them all again, especially Simone and Kai. I didn't much like the notion of Kai—like me—growing up without a father in his life. I took a few more deep breaths of the fresh ocean air.

A short time later, Nancy took a picture of me with *Discovery* in the background. It became a tradition for me to do that before each launch and those pictures became very special to me and serve as a reminder of the final moments before boarding the space shuttle.

Finally I heard someone yell out from the White Room, the last stop for the astronauts before climbing aboard, "Hey, Don. We're ready for you."

It was my turn to enter *Discovery*. After one final huge gulp of fresh air, I told Nancy, "See you inside," and walked across the access arm into the White Room. There, closeout crew members Chris Meinert and Jim Davis greeted me, helped me into my parachute harness, and made a final check of my equipment. When all looked good and after giving my camera to the closeout crew, I got on my hands and knees and crawled into *Discovery*.

The countdown clock read T-2 hours, and Steve Smith, our astronaut support person, and Jean Alexander helped those of us on the flight deck get into our seats, strapped us in with our helmets on, and got everything ready for hatch closure. On the middeck, Max Kandler was doing the same for Mary Ellen.

As they finished getting me settled, my heart was racing. I was seated directly behind Kevin and looking out the forward windows. I could see blue sky and a few puffy white clouds. To the left of Tom's left shoulder, I could see part of the launch tower. I pulled out a mirror from my kneeboard and took my first look out the overhead windows. Looking back into the flame pit, I was humbled how close I was sitting to the shuttle's main engines.

There were no windows in the middeck, where I experienced my first launch. It seemed as though I was sitting in one of the simulators in Houston. But the view on the flight deck—looking at the blue sky, the flame pit, and the Atlantic Ocean in the distance—made it clear where I was. This was no simulator. It was the real thing.

I put away the mirror, organized my flight data file books, and reviewed our emergency egress procedures. I went over the location of my emergency oxygen activation device, a bright green ball that I needed to pull in an emergency. I also felt around to find the locations of all my suit connections—the water cooling and the communications lines and the oxygen and the parachute harness connectors. In the event of an emergency, there wouldn't be time to find them, so I spent a few minutes making sure I could readily locate them just in case I needed to. I sure hoped I wouldn't, but it pays to be prepared. Soon I was oriented, comfortable with my environment, and ready to go.

About the same time, the Final Inspection Team, also known as the Ice Team, was crawling around the launch pad making detailed inspections of Discovery. Theirs was the final look to verify that Discovery was ready for launch. Typically the team's main task was to look for ice buildup on the external tank. Filled with a half-million gallons of ultra-cold liquid oxygen and liquid nitrogen (-300°F and -423°F, respectively), it was easy for ice to form due to the high humidity conditions of a typical summer day in Florida. The extremely cold temperatures caused moisture to condense, freeze into ice, and become dangerous projectiles if they broke off during launch, possibly damaging the shuttle's fragile thermal tile protection system.

Besides visually inspecting the shuttle and the launch pad, the team carried portable thermal imagers to search for unusual temperatures that could cause problems. For example, unusually low temperatures could indicate problems with the foam insulation on the external tank.

The team also took one last look at 174 of the woodpecker damage site repairs, but no ice was seen in those areas and no debonding of the repaired foam was noted. So far, so good—the repairs appeared to be holding. After reporting to the Launch Control Center that everything outside Discovery was looking good, the team left Pad 39B for safe cover three and a half miles away. We were still go for launch.

Our families arrived at the LCC and were escorted to the office of Jim Harrington, the STS-70 launch director, where NASA provided food and beverages for them. A parade of NASA managers stopped by to meet them. Our family escorts were fellow astronauts Dan Bursch and Carl Walz, another Ohioan, and from Cleveland no less. They assisted our families any way they could. They were a tremendous help during our mission.

While waiting for the launch, there was a LCC tradition for the crew's children to create artwork of their choice relating to the mission. NASA supplied a white board and some markers and let the kids be creative. By diverting their attention a bit from the impending launch, it served as a stress reliever for both the children and our spouses. After the mission, NASA framed the artwork and hung it in the LCC building.

Included in the STS-70 artwork was a great image of Woody Woodpecker drawn by Kevin's daughter, Frances; and a few versions of our official crew patch drawn by Nancy's daughter, Stephanie; Tom's children, Heather, Katie, and Tommy, and Kevin's other children, Carolyn, Nick, and Kate. At five months old, Kai, wearing a Cleveland Indians outfit that morning, added a few touches of his own. In addition, the messages "GO STS-70" and "Go *Discovery!*" were a part of the artwork. NASA tried to keep the families as busy as they could to pass the time while we worked through the countdown.

The closeout crew checked *Discovery*'s flight deck and middeck one last time to make sure everything was in order and closed the hatch at 8:11 a.m. We then conducted a leak test to be certain the cabin was sealed properly. We were done with most of the major activities before launch with about an hour and thirty minutes to go. We spent the time joking around a bit, which was pretty normal. It was another way to burn up some nervous energy and try to settle the butterflies.

We were on our backs watching a few buzzards soaring just above the shuttle. Tom noticed a few others perched nearby, apparently watching us. A short time later, Tom gave us a buzzard update. "There's two of them. One of them's sitting right on top of the lightning rod," at the very top of the launch pad's service structure just to the left of the shuttle.

"This is going to be classic," Kevin commented, anticipating the surprised look on the buzzards' faces once the engines ignited.

"Not for long," Nancy shot back.

"There's one orbiting [above] and there's one that's sitting on top of the structure just watching us," Tom said, with yet another update.

Kevin said they probably were thinking, *Hey, lunch in a can!*

That notion reminded me of an old *Looney Tunes* cartoon with two buzzards sitting in a tree wearing napkins around their necks and holding knives and forks in anticipation of a meal as they looked down on their prey. But this wasn't a cartoon. We had two real buzzards looking down at us from above, practically smacking their beaks. Was it some kind of organized woodpecker revenge?

The countdown proceeded smoothly, and at T-9 minutes we went into a preplanned twenty minute hold. During that time, the launch team checked all the systems to make sure space shuttle *Discovery* was ready to launch. It was also a time to review the weather.

There wasn't much else for the crew to do, so we just tried to relax. Tom and Kevin periodically cycled through a few of the computer displays to look at various temperatures, pressures, and other items to make sure everything was in order. At that point, we were all starting to get a little uncomfortable from being on our backs on top of our lumpy parachutes, oxygen bottles, and survival gear. We all just wanted to get going. Tom reported that we had two buzzards roosting above us staring right at him through *Discovery*'s front windows.

Right on schedule, we came out of the hold and the countdown clock once again ticked away—8 minutes, 59 seconds and counting...8:58...8:57...8:56...Hopefully, it would march uneventfully to zero. We came alive in the cockpit, monitoring systems, checking the launch procedure books, and keeping a keen eye on that clock. The last minutes before launch flew by. Tick. Tick. Tick.

At T-7 minutes, 30 seconds, the orbiter (crew) access arm began to retract. The White Room that butts up against the shuttle hatch was at the end of the arm. It was kept in place up until this point in case there was an emergency that necessitated our quick exit from the shuttle. But now it was moved out of the way for launch. If an emergency developed, it could be returned within seconds to allow us to escape. Tom mentioned that the cameras mounted inside the White Room probably were still operating, so Kevin gave a final wave as it swept out of *Discovery*'s way.

All was quiet for a moment inside *Discovery*. Then Tom said, "It's time to get out of town."

We all silently concurred. We were more than ready to go. The tension continued to build. Underneath my gloves, I could feel Simone's wedding band, which I was wearing on my little finger. I gave it a little rub and thought about her three and a half miles away.

The countdown clock passed five minutes. Everything was looking good and things remained quiet inside *Discovery*. Nancy broke the silence at one point by saying: "Okay, this one's a counter," meaning it wasn't just another simulator exercise.

About two and a half minutes before launch, the external tank's sixty-five-foot-long gaseous oxygen vent arm with its thirteen-foot-in-diameter vent hood, affectionately known as the beanie cap, began to retract. Raising the vent hood took about twenty-five seconds. Retraction of the entire arm that swung toward the fixed service structure was completed by T-1 minute, 45 seconds.

Two minutes to go. I took a few deep breaths to relax. I had no idea what my heartbeat was, but I can only imagine that it must have been pounding pretty good. One after another the seconds fell away. Sometimes when I am in boring meetings, I stare at the clock, and it seems like the second hand just stops. It's as if each tick of the second hand lasted five to ten seconds. Not so for the final minutes before launch. For me the time seemed to go faster and faster. It was a little like falling through the air, accelerating every second and with no way to slow yourself down. You knew at that point you were just along for the ride and there was nothing you could do to stop it. We were going to launch.

Our next call was from Ray Probst, our orbiter test conductor (OTC): "Flight crew, OTC. Close and lock your visors and initiate O_2 flow and on behalf of everyone in the control room have a great mission."

Tom replied, "We're looking forward to it. Thanks for your help."

We all closed our helmet visors and turned on oxygen flow to our suits. We were each sealed in our suits now. If there was a cabin leak during launch and all our air escaped, we would be protected by our suits.

The sun was streaming in through Kevin's front and side windows. He commented on how bad the glare was. Tom told him not to worry because the sun angle was about to change.

One minute until launch. Kevin produced a big yawn and said, "Don, I think I'm starting to get excited."

I chuckled. As for me, the butterflies took one more lap around my stomach.

The seconds ticked by until about forty seconds from launch. At that point we were told there was some sort of a problem with communications data, and they called for a hold in the countdown; there was some sort of multipath error. I initially thought the problem was with TDRS in our payload bay. My ears strained to hear any hint of what the problem was. The launch director notified everyone: "We'll hold at T-31 seconds."

Oh, rats! I thought. *Thirty-one seconds from launch. What's the problem?* The countdown clock froze at T-31 seconds.

But it turned out that the problem had nothing to do with our satellite. The hold was initiated by the range safety officer, who saw some fluctuations in the signal for the external tank range safety receiver. If the space shuttle began veering off course during launch and possibly threatening populated areas, the range safety officer could send a signal to blow up *Discovery* and her crew. It's one of those things that makes you a little nervous knowing about. Each of the SRBs contained explosive charges, as did the external tank. It was one of those detonator receivers on the ET that was sending erratic readings back to the range safety officer. That wasn't something you wanted to mess with. Inside *Discovery*, we had no clue what the problem was at the time. We only knew it was communications related.

There was nothing we could do but wait. My first thought was that we would be scrubbing the launch, and I felt a momentary letdown. We held for fifty-five seconds until some contingency procedures were worked out.

Suddenly we heard from the launch director. "All personnel, the countdown clock will begin momentarily," followed almost immediately by, "The clock will resume on my mark—two, one, mark." They didn't waste a second getting things going; apparently they wanted us out of town as much as we wanted to go.

The countdown clock started ticking down from thirty-one seconds. *Discovery*'s five onboard computers were now controlling. We were go for launch—thirty seconds...twenty-nine seconds...twenty-eight seconds...

3...2...1...Liftoff!

Two things happened simultaneously at T-6.6 seconds. The launch pad's water sound suppression system activated and the shuttle's three main engines ignited.

Water flowed into the flame trench below the space shuttle to protect both the pad and the vehicle, its crew, and payload from the intense heat and acoustic energy generated by the main engines and the solid rocket boosters.

The water absorbed much of the engine noise, thus preventing the shuttle from shaking itself apart in the first seconds of flight. The acoustic energy peaked nine seconds after liftoff when the shuttle was about three hundred feet high. At that point, the water flow rate was an amazing nine hundred thousand gallons per minute.

Also at the T-6.6 second mark, *Discovery*'s five main computers sent the engines start commands, and immediately, one by one in rapid succession, they ignited. The engines reached 90 percent of their maximum thrust within 3.6 seconds. The top of the external tank deflected 31½ inches as the main engines achieved full power. For the next 3 seconds, they continued to burn while the space shuttle stack swayed back to perfect vertical. That movement was called the twang. That's when the solid rocket boosters ignited and, simultaneously, the eight large bolts holding us to the launch platform were severed by explosive charges, breaking us free of our final hold to Earth.

Sitting in my seat on *Discovery*'s flight deck, I used the mirror strapped to my right knee to look out the overhead windows. I was able to see directly down into the flame trench about two hundred feet below. First I saw the water start to flow; then a bright white steam created when the main engines ignited and their heat collided with the water. Billows of beautiful white steam poured out of the trench. When the SRBs ignited, there was a new, incredibly impressive wave of intense orange flames and brownish smoke that shot from the trench in waves of unbelievable speed and turbulence. It was an awesome sight. It was nothing like the gentle billows of smoke I grew up seeing at the steel plants in Cleveland where tall smokestacks steadily and gracefully spewed their puffy discharges skyward. The detail I was able to see from only 183 feet away was unlike any of the launch films I had seen. It was truly a jaw-dropping experience.

NASA Launch Commentator George Diller counted down the final seconds: "10...9...8...7...6...main engine ignition...5...4...3...2...1...and liftoff of space shuttle *Discovery* to complete NASA's constellation of tracking stations in the sky!"

We were on our way and, boy, were we ever! As we vaulted from the pad, the space shuttle's three main engines consumed 1,000 gallons of fuel each second—750 gallons of liquid hydrogen; 250 gallons of liquid oxygen. Our two solid rocket boosters were burning 20,000 pounds of solid fuel a second. Those five engines gave us a total liftoff thrust of nearly 7.5 million pounds. Once the SRBs lit, there was no turning back. Theirs was an uncontrolled burn. No matter what, we were going somewhere that day.

Liftoff officially occurred at 9:41:55:020 EDT. NASA could calculate the exact launch time to the nearest $1/1000$ of a second, so they did.

As we left the pad, a large piece of foam near the bottom of the right-side SRB broke off. That chunk was eighteen inches long by ten inches wide by ten inches thick and came off the booster's flared skirt area. It harmlessly fell away into the flame pit and probably was shattered to bits by the violent SRB plume as we lifted off. Thankfully it had no affect on our mission.

I could feel the push in the middle of my back from the acceleration that was immediate at liftoff. It felt like someone was pushing me into the air. And the vibrations at liftoff were tremendous. On my first flight, I was particularly impressed with the vibrations when the main engines started. That immediately was eclipsed by the vibrations that occurred at SRB ignition. They were intense. It was like driving down a very bumpy road at high speed in the back of an old pickup truck, rocking and rolling from side to side and bouncing up and down all at the same time. That was why it was so important to be strapped in tight.

I continued watching out the overhead window using my mirror as the shuttle shot skyward. I got no sense of speed looking straight ahead out the forward windows at the blue sky, but looking back at the launch pad gave me a great sense of speed. We were really out of there, and the view out the overhead windows provided an incredible perspective. As we cleared the launch tower's lightning mast only four seconds after liftoff, we were traveling one hundred and twenty miles per hour and accelerating like a bat out of hell.

Shortly after we cleared the tower, we began our roll program to get on the right trajectory for orbit. That put us in a 'heads down' orientation, although we had no sensation of being heads down. "Houston, roll program," Tom reported to Mission Control.

"Roger, roll *Discovery*," was their reply. Those initial calls, the first ones after liftoff, were used as a communications check between the shuttle and Mission Control.

As the shuttle rolled, I got a great view out the overhead windows of the beach at Cape Canaveral and the Atlantic Ocean. I could see waves rolling in and again was overwhelmed by the sense of speed I got as the view quickly faded.

On STS-65, I was seated on the middeck and had no view. As a result, the launch experience was something I felt and heard and sensed but didn't experience visually. Sitting on the flight deck for STS-70 looking out the windows at the flames and smoke, I don't remember hearing or feeling much of anything during launch. It was a totally visual experience, which overrode all other senses.

The two launches were totally different experiences, yet equally thrilling. I was amazed how the same event could be experienced so differently depending on whether one was sitting on the flight deck or the middeck. Down in the 'basement' you got a much keener feeling for the launch because you were deprived of any visual stimulation.

When I put the mirror down and looked straight ahead through the forward windows I finally got a good sense of our speed as we rapidly accelerated toward a thin layer of clouds and, in the briefest moment, punched a hole in them and continued our upward climb. As we did, I used my mirror to look at them again as we climbed skyward. I got another excellent sense of our speed watching the clouds rapidly recede.

A few seconds later we approached, passed through, and quickly flew away from a second thin layer of clouds. Again I watched through the forward windows as we approached and saw them fall away as I looked out the overhead windows. We were really moving.

I never experienced that sense of speed before. Looking straight up at the blue sky with no clouds or other frame of reference made it impossible to determine how fast we were going. Only by watching the ground seemingly fall away or watching clouds approach and then recede did I get any sense of speed. We were probably still going less than Mach 1, but from where I was sitting we were going freakin' fast.

A moment later, Tom let us know we had just passed Mach 1, the sound barrier.

Our families and friends on the ground watching the launch experienced some moments of concern about forty-five seconds into the flight. What appeared to be a white cloud suddenly enveloped *Discovery*. It lasted only a few seconds before disappearing. While momentarily alarming, it turned out to be a simple condensation cloud that formed because of supersonic shock waves and the passage of the shuttle through layers of high humidity. Initially the cloud formed as a small ring around the vehicle, which is commonly referred to as a condensation collar. Within a few seconds, it rapidly expanded into an impressive cloud surrounding the entire shuttle.

And seconds later, it was gone. While both alarming and spectacular, the phenomenon was not a totally unexpected occurrence when launching in July in the hot, humid conditions typical in Florida. But the crew aboard *Discovery* saw none of that.

I had always wondered precisely when the sky transitions from blue to black, so I tried to take notice of the color every few seconds in an attempt to pinpoint that moment. The sky remained blue for quite some time, gradually getting darker and darker blue. It was only as we came up on SRB separation that I first noticed that the sky was turning black. That occurred twenty to twenty-five miles above Earth. As we were approaching the time for SRB separation—two minutes into the flight, I was prepared for the reduction of thrust just before burnout and separation. That caught me by surprise on STS-65. Not that I was feeling any accelerations this time—as I said, STS-70 was mainly a visual experience. When the SRB thrust tailed off, I looked straight ahead out the front windows. I didn't want to blink and miss the flash that occurred at SRB separation. Finally, the thrust tapered off, and I heard Nancy report "Pc less than fifty," indicating that the SRB engine chamber pressures were less than fifty pounds per square inch which allowed the SRBs to separate.

So with eyes wide like saucers, I froze for a few seconds and waited for the flash. Finally it happened. An instantaneous flash for the briefest fraction of a second illuminated the flight deck—so bright that it reminded me of a quick glance at the sun or the flash of a camera.

Simultaneous with the flash was a loud and strong "ka-thunk" from the explosive bolts firing to separate us from the SRBs. I so anticipated the flash that I almost forgot about the noise the explosive bolts made, but there it was. My senses were getting overloaded. Seeing and hearing at the same time was just too much in that environment.

The separation motors on the SRBs, which would push them off to the side, were directed toward the shuttle and always had the habit of 'dirtying' the windows on the sides of the orbiter. After SRB separation, Kevin noted, "The windows are dirty just like advertised."

We then got a call from Curt Brown in Mission Control that we were waiting for. "*Discovery*, Houston. Performance nominal." That meant that the solid rocket boosters had done their job and provided us with the anticipated amount of thrust. Everything was looking good.

With the SRBs gone and the sky ever blackening, we continued uphill. We were now being pushed into the sky with just the shuttle's three main engines, and the ride was totally smooth, just like it was on STS-65. Electric Glide all the way.

"Smooth ride," Kevin said as we passed Mach 6. The acceleration slowly began to build and, as it did, I began to feel like I was slipping off my seat to the right. So with my right arm, I attempted to push myself back into position, which was no easy task with the shuttle accelerating up to a maximum of 3 gs, three times the force of gravity on Earth.

When Steve Smith helped strap me into my seat, I must have been slightly off to the right side instead of being perfectly centered. Being off center on my seat caused a bit of uneasiness. I had the feeling I was going to be pulled out of my seat and get wedged in between it and the flight deck's aft starboard wall. During the rest of our time headed to orbit, I pushed hard with my right arm against the panels to my right to keep myself from sliding even farther.

I looked out the overhead window a few more times during ascent but never saw too much after SRB separation. I couldn't see the ground because the shuttle was not climbing straight up but was arching over more and more to gain orbital speed as we climbed. The view was more of a very dark blue behind us, and I could just catch a glimpse of a few wisps of flames from the three roaring main engines that were pushing us ever closer to orbit. At higher altitudes under the near-vacuum conditions, the engines' flames expanded, and that was what I saw. Definitely not any bright flames. Just faint wisps visible out the overhead windows.

As we passed the fifty-mile mark, Kevin called it out, "Fifty nautical miles. M. E. [Mary Ellen], it's a counter."

To which Mary Ellen enthusiastically responded, "All right!" NASA designates that a person officially becomes an astronaut once they cross the fifty mile mark. For Kevin and Mary Ellen, the two rookies on the flight, it was a special, if brief, moment for celebration. They were now officially astronauts—real honest-to-goodness flown-in-space astronauts.

Discovery was operating magnificently and we received many of the standard calls from Mission Control, such as "*Discovery*, negative return." That meant we could no longer return to KSC in the event of any problems. We would head across the Atlantic from that point on.

"Here come the gs," Kevin noted as we passed Mach 15. The last three minutes of powered flight were pretty uncomfortable for me. I knew to expect the increase in gs from the normal 1 g on Earth to the maximum of 3 gs as we accelerated until the computers throttled the engines back to maintain a maximum of 3 gs. Breathing became a little more strained and it felt like someone was sitting on my chest pinning me down. But what made the last three minutes even more uncomfortable was being off center on my seat. It was a little unnerving as I fought hard with my right arm to push myself back to the left to keep myself in my seat.

Kevin noted when we were passing Mach 20. Tom followed up with, "One minute to go." Everything was still looking good, and we were almost in orbit.

"Half minute to go. Mach 23," Kevin said. "Ten seconds [to go]."

I was more than a bit relieved when, with a sudden jolt, the main engines shut down right on schedule. A few seconds later, Tom and Kevin simultaneously made the main engine cut off call, "MECO."

Nancy followed with, "MECO confirmed," indicating the computers recognized that all the engines had shut down.

Once again it was not a smooth and gradual transition from 3 gs to weightlessness. It happened in a fraction of a second. One second you are being pushed back in your seat, and a microsecond later you are weightless. It is one of the most dramatic transitions in spaceflight, maybe second only to the vibrations and acceleration surge that occurred at liftoff.

"Welcome to Space"

Everything was 'nominal' as we say in the space business, meaning there were no problems during launch, all systems had worked properly, and everything was fine. Nominal sounds a bit bland, but in the shuttle world nominal meant excellent. That was as good as it got as far as launches were concerned.

"[Mach] 25.8. Welcome to space," Kevin said.

With the shutdown of the three main engines, I waited the few seconds for the external tank to separate. "Ka-thunk" went the charges exploding the bolts that held the tank to *Discovery*. The bang was unmistakable and indicated a clean separation, after which Kevin noticed lots of ice crystals floating outside his windows. Tom explained that the ice was coming from *Discovery*'s maneuvering thrusters as they fired for the first time.

Kevin was excited to see things floating inside the shuttle, too. "Pretty cool," he said.

A few moments later, Nancy said, "It's great to be back."

Tom concurred, and Kevin remarked, "In the words of Dudley Moore... 'This doesn't suck.'"

It was great to be back, and Kevin was right—it didn't suck being in space again.

While Tom, Kevin, and Nancy looked over our orbital parameters and did a quick systems check to make sure all was well with *Discovery*, I began taking off my gloves and helmet and removing the mirror on my knee along with my kneeboard. I disconnected my oxygen line, communications line, and the parachute from my parachute harness. I tucked my gloves into my mesh helmet bag, added my helmet, kneeboard and mirror, and floated out of my seat.

Down in the middeck, Mary Ellen was doing the same—unstrapping from her seat and getting her first taste of weightlessness.

The first few seconds floating around on the cramped flight deck with my launch and entry suit and boots on was a little rough. It had been a year since I had been in space, and while the initial moment of floating in weightlessness felt a little uncoordinated, I quickly adapted as I made my way over to the lower aft corner of the flight deck to retrieve a camcorder for myself and a 35 mm camera with a huge 400 mm lens for Mary Ellen. We were to document the external tank as we separated from it. I floated to the starboard side overhead window with the camcorder, while Mary Ellen took her position at the neighboring overhead window.

I saw the external tank off in the distance. At that point, we already were about two-thirds of a mile from the huge tank that appeared to be about the size of my little finger when held out at arm's length. I went through the camcorder's quick checklist but had difficulty getting an image through it. Either I somehow screwed it up or the settings were changed accidentally when it was loaded on board *Discovery*. But something wasn't right. I worked methodically through the cue card to verify all the settings and tried again. Finally, I was getting an image, but by that time *Discovery* had pulled far enough away from the tank that it appeared quite small in the viewfinder. I videotaped a bit, but it became pointless as the tank moved farther from *Discovery*. I watched as the tank faded away.

Soon we were coming up on the west coast of Africa, and I could see a dust storm. Tom pointed out the Cape Verde Islands as we passed overhead. Both were spectacular views.

As we passed over Banjul, the capital of Gambia in western Africa, Tom noted the runways at Banjul International Airport. That would have been one of our emergency landing sites if we had had engine problems or another emergency during our launch. "Not a good day to go to Banjul," Kevin noted, as clouds covered much of the city.

With the ET Photo DTO (external tank photograph detailed test objective) finished up, Mary Ellen went back down to the middeck while I finished securing my items that I had quickly doffed after MECO. I took my helmet bag with me as I floated down to the middeck. There I found Mary Ellen using one of the airsickness bags. About a third of the astronauts experienced some form of space sickness during their missions, and it just happened to hit Mary Ellen a bit sooner than normal. After poking my head up to the flight deck to apprise Tom of the situation, I floated back down to help her as best I could. There isn't a lot that you can do in those situations besides helping anchor them in a stable position and provide them with a towel and fresh bag as needed.

We were taught to take it easy the first day in space—not do any flips and try to maintain a normal heads up, feet down orientation. That was to keep from getting disoriented and sick and, equally as important, to help prevent others from getting sick. For someone right on the edge, the sight of someone else floating upside down or doing somersaults could be enough to induce vomiting.

One thing that usually made people feel better was to get out of the spacesuit, so I helped Mary Ellen get out of her suit and cooling garment. There was almost instant relief because the suits became quite hot after disconnecting the cooling lines. Astronauts usually were soaked with sweat by the time they wrestled themselves out of the suit, and the cool cabin air felt great and brought quick relief.

Then, with her help, I started getting out of mine. I stuffed our suits into large mesh bags and temporarily stowed them out of the way. I activated the waste containment system (WCS), better known as the toilet. Shortly after reaching orbit, there is a pretty dramatic fluid shift in the body from your lower extremities to your head. Your legs get thin ('chicken legs' was a common term for them), and the face frequently gets quite puffy. The eyes can swell up a bit, and some crew members I have seen looked almost unrecognizable as if they had been in a fight or something. The brain senses all this excess fluid in the head and attempts to compensate by ridding the body of it. As a result, astronauts have a strong urge to urinate quite a bit after they get to space. It usually takes an hour or so after launch for the urge to kick in, but soon everyone feels it. That's one reason why we activate the WCS as quickly as possible.

Another was because that was where we stowed our first day's set of clothes. They were rolled tightly, placed in plastic bags, and stowed behind a panel inside the WCS in what was call the wet trash volume. So after opening the door, connecting a few hoses, deploying a mirror and some other components as well as the vitally important procedure cue card (step-by-step instructions for going to the bathroom in space), the WCS was open and ready for business.

The first woodpecker reference wasted no time in presenting itself. During our first orbit, launch Capcom Curt Brown radioed: "Tom and Kevin, if you have a chance, we are ready to copy your debris report if you have one." Mission Control was inquiring whether we noted any unusual debris during the external tank separation or if we noted any unusual or significant damage to the tank itself.

In a serious and business-like manner, Tom asked all of us if there was anything to report. Mary Ellen and I saw nothing unusual when we photographed the ET.

Kevin added: "I saw a couple of white and red feathers," suggesting we might have hit a woodpecker or two during launch.

Tom, in a strictly serious manner, asked Kevin, "Anything else?"

Kevin replied with a simple and rather quiet, "No."

Tom paused for a second and then, in a straightforward and serious manner, told Brown, "No debris, especially no feathers. Just the normal hazing [on the windows] from SRB separation."

We hadn't actually mentioned the word woodpecker, but it would be the first of many references to them during our flight. It would remain one of the dominant themes of our mission, and Tom, with an assist from Kevin, was the first to make reference to the pesky bird.

Time for TDRS

On a typical space shuttle mission, the crew would spend the first five or six hours of their mission getting out of their spacesuits, activating the toilet and galley, setting up the exercise equipment, unpacking things they need such as cameras and flight data file procedure books, and stowing items they wouldn't need for a while, like launch and entry suits and seats. The first day in space usually was spent getting organized.

But because we were scheduled to deploy TDRS so early in the mission—a mere six hours after launch, we did not have the luxury

Discovery to Orbit

of spending that first day setting up housekeeping. After being given a "Go for orbit ops" from Mission Control, we started setting things up as fast as we could before Mary Ellen and I were scheduled to start the TDRS/IUS checkout only ninety minutes after launch. The idea was to make sure TDRS/IUS was working properly prior to deployment. If something didn't look right or if problems were detected, we could always bring the payload home with us. But once we hit the deploy switch and the satellite started moving out of *Discovery*'s payload bay, there was no turning back. The checkout prior to deployment was critical.

While Tom, Kevin, and Nancy were on the flight deck getting *Discovery*'s systems set up, Mary Ellen and I were on the middeck getting the toilet and galley operational, setting up handheld microphones for communicating with Mission Control, and temporarily stowing our suits and other equipment. Soon it was time to begin the TDRS/IUS checkout, so Mary Ellen and I broke off from configuring things for the flight and jumped into the procedures powering up and checking out the satellite and its booster rocket. I say Mary Ellen and I because that is the way we planned and trained. But because of her bout of space sickness, she wasn't available very much during checkout. Unfortunately, we couldn't just stop the mission and wait until she was feeling better. So right on time per the timeline, I started powering up and checking out TDRS/IUS. I had trained for that moment so many times in the simulator, but now I was doing the real thing. It was exciting and fun. I took my time so I didn't miss anything or make any mistakes.

Periodically, Mary Ellen would float up to the flight deck to see what and how I was doing. She'd pick up a procedure book and ask what step I was on. I'd tell her where I was and minutes later she would head back to the middeck. She definitely wasn't feeling well.

As I worked through the procedures, Nancy was on the flight deck setting up cameras to capture the deployment and document our activities on *Discovery*. She didn't hesitate to offer her help. She did a great job pitching in as my backup, helping to make sure everything went smoothly. Tom and Kevin were getting things set up on the middeck and taking care of orbiter attitudes and other shuttle systems. But Tom was also watching over our shoulders to make sure everything was going well. It was. We accomplished the procedures right on time. Everything was working nominally that day. There were no error messages or malfunctions. After being trained and prepared for the worst, it was a pleasure to have everything working properly—well, except for under-the-weather Mary Ellen.

One of the last steps before we released the satellite was to increase the tilt of the cradle, elevating it from twenty-nine degrees during the checkout phase to the full sixty degree position for deploy. That angle guaranteed that the satellite would move out of the payload with plenty of clearance to avoid hitting *Discovery*'s crew compartment.

The deploy was a complicated but extremely well-coordinated event that involved three control centers to manage the operation. Mission Control at the Johnson Space Center was responsible for control of the shuttle and was the lead center for the deploy. The White Sands ground station in New Mexico was responsible for control of the TDRS, while control of the IUS was managed at Onizuka Air Force Base in Sunnyvale, California.

After months of training, countless pages of procedures, and hours of checkout, the final step in the deploy sequence was to move a switch to the deploy position. It was childishly simple. I pushed the switch right on time, and we watched TDRS/IUS inch its way out of the cradle ever so slowly and float from *Discovery*'s payload bay. There was no loud "ka-boom" as the explosive charges fired, allowing springs to push the satellite out into space. It just silently started moving and ever-so-smoothly inched its way from the back of the payload bay upward and forward, flying right above the aft flight deck's two large overhead windows. All five of our faces were pressed against the windows straining to get a good view. The satellite was huge, and it appeared to me as though it would barely miss hitting the crew compartment. In reality, there was plenty of clearance.

The Tracking and Data Relay Satellite-G was successfully deployed from *Discovery*'s payload bay exactly on time at 2:55 p.m. CDT—six hours and thirteen minutes after our launch.

"Houston, *Discovery*. It's a deploy," I radioed to Mission Control.

"Excellent job. Happy faces here," replied astronaut Tom Jones, our Capcom and my good friend.

All during the deploy, Kevin and Nancy were taking pictures and recording the event. I could hear the cameras whirring and the shutters clicking. Tom kept an eye on everything and was ready to do a rapid separation maneuver in case it was needed. But everything was looking good. I don't remember anyone cheering or applauding. We were focused on what we were doing and making sure no problems developed. But I do remember a series of quiet gasps, such as "Wow! Look at that!" and "That thing is huge!" It was an impressive sight.

All the time and effort we had put into training the past year paid off. The satellite was on its way, getting smaller and smaller as we moved apart.

About fifteen minutes later, Tom fired the shuttle's engines to raise our orbit and to move us away from the satellite and its soon-to-be firing Inertial Upper Stage booster. We never saw TDRS again. All that work and we only got to see it fly for a brief ten minutes. About an hour later, the IUS booster fired the first of two burns that put TDRS into its proper geostationary orbit over the central Pacific Ocean 22,300 miles above Earth.

The final TDRS-related task was to configure the cradle for landing. That was a relatively simple procedure. But when I tried to lower the cradle, it wouldn't move. I stopped and reported the problem to Mission Control.

After a few seconds, Jones replied, "Stand by. We're evaluating." After ten or fifteen minutes and no word, I called Jones and asked how to proceed. He told me to leave everything as it was and that ground teams would look into the issue over the next few days and try to pinpoint the problem. So for now, we were done.

I was kind of hopeful that Mission Control might ask us to do a spacewalk to manually lower the cradle. After all, Mary Ellen and I trained for that. But after a few days of analysis and discussions, we were told we should land with the IUS cradle as it was. It did not pose any problem for our return and landing and would not interfere with our closing of the payload bay doors at the end of the mission.

I was a little sad as I powered down the cradle because that meant there would be no spacewalk. But I had a strong sense of pride and feeling of satisfaction in accomplishing the mission's primary objective. Whether it was a software issue or some other problem didn't really matter. The deployment of the final TRDS from a space shuttle was 100 percent successful, and the satellite was on its way to its final destination. Mission accomplished. The entire crew was elated. It had been a great team effort.

After a relaxing dinner, I took a few minutes to write some thoughts in my journal, a tradition I started on my first flight and continued nearly every day I spent in space.

Mission Elapsed Time (MET) 0 days 8 hours 44 minutes
Just winding down from a busy first day in orbit. I feel great!! We had a successful deploy—on time with no problems.

Launch was spectacular!!! I watched the engines light looking out the overhead windows with my mirror. First I could see the spray of water, followed by the white steam (very bright white) followed by the brownish/orange-yellow smoke from the SRBs. I watched as the ground fell away. As we did the roll program I could see the beach and later the VAB. What a rush! After liftoff I watched us racing up to clouds, pass thru them and leave them behind. You could really sense the speed!

The next big event was SRB sep—What a bright flash, much brighter than I thought it would be. It was like looking into the sun for a second. It was right about this point that the sky went from blue to black. The next great event was flashes in the overhead window due to the main engines. It looked like the flashes I've seen on tape from reentry. Pulsating quick flashes. They lit up the back of Tom's and Kevin's suits and helmets. The last minute of flight was a little uncomfortable with the 3 gs—more so than I remembered and MECO was more dramatic. Boom! Zero-gravity instantly!!

Overall today was a great day. I am relieved that it went so well and the deploy is behind us and successful.

I seemed to adapt very quickly this second flight—just like everyone has said happens. I didn't eat lunch but did have an appetite for dinner—four tortillas, an oat bran bar, and some crackers. I could have eaten more but why push it.

The New Mission Control

Near the end of our rather busy first day in space, Capcom Story Musgrave let us know that for the first time since the Gemini 4 flight in June 1965, a manned spaceflight was being controlled from a different facility in the Mission Control Center in Houston.

"That old Mission Control Center is an icon. It's been a symbol for billions of people of humanity's pursuit in space," Story said.

Looking to the future, Tom replied that he hoped "the next thirty years [would] carry us as far as the last thirty years [had]."

For all six Mercury missions and for the Gemini 3 flight, NASA used the Mercury Control Center at Cape Canaveral, Florida. NASA commissioned its new control rooms in 1965 on the second and third floors of a building at the newly opened Manned Spacecraft Center, later renamed the Johnson Space Center. Since that time, all subsequent Gemini missions, Apollo flights (including those to the Moon), Skylab, Apollo-Soyuz, and the previous sixty-nine space shuttle missions were controlled from one of the two identical control rooms.

The new control room was just down the hall from the old control rooms and was located in a recently built addition to the Mission Control building. NASA was nervous about using the $250 million center for the critical launch and landing phases of our mission without first doing some checkout and testing during the mission's less-critical orbit phase. If something didn't work correctly or if there was a problem in the new control room, it was much better for it to happen as we orbited the Earth as opposed to during launch or landing when fractions of a second could mean the difference between mission success and failure.

So they used one of the old mission control rooms for our launch and landing and also for the critical operations associated with the checkout and deployment of TDRS. Once the satellite was deployed and we performed our separation burn to move away from the TDRS, the Mission Control team moved into the new Flight Control Room. The transfer to the new control room was totally transparent to us on orbit because we only heard the Capcom's voice. That the voice was coming from a different room hardly mattered to us as long as we kept hearing it. But for the record, the STS-70 mission will go down in history not only as the 'All-Ohio Space Shuttle Mission' and the 'Woodpecker Flight,' but also as the first mission to be controlled from the new Mission Control at JSC.

Sleeping in Space

Each of us claimed a small piece of real estate inside *Discovery* where we would put our sleeping bags. The flight deck and middeck areas were pretty small, but we tried to separate as best we could to have a little sense of privacy and to have a small corner of the shuttle that we could call our own.

Tom chose to sleep on the flight deck in the commander's seat. Most shuttle commanders prefer sleeping there to be near the controls in the event of a problem. If an alarm went off, they would be nearest to the computer displays and the caution and warning alarm system control panel. In addition, the front cockpit windows provided a great view before falling asleep.

Kevin was a tall guy and chose to sleep on the middeck where he had a bit more room and could stretch out. He put his sleeping bag on the lower starboard side of the middeck just above the middeck floor. Each sleeping bag had six carabineer hooks that could be attached to the wall to keep them stretched out and prevent them from floating around. Sometimes astronauts slept floating free, not tethered to anything, but usually not. There were fans throughout the shuttle that circulated the air, absorbed the carbon dioxide exhaled by the astronauts, and filtered out the dust in the air. The fans created small air currents that slowly moved floating objects. For untethered sleeping astronauts, the air currents would move them until they bumped into something or someone, possibly waking up some of their crew mates. So typically we were tethered in such a way that we were able to float somewhat free, but not drift too far away from our starting position.

The specially made sleeping bags were lightweight with a zipper down the center just like regular ones. But unlike conventional sleeping bags, ours had holes we could put our arms through, almost like wearing a vest. By our heads were small pillows with headbands velcroed to the sleeping bag. We slipped on the headband and placed it across our foreheads like a sweatband. This pressed our head to the pillow to create a feeling more like what we were accustomed to on Earth. If we didn't use the headband, our heads floated free, which felt a little strange. It generally helped to use the headband for a few days until we got adapted to sleeping in zero gravity.

There were two large straps around the waist area of the sleeping bag, similar to wide belts or a tuxedo cummerbund, which we used to pull ourselves toward the wall where our sleeping bag was attached. That put some pressure on our backs to make us feel a little like we were sleeping on a mattress on Earth. Even with the softest pillow top mattress in the world, there's still pressure on your body when you sleep on Earth. Take that away in space and it feels a little strange at first. Most astronauts use the waist straps the first few days for a more comfortable Earth-like sensation. Once we adapted to weightless sleeping, we no longer needed to tighten down those straps because we found it comfortable just floating inside the sleeping bag.

Many astronauts tried to cross their arms across their chest when they slept, otherwise their arms would float out in front of them creating the possibility of inadvertently bumping into cables or switches. But being nice and snug in the sleeping bag up against the wall was a comfortable and secure feeling. That was what Kevin chose to do using a middeck wall.

Nancy placed her sleeping bag directly above him on the wall, creating a sort of bunk bed. Hers was close to the ceiling.

Mary Ellen chose to sleep on the flight deck. With ten flight deck windows, it was important to install the window shades before calling it a night. During our ninety-minute orbits, we experienced forty-five minutes of daylight followed by forty-five minutes of darkness. So if you didn't want the sun shining in your eyes every forty-five minutes while you slept, the shades were the perfect deterrent.

I chose to sleep in the middeck airlock, a cylinder 63 inches in diameter that ran from the floor to the ceiling of the middeck (83 inches). It had an internal volume of 180 cubic feet, so it was about the size of a small hallway closet at home. On the back side was a hatch that opened to the payload bay and the vacuum of space. But that was only opened during spacewalks. On the front side was another hatch that opened to the middeck. It was closed for launch and landing but typically stayed open during the flight except when a spacewalk was being performed. Each of the airlock hatches were 36 inches in diameter.

The airlock was used for stowing equipment during the mission and also contained two extravehicular mobility units, more commonly referred to as spacesuits, which were used in the event of a spacewalk. The suits were attached to the wall of the airlock and folded up to minimize the volume they took up. Attached to the airlock's ceiling and floor were large canvas stowage bags loaded with food, clothes, and other equipment we would need during the mission.

No air was fed directly into the airlock, so we hooked up a three-inch diameter hose from the middeck into the airlock, which provided fresh oxygen and kept carbon dioxide from accumulating. We called it the elephant trunk.

So with two spacesuits, one large stowage bag on the floor and another up on the ceiling, the airlock's actual useful volume was quite limited. But I liked the privacy and thought it would be a little quieter in there.

Because there wasn't a long open wall to which I could attach my sleeping bag like Kevin and Nancy on the middeck, I clipped mine to one of the stowage bags in the airlock. There was just enough room for me to stretch out. I felt like I was sleeping inside a closet nearly full with clutter. But it was cozy and I made it my home. Nancy called it 'Don's Condo.' I put up pictures of Simone and Kai and a plastic bag containing some personal items. When it was time for bed I slipped inside my sleeping bag, stuck my arms through the holes, and zipped it up.

It had been quite a day. I was tired but still had lots of adrenaline running through me from the excitement of launch and the successful deployment of the satellite. But there also were many other things that could keep you from sleeping well in space, and none of them had anything to do with weightlessness.

The toilet was extremely loud when it was operating. The fans sounded like loud vacuum cleaners, so it was nearly impossible not to wake up the rest of the crew when one of us used the toilet in the middle of the night. There were a lot of other noises inside the shuttle, from fans and pumps operating to valves opening and closing under the floor. It was not the silent world on orbit one might imagine as we coasted around the Earth. In reality it was quite noisy, but after a period of time we adapted and the noise faded into the background.

We also frequently shifted our sleep time a bit every day to make sure that lighting conditions would be optimal for our landing at the end of the mission. We would go to sleep about an hour earlier each night to shift to the landing day schedule. After a few days of that, most of us just weren't tired at our scheduled bedtime.

So we also carried some items we considered vital to help us get to sleep and allow us to stay asleep—earplugs to drown out unwanted noise, eyeshades to make sure the sun or any other light source didn't invade our eyes if the window shades weren't in place, and the hammer on the anvil, so to speak—sleeping pills.

One of the most amazing things I experienced while sleeping in space was the occasional flashes in my eyes caused by cosmic rays and other particles hitting your optic nerve and creating a sensation of bright flashes of light. I was just falling asleep in the dark one night during my first mission when a bright flash suddenly occurred as if someone set off a camera flash right in front of my eyes. It startled me and was alarming for a moment, but I quickly realized that it was a cosmic ray flash.

Sometimes they appeared as a small line across the eye, almost like seeing a shooting star quickly traveling in the night sky. Other times they seemed to hit straight on and would lead to the intense bright bursts of light. On one occasion, I saw the light streak pass across my left eye and then enter my right and go across it. Whatever that was, it passed through both eyeballs and right through my head.

They were fascinating to see. Each night as I tried to doze off, I would wait for the next flash to occur. Some nights I saw a couple of them; other nights none. It had a lot to do with altitude, orbit, solar activity, etc. They were known to be more active as we passed through the South Atlantic Anomaly where the Earth's magnetic field dips down a bit, exposing us in space to slightly higher levels of radiation. About 80 percent of crew members reported seeing light flashes, and about 20 percent reported that the flashes may have disturbed their sleep.

One of the last things I did before turning in was send Simone an email to let her know all was well.

Hi Simone and Kai!!

Boy, am I ever glad to be here as I know you probably are!! What a ride it was on the flight deck. Watching the ground rush away through the overhead window was fantastic. I've felt totally great all day and even had an appetite for dinner.

The deploy was smooth and went so quickly. It was beautiful watching it sail out of the bay. I have a big smile on my face because I know you have one also. I'm missing you and can't wait to see you both. Look up in the sky in your dreams tonight. Look for that special constellation!!
All my love,
Don on *Discovery*

With a great sense of accomplishment and sense of making history, the STS-70 crew began its first sleep period after a rather long and busy first day in space. With the occasional whining of the toilet, I faded to sleep. I felt safe and secure inside *Discovery*, and I was looking forward to the rest of the mission.

Chapter Five
Open Season on Woodpeckers

Rob Kelso, our lead flight director, declared the STS-70 mission a woodpecker humor-free zone—at least until TDRS was deployed. He told everyone to hold off on the woodpecker jokes, and rightly so because he wanted all of us focused on the primary and most critical aspect of the mission. But once the satellite floated from the shuttle's payload bay and was placed successfully in its own orbit, things got much more laid-back, not only aboard *Discovery* but also in Mission Control.

And it soon was open season on woodpecker references—something that would continue unabated until long after our flight was over. The folks in Mission Control wasted no time in leading the way. To mark the dawning of Flight Day 2, a large stuffed Woody Woodpecker showed up on the flight director's console, apparently to help keep an eye on things. Woody would circulate among the various consoles in Mission Control during the flight so everyone got a chance to enjoy his company.

Aboard *Discovery*, the day got off to a great start. Shortly after 3:00 a.m. CDT, we were awakened as astronauts usually are—with music, a tradition that dates back to the Apollo Program. The mission-specific musical selections were chosen to promote camaraderie and esprit de corps among the astronauts and ground support personnel.

To appropriately mark our first full day in space and acknowledge the travails we experienced getting there, Mission Control chose the theme from the Woody Woodpecker cartoons, which ended with Woody's staccato laugh. Our second shift Capcom, veteran astronaut Story Musgrave, followed up with a laugh of his owns as he said, "Good morning, *Discovery*."

Nancy answered for the crew, "Good morning, Houston. Great wake-up music."

To which Story replied, "Just because you're one hundred and seventy-nine miles up doesn't mean you can get away from that little guy. He's after you."

Nancy responded in kind, "We had a feeling we would hear more about that."

That was the first time I was able to enjoy the wake-up music tradition while aboard a shuttle. I played songs on many occasions for other shuttle crews when I was a Capcom but was not on the receiving end until STS-70. During my first flight on space shuttle *Columbia* for STS-65, we were a two-shift operation with half the crew working while the other half slept. Because there was no single time when the entire crew was waking up, Mission Control had a policy of not playing wake-up music on those flights. So I had been looking forward to enjoying the wake-up music on STS-70 and got a kick out of the Woody Woodpecker theme song as the first such music I experienced in space.

One of the first things I learned about wake-up music was that the speakers used on the shuttles were not exactly ideal for listening to music. There was one speaker on the flight deck and a second in the ceiling on the middeck, and both had the same fine acoustic qualities familiar to all on Earth when placing orders at fast-food take-out windows. If you ever struggled to understand what the person taking your order was saying, you have a pretty good idea of what the sound quality was like for our wake-up music.

The second woodpecker assault—the first, of course, being the actual attack on the external tank—didn't end with the wake-up music. As part of our morning mail and messages from the Mission Control team, the so-called STS-70 Execute Package, we received the following imitation of David Letterman's Top Ten List:

> OK, you knew it was coming. From the home office in Room 3308. The Top 10 STS-70 crew-requested galley items.
> #10 Fettuccini Alfredo with blackened woodpecker
> #9 Woodpecker and Swiss Lunchables
> #8 Woodpecker Boudin
> #7 Caffeine-free Woodpecker cola
> #6 International Coffee "Woodpecker Mocha"
> #5 Woodpecker on a shingle
> #4 Woodpecker Picante (medium)
> #3 Slimfast Woodpecker shake
> #2 Szechwan Woodpecker (spicy)
> #1 More astronaut ice cream

That brought some big laughs from all of us as we passed the list around.

Along with this top ten list, Mission Control sent a picture titled "STS-70 Prelaunch Photo," which showed the shuttle on the launch pad ready to go. Prominently featured in the photo was a monstrously large woodpecker, half the size of our external tank, poking holes once again. The name on the shuttle had been changed from *Discovery* to 'Spruce Goose.' A caption at the bottom of the picture claimed, "We promise this is the last woodpecker joke."

The people in Mission Control were terrible liars. We knew full well that the woodpecker jokes would continue. And truth be told, we were looking forward to the next round.

We were one relaxed crew and thoroughly enjoying our flight.

The Morning Routine

Each day after the wake-up music, we got involved in what NASA called our postsleep activities. Regular people call it getting ready to go to work. Just like at home, we used the waste containment system (okay, at home it's known as the toilet), washed up, brushed our teeth, tried to comb our hair, shaved, prepared and ate breakfast, and reviewed the morning mail and other messages Mission Control sent us while we slept. There was a lot to be done in the allotted hour, so the shuttle—particularly the middeck—was a beehive of activity with bodies floating past one another as we got ready for the day ahead. Typically, Mission Control didn't disturb us unless it was something pretty important.

One of the most often-asked questions is how astronauts go to the bathroom in space. Well, it starts at the WCS, located on the middeck right inside the shuttle's hatch. It's comparable to a small closet about four feet wide and maybe six feet high. Think of it in terms of a porta-potty you might see at sporting or entertainment events or construction sites. Another good example is the restroom facilities on commercial airplanes. The main functional difference is that on Earth we utilize gravity to collect the waste material in the toilet. In the weightless environment, everything floats, and that was a potentially huge problem that NASA solved with the WCS.

Velcroed to one of the WCS's walls was an eight-by-ten-inch cue card that instructed us on how to go to the bathroom. You might think astronauts were smarter than that and that we shouldn't need instructions on how to go to the bathroom, but we did. It sounds silly, but we had procedures for everything we did on the shuttle. About the only thing we memorized was what to do in case of a fire or if there was a cabin leak and we were losing all our oxygen to space. For everything else we had detailed procedures, even for going to the bathroom. The procedures detailed each step and each switch and dial needed to operate the toilet. There were separate procedures for urination only, defecation only, and the granddaddy of them all, urination and defecation.

Once we were procedure compliant, we closed the accordion-like privacy curtain and fastened it with Velcro. There was a second privacy curtain up near the ceiling that closed so there could be no public viewings from the flight deck. With the two curtains in place, we had a fair amount of privacy.

For urination, we attached a small clear plastic funnel about the size and shape of a Dixie Cup—the female urine funnels had elongated openings, while the male versions were perfectly round—onto the end of a hose connected to a vacuum. We urinated into the funnel and it got sucked into a waste tank.

The WCS's toilet bowl contained fans that created a downdraft that functioned much like gravity on Earth to keep the solid waste from floating up, up, and away. Our derrieres were lightly held down on the toilet seat by that slight suction. The WCS also had two foot restraints and two thigh restraint bars that could be swiveled over our thighs when seated to further anchor us and prevent any unacceptable floating around while doing our business. Once in place, we pushed a large white knob forward to activate the fans and open a sliding door that normally covered the toilet bowl when not in use. After that, well, let's just say the rest was history.

When done, tissues and wet wipes were readily available, but none of those were discarded in the toilet like we do on Earth. Instead, all used materials were put into a liner inside a small, vented container affectionately known as 'the coffee can.' After we were finished using the WCS, we pulled the liner's drawstring to close it and then deposited it into the wet-trash compartment behind a small door at the back of the WCS. By not throwing tissue, wipes, and other material into the toilet, its storage unit did not fill up as rapidly, an important consideration for missions lasting more than a week.

Proper WCS etiquette dictated that users install a new coffee-can liner and make sure the toilet was as clean when they left as it was when they arrived.

Both the WCS and the urine waste tanks were not emptied until after our return to Earth. None of that was dumped into space.

Brushing teeth was pretty much the same in space as on Earth. We each had one or two toothbrushes (always a backup in case one of them floated away) and a tube of toothpaste. Crest was one of NASA's standard-issue items, but the space agency pretty much let us select our own brand if we cared that much. The big problem with space-borne teeth brushing was that there was no sink to spit in or use to rinse our toothbrushes. Before I started brushing, I simply took a small sip of water and then brushed as I normally did. I ended up with a mouth full of toothpaste that I needed to expectorate. Our options were fairly limited because we weren't allowed to spit into the toilet. We could either swallow the toothpaste or spit it into a towel, napkin, or tissue. That second option unfortunately left us with wet towels that later would begin to smell. The same was true with a napkin or tissue if they got thrown in the trash. They soon emitted odors all their own. I always chose to swallow my toothpaste, after which I would take another drink of water to rinse my mouth. I'd then swallow that—not a practice recommended by dentists and doctors, but it was a practical solution when in space. With that accomplished, I'd fill my mouth with water and insert the toothbrush to rinse it off. Then I would put it and the tube of toothpaste back into my hygiene kit, much like a travel kit but with plenty of Velcro inside to keep the items from floating away when opened. Toothbrush, toothpaste, floss—everything had a small blue dot of Velcro on it to anchor it down.

Next I'd put in a new pair of contact lenses. Most of the astronauts used disposable contacts to minimize eye infections while in space. After carefully peeling back the aluminum foil sealing the contacts, I used a cloth towel to absorb most of the solution packaged with the contacts and then carefully scooped up the contact with my index finger. The surface tension of the solution on my fingertip prevented the contact from floating away, so inserting contacts in space was easier than on Earth.

We flew a small eight-by-ten-inch mirror that could be velcroed to any surface. That was handy for putting in contacts or when shaving. It wasn't glass or metal, but rather a piece of shiny silver Mylar pulled taut around a frame. It was lightweight and unbreakable, making it an excellent space mirror.

Next came shaving, and NASA offered us two methods to choose from. One was to wet shave using a safety razor and shaving gel (appropriately called Astro Gel). I'd squeeze some of the gel from a tube and it would begin to foam as I spread it on my face. The difficult part was cleanup. Since we had no sink with running water, it wasn't possible to rinse the razor. The only option was to wipe it off with a towel or tissue and we'd be back to square one with the smell issue. On my first flight, I wet shaved not realizing the ensuing mess. (Hey, I was a rookie. What did I know?) I learned my lesson.

So on STS-70, I went with the other method—an electric razor. I don't use one on Earth, which is why I didn't choose this option on STS-65. But the messy cleanup made it clear to me that an electric razor was the way to go. I could shave in a few minutes and there was no messy cleanup. But where did all the whiskers go? They appeared to be magically vacuumed up inside the razor. I never saw small whiskers floating around in the cabin after someone shaved with an electric razor, but many of them must have ended up doing exactly that. Sooner or later, the cabin fans that circulated our air would cause the floating whiskers to make their way to an air inlet filter where they would be trapped.

The notion that every day in space is a bad hair day was fairly accurate. My hair was always a mess when I woke up because there was no gravity to help keep it in place. I always flew a small aluminum comb and a hairbrush to bring order to my unruly hair. I could never get it perfect, so I just gave it my best shot.

In space as on Earth, it always felt great to put on clean clothes. The postsleep activity period right after wake-up was a good time to gather up a set of clean clothes for the day, and I'd change into them when it was my turn to use the WCS. Changing clothes was quite easy in weightlessness. No need to sit on the edge of the bed to put on a clean pair of socks; you could do so simply floating in the air. You could even put on a pair of pants both legs at one time! All our dirty clothes and towels went into large brown mesh drawstring bags that we filled until we couldn't force any more in. Then we pulled the drawstring tight, stowed the bag out of the way, and began filling a new one.

We had a clean shirt, clean socks, and clean underwear available for each day of the mission. By clean underwear I don't mean brand new clean. Men's underwear was reused from flight to flight, unlike the women who got to purchase their own underwear with a government credit card and then got to keep the underwear after the flight.

We were permitted to choose our own shirts, most of which came from places like Lands' End, L.L. Bean, or J. Crew. Typically we would have them embroidered with our name, mission number, or mission patch. Two of the shirts were selected to be matching crew shirts, which all of us would wear on a selected day or for a special event during the flight. One of those was our on-orbit crew press conference, which also made it an ideal day to take some official crew pictures.

One of our crew shirts was a black short-sleeved polo with the STS-70 mission patch embroidered on it. Our second matching shirt was a grey-and-blue multi-colored long sleeved polo also bearing the mission patch.

NASA provided the long or short pants we wore, and we were expected to make them last two to three days each.

Settling In

So with clean clothes, brushed teeth, a close shave, a fresh pair of contacts, and my hair combed to the best of my ability, I was ready to begin my second day in space—and a busy one it would be, conducting lots of experiments while setting up housekeeping.

Soon after we awoke I learned that our TDRS satellite had successfully achieved its orbit 22,300 miles above the Equator over the Pacific Ocean. All its solar panels and antennas deployed, and the controllers were busy with the command and checkout of the spacecraft. That was great news—a nice way for all of us aboard *Discovery* to start the day. It would take a few more weeks for the ground teams to complete the testing and checkout of the satellite, but everything was working properly so far.

We began focusing on the wide variety of experiments that were conducted on the flight deck and the middeck. Tom and Kevin were busy taking pictures of targets on the Earth using a Department of Defense developed camera called HERCULES. It was capable of imprinting the latitude and longitude of areas photographed so that researchers could easily determine what locations on Earth were photographed by the crew. While most pictures taken nowadays with cell phones have GPS-location data imprinted on them, back in 1995 that was totally new technology with which we were experimenting.

While Tom and Kevin were busy with the HERCULES camera at the overhead flight deck windows, Nancy set up another camera called WINDEX looking out the payload bay windows toward *Discovery*'s tail end. That experiment was studying the effects of space shuttle engine firings on the glowing phenomenon reported on many previous missions. With the WINDEX camera set up to record the event, Tom fired *Discovery*'s engines to lower our orbit and enhance our landing opportunities at the end of the mission. That allowed Nancy to observe and record any glowing on the shuttle's surfaces created by the firings. A better understanding of the glow phenomenon would be useful to designers of future orbiting astronomy satellites with which such glowing could interfere.

We also checked on some of *Discovery*'s other passengers that were part of a National Institute of Health-Rats experiment, also referred to as 'Rodents in Space,' or more affectionately by us crew members as 'Rats in Space.'

The rodent experiment was studying the musculoskeletal system development of unborn baby rats in microgravity. Astronauts experience quite a bit of muscle and bone loss while in weightlessness. Our rat experiment was being flown to help study those changes with the hope of helping future astronauts on long duration missions and possibly shed some light on osteoporosis and muscle deterioration in senior citizens on Earth. The experiment required a minimum of eight days in space, which influenced NASA's decision to extend the STS-70 mission from five days to eight. We were happy to share *Discovery* with those fellow passengers because it gave us the opportunity to spend a few additional days on orbit.

The dozen or so pregnant rats each carrying four or five pups were flown in a small chamber roughly as big as a medium-size suitcase called an animal enclosure module. It took the place of one middeck locker. The rats lived in darkness inside their module for most of the mission. But periodically we would pull out their drawer and install a bright fluorescent light above their chamber to expose them to light. That also allowed us an opportunity to observe them. While rats are considered by some to be relatively clean animals, I can assure you their module was not. Any feces and urine produced by the rats seemed to end up on the glass viewing window. During the mission's second day, we were still able to see through the window and observe the rats, but we could tell the window was starting to be covered with, well, let's just call it debris. It wasn't a pretty sight.

When I checked on the rats a few days later, I could no longer see through the window. A few days later, a portion of the window suddenly was clear enough to see through. Without a way to clean the inside of the window, I wondered how it suddenly got so. Upon closer examination, the clear area was where one of our floating rats urinated on the window.

I told Nancy: "That pretty much defines how dirty the windows are...that when you pee on them, they suddenly look clean again." So from the midpoint of our mission to the end, we never got a good look at the rats unless one of them obliged by urinating on the window for us.

I set up the Shuttle Amateur Radio Experiment (SAREX), a handheld amateur radio with an antenna that fit over one of the shuttle's forward windows to communicate with amateur radio operators around the world. More significantly, we used it as an important part of our school outreach. We had nine schools around the world scheduled to talk with us. During those relatively short communication opportunities—typically five to eight minutes—students would ask questions about our mission and living in space as we passed overhead. Nancy and I both had our amateur radio licenses (KC5OZX and KC5FVF), and the entire crew would be involved with that experiment during the flight.

As part of its capability, the SAREX equipment was configured to broadcast a short message to amateur radio operators. Nancy and I wrote the following for one of those transmissions:

> Greetings from the STS-70 crew aboard the Space Shuttle *DISCOVERY*!! We had a fantastic launch yesterday and a successful deploy of the Tracking and Data Relay Satellite. It was spectacular watching it move out of the payload bay. We look forward to making contact with as many people as possible during the next week. Thanks for your interest in our mission.

Meanwhile, Tom, Kevin, and Mary Ellen participated in our first scheduled press event with the *New York Times* and *American Online* (AOL). It was the first ever interview conducted from space using an electronic online news service. Our answers to their questions were transcribed and posted online in near real-time, something quite common today via instant messaging, Twitter, Facebook, and other social media.

One notable question from someone named George in Connecticut asked what the difference was between the commander and the pilot on a shuttle crew. "The biggest difference is that the commander is much, much older," Kevin answered. That brought a big smile to Tom's face.

After that brief live interview, we received other questions about our mission and living in space from the general public via the *New York Times* and *AOL* throughout the rest of the flight. We took turns answering them whenever we had a few minutes. Our posted responses were pioneering first steps in using the Internet to conduct interviews and make information about NASA and its various missions more accessible to the public.

Tom, Kevin, and Nancy also did a special downlink to commemorate the opening of the new Mission Control room being used for the first time. Tom held a plaque that would be later placed in the new facility as he thanked the flight controllers working there in support of our mission as well as all those who would be working there supporting future missions. Tom brilliantly summarized how we all felt when he told Mission Control, "It's not the rooms that make the difference, but the people who man those rooms."

Exercising in Space

Exercise, while important for good health on Earth, is even more critical in space. There was little need or opportunity for us to use our muscles because of weightlessness. As a result, muscles atrophy quickly in space, much as they do after being in a hospital bed for an extended period of time.

So we wasted very little time setting up our exercise equipment, a bicycle-like device without the wheels or frame known as an ergometer. It was a relatively simple piece of equipment, essentially a rectangular box with pedals sticking out the sides, mounted on the shuttle's floor. We flew special cycling shoes that clipped into the pedals. A unique feature was that it had no seat because there was no need to sit down in space.

The ergometer had a small white strap-like belt that went around your waist to pull you down near the floor. With our legs extended, we could pedal away as we floated weightlessly. The ergometer had an adjustment to change the friction to make pedaling easier or more challenging, so it could accommodate a wide range of exercise intensity.

The ergometer was particularly important for the commander and the pilot because they would need to use their legs quite a bit during the landing, steering with the rudder pedals and applying the brakes to stop. It also was important to make sure their hearts were still strong and kept the blood pumping to their brains so they wouldn't get dizzy and possibly pass out while returning to Earth.

In order to keep the flight deck clear for our Earth observation experiments, we mounted the ergometer in the middeck where more room was available. The only downside was there were no windows, so we weren't able to watch the world go by as we exercised, something that I had greatly enjoyed on my first flight.

The flight planners allotted each of us one hour a day for exercise. During that time, we would change into exercise clothes, find and put on the special shoes, exercise, stow the shoes, clean up a bit, and then get into our work clothes. So we were lucky to get in forty-five minutes of actual exercise.

In spite of that, I always looked forward to and enjoyed the exercise periods because they provided a break from working on the numerous experiments and gave my brain a chance to relax a bit. It was a great stress reliever and an opportunity to get away from the normal activities.

The entire STS-70 crew enjoyed exercising, but nobody as much as Nancy. She was an avid runner and liked weight training and a host of other activities. During one of the first days of our mission, I floated down to the middeck and saw an unusual sight—Nancy floating upside down pedaling with her arms, her feet floating up above her almost touching the ceiling.

The ergometer made a lot of noise when in use, but it just blended in with all the other noises inside the shuttle—from the fans, pumps, valves opening and closing, and toilets operating to normal conversations and radio communications from Mission Control. With no shower, bathtub, or sink on the shuttle, the best you could do post-exercise was take a sponge bath. We flew a few of our drink bags with powdered soap instead of powdered beverages inside which we'd fill with hot water from the galley to make hot, soapy water. After squeezing some onto a washcloth, we'd give ourselves a refreshing sponge bath. While less satisfying than a hot shower on Earth, it still felt great to clean up a bit.

We flew a special shampoo originally developed for people in the hospital who couldn't get out of bed to take a shower, called No Rinse Shampoo; it required no water. But I found it best to prewet my hair using a little water from the personal hygiene hose attached to the galley. After squirting some of the shampoo onto my hair, I'd then work it into a lather. It was important to do that slowly, because if you were too vigorous in your lathering operation, water and shampoo could go flying everywhere, splashing your crew mates. When I was done, I patted my hair with a towel to absorb any water and lather. After some work with a comb or brush, I was finished.

Once cleaned up and back in my work clothes, I'd look for a place to hang the towel, washcloth, and exercise clothes so they could dry. There were color-coded grommets inside the WCS for each crew member to put such items. I chose not to do that because it was disturbingly close to the toilet seat. I thought it prudent to put more space between my towel and, well, you get the picture.

Music Flown on STS-70

Long before iPods and MP3 players, we flew portable CD players, and each astronaut was allowed to fly five CDs. So while I exercised on the ergometer, I'd put on my headphones and crank up the music. I felt like I was in my own little world. It gave me a chance to get away from the rest of the crew for a while to enjoy some private time—or as private as it got in the shuttle's cramped quarters with four other human beings.

My STS-70 CD selections were:

1. *Ode an die Freiheit* (Ode to Freedom)—*Bernstein in Berlin* which featured Beethoven's Symphony No. 9. This was one of my father-in-law's favorite CDs, and he asked me to fly it on the shuttle for him.
2. *Dark Side of the Moon* by Pink Floyd. This became one of my all-time favorite albums during the summer of 1973 when

my brother Dennis and I saw them perform it in Cleveland shortly after it was released. Staring down at city lights passing by at five miles a second, or watching thunderstorms lighting up the cloud tops with thirty to forty flashes per second with beautiful subtle shades of blue, purple, and white, or the occasional shooting star creating a momentary trail of light one hundred and thirty miles below us as it entered Earth's atmosphere, *Dark Side of the Moon* made the experience that much more intense. Some of the screams and background effects on the album were dramatic when just staring out the window at the pitch-black sky filled with millions upon millions of stars—not a single one of them twinkling, each a steady pinpoint of light. If you fly in space and you don't take that one along with you, you are missing one incredible experience. I flew that album on all four of my missions. "Don't leave home without it."

3. *Bat Out of Hell* by Meatloaf. This album was a great one to listen to while riding the ergometer. It always got me going.

4. *Wish You Were Here* by Pink Floyd. That album was equally good for exercising or looking out the window at the Earth and the blackness of space.

5. *Little Rascal Music* by The Beau Hunks. That album by a group from The Netherlands recreated the background music used in the old Hal Roach *Little Rascal* and *Laurel & Hardy* shorts and was filled with a selection of cheery tunes.

Tom brought along Tom Petty's *Full Moon* album as one of his selections. One evening we were all up on the flight deck after dinner and had an hour or so before our scheduled bedtime. Tom plugged in a small set of speakers he brought along, and while the speaker quality left something to be desired (there were no decent speakers on the space shuttle, trust me), they allowed all of us to share in listening to the music together. It was the only time on my four flights that the entire crew shared such a moment and listened together. I absolutely loved it when "Free Fallin'" started playing. Can you think of a better place to listen to that song? Floating in a state of perpetual free fall 175 miles above the Earth traveling at 17,500 miles per hour. I'm convinced Tom Petty wrote that song for just such a moment.

Flight Data File—Official and Special

Flight Data File (FDF) books are what NASA called all the official procedures and instructions for how things were to be done on the shuttle. We carried approximately one hundred different books weighing a total of nearly one hundred and twenty pounds. Most of them were stored in the middeck lockers and brought out depending on what phase of the mission we were in—orbit operations, deorbit preparation, entry, etc.

There were procedures and cue cards for everything we did in space. Each procedure had detailed step-by-step instructions for such things as which switch to throw or for how to enter specific commands into one of the shuttle's computers. The FDF told us exactly what we needed to do and precisely when. The idea was to keep crew errors to a minimum.

There was another class of Flight Data File that had absolutely nothing to do with shuttle procedures, was not critical to the safe operation of the shuttle, and was not controlled by any official NASA board or organization. It was called the Special Flight Data File and was usually stored in one of the middeck lockers. That is where astronauts were permitted to fly school banners, baseball caps, or tee shirts to have access to them during the mission so they could enjoy them and take a few pictures with them.

The Special Flight Data File locker was also where astronaut spouses could send along a special message or a few pictures, where the training team could put in a few funny surprises, and where the Capcoms sometimes inserted a few items. While the spouses typically sent sentimental pictures and messages to remind us of home, most of the other items were intended to make us laugh.

The commander usually opened the Special Flight Data File locker on the mission's second day, once we were safely on orbit and everything was under control. Our STS-70 training team had put in a picture of our crew with them to remind us of all the great folks who got us ready for our mission. Also flown by a person unknown was a small stuffed Woody Woodpecker. Tom handed me something that Simone put in there for me, a great picture of her and Kai with "LIMA YANKEE!!!" written on the front of it. 'Lima' is the official aviator designation for the letter 'L' and 'Yankee' is the same for the letter 'Y.' That was our a special code for 'Love You.' Simone laminated the eight-by-ten card and added Velcro to the back so I could hang it up. Floating in the middeck, I stared at the picture for a few minutes after Tom gave it to me. It made me homesick to see the big smiles on their faces. It made me a bit lonely, and I felt detached from them. As much as I enjoyed seeing that picture, I somehow wished it had included me as well just to show us together as a family—all three of us. I never would have thought that I would have felt quite like that. But during training and as you got closer and closer to launch, you got busier and busier and tended to separate more and more from your family. That picture somehow reinforced that separation for me. Still, I hung it next to my sleeping bag inside the airlock, which made it a bit more like home.

Each crew member was allowed to fly three or four items in the Special Flight Data File locker as well, depending on how much room was available. To commemorate John Glenn, who flew the first All-Ohio mission and who played such an important role in inspiring me as a young boy, I flew two of his books. The first was *P.S. I Listened to Your Heartbeat*, a collection of letters written to Glenn immediately following his historic flight on February 20, 1962. I had that book since childhood. The other was *We Seven*, written by Glenn and the other six Mercury astronauts. That, too, was one of the books I had as a child. Over the years, I had it autographed by Glenn, Alan Shepard, Scott Carpenter, Wally Schirra, Gordon Cooper, and Deke Slayton. Betty Grissom signed it for me on behalf of her husband, Gus Grissom, who died in the Apollo fire on January 27, 1967.

Both books helped in inspiring me to become an astronaut, which is why I chose to fly them. On the back cover of *P.S. I Listened to Your Heartbeat* was a great color photograph of Glenn. I had Nancy take my picture with the book showing that photo of Glenn. I liked to tell people I had a picture of John Glenn in space aboard the shuttle, which was quite a novelty at the time.

Food aboard the Space Shuttle

Space shuttle food was just okay. The selections were somewhat limited and few menu items were overly appealing. When you throw in the fact that they're either freeze-dried or heavily irradiated, the desirability level dropped even further.

We sampled most of the food during our training sessions in the simulators in Houston. For any simulation that lasted more than four hours, we were provided with a limited shuttle food selection depending on what was available or what was about to go past its expiration date. That was fun and exciting when I first became an astronaut—eating real astronaut food just like the real astronauts. Wow, what a treat! But I soon grew weary of it and wouldn't eat that much of it when we were in the simulators. I figured the less of it I ate before a flight, the less sick of it I would be during the flight. So I minimized my consumption in hopes that it would be more appealing once I made it to orbit. That helped only to a limited extent.

About eight or nine months before our flight, as part of our crew training, we conducted an official taste testing session at lunchtime in the food lab at the Johnson Space Center. The entire crew was scheduled for lunch at the same time, and we invited our crew secretary and a few folks from our training team to join us because it was usually a fun event. The food lab personnel would prepare a number of items for us to sample, working hard to highlight anything recently added to the space food inventory. We sat on stools in a line along a typical kitchen counter and each took a small taste and then passed the item on to the next person. Only a small taste of each item was required. You'd know pretty quickly whether you liked it, plus we had a large number of items to sample. We then rated the food on a scorecard so we had a record of what we liked and didn't like. Extra food was usually included on each mission, and the scorecards were used to stock the backup pantry with items most liked by the crew.

After the sampling session, we received a list of all the available food items and a schedule for our meals for our upcoming mission. We were asked to prepare our own menus for each meal of the flight. We planned each meal for the entire flight. It sounded relatively easy, but trying to decide what you want for dinner six months down the road was a bit challenging. At home, I barely planned my meals a day in advance, let alone six months in the future. So I gave them my best guess.

For my first flight, I selected only my top favorite items. That was a big mistake. They had just developed a new powdered drink that was orange-mango flavored, which I liked quite a bit. So I requested it for breakfast, lunch, and dinner for nearly every day of the fifteen-day mission. Many of my crew mates did the same. It wasn't too long after our launch that nearly all of us were sick of orange-mango. Suddenly a lemonade or grape drink seemed so exotic. We were trying to trade our selections away, with a lemonade going for three or four orange-mango drinks. As a result, I learned to select the food I liked best and add a few items that I only moderately liked. That way I had access to more of a selection.

But I never flew anything that I definitely didn't like on the ground, assuming that nothing was going to happen to me in space that suddenly would cause me to crave creamed spinach or freeze-dried Brussels sprouts in some mystery sauce. Some astronauts have claimed their taste buds changed while they were in space and that they started to crave different foods. That never happened to me. If I liked it on the Earth, then typically I liked it in space. I didn't eat grits at home, and I was sure I wasn't going to start craving them in space.

One of the most popular food items requested by astronauts was shrimp cocktail. That was nearly legendary for shuttle crews, much like the classic steak and egg breakfasts the Mercury, Gemini, and Apollo astronauts ate just before their launches. The shrimp cocktail had a great flavor and was a bit spicy, which made it appealing to many astronauts.

Once we made selections, our menus were given to NASA dieticians for careful review and analysis. Diets were designed to supply each shuttle crewmember with all the recommended dietary allowances (RDA) of vitamins and minerals necessary to perform in the environment of space. More than once I had the dietician police inform me that my food selections did not contain enough zinc, for example. They suggested that I consider adding some freeze-dried lima beans. I wouldn't fight them. I'd simply respond with "Great idea. Let's add a few packages" even though I had no intention of ever eating them. Including them was a gesture aimed at keeping everybody happy, while giving a few hundred freeze-dried lima beans a free round-trip ride aboard the shuttle.

To determine exactly how much food we needed, NASA used a scientific formula that determined the caloric intake we needed while in space. Those caloric requirements were based on a National Research Council formula for something called basal energy expenditure (BEE). For women, BEE = 655 + (9.6 x W) + (1.7 x H) − (4.7 x A), and for men, BEE = 66 + (13.7 x W) + (5 x H) − (6.8 x A), where W = weight in kilograms, H = height in centimeters, and A = age in years. Not too complicated, eh? For my age, weight, and height at the time, my own BEE was 1739.27 calories, or the equivalent of eating three Big Macs a day.

In spite of all that careful and detailed analysis, we always flew much more food than we could ever consume, with the food lab people packing nearly three thousand calories a day for each astronaut. Much of that returned with us at the end of the flight, including those lima beans, and later ended up being consumed in the simulators during training sessions. I thought it was a good idea to fly more rather than less food because it allowed for a bit more of a selection in case something suddenly became more appealing.

I'm frequently asked how I liked the astronaut ice cream. While it is sold at nearly every science center and museum gift shop as the quintessential astronaut space food, I almost hate having to confess that we didn't fly it. It was too dry and powdery and produced too many fine particles and crumbs that would float in space and get in astronauts' eyes or be breathed into our lungs. Because of that, we shied away from most dry foods that created crumbs, such as bread, potato chips, and the like. We did fly a few packages of shortbread cookies that could produce crumbs. But those were bite-sized, so crumbs were kept to a minimum. Most of the food we ate was designed to be a bit wet and pasty so it adhered to a fork or spoon and made it easier and neater to consume.

Much of what we flew was classic freeze-dried food prepared in the food labs by Boeing, which had the contract for shuttle food preparation. They cooked and then freeze-dried the food and packed it in small plastic containers for flight. They also took great care to track the shelf life of each item to make sure no food was launched that went beyond its recommended use by date. There were freeze-dried scrambled eggs—the spicy Mexican scrambled eggs were one of my favorites—and freeze-dried breakfast sausages. There were packages of Rice Krispies and other cereals available that contained powdered milk. There also were freeze-dried fruits. The freeze-dried strawberries turned bright red once water was added and were delicious although a bit mushy and not crispy like normal fresh strawberries on Earth. But their taste was outstanding. Freeze-dried blueberries were added to some of the cereal packs, and that was quite a treat. I also flew a few packs of instant breakfast drink, which was like a thick chocolate milk drink. It was quick to prepare and something you could easily drink during the commotion of everyone getting up and ready for another day of work in space.

We also had freeze-dried spaghetti and meat sauce. I learned during my later flights to put that onto a tortilla to create a space pizza. I liked the sound of it. While it turned out not to be my favorite pizza in the world, it was a fun substitute in space. We also flew freeze-dried hamburger patties. Those were hard as a rock until hot water was added, which magically softened them and made them nearly edible. They were every bit as good as you can imagine a freeze-dried hamburger would be.

We flew a selection of military rations called 'Meals Ready to Eat' (MREs)—and they're just that. They weren't freeze-dried and came in dark green foil pouches. Before packaging the food, the items were bombarded with radiation to kill off any germs or bacteria, which gave them a long shelf life without the need for any type of refrigeration. Many of them were my favorites, including beef in barbeque sauce, sweet and sour chicken, grilled steak, and grilled chicken breasts. I also found that MRE applesauce tasted Earth-like to me. In my opinion, MREs were some of the most appetizing foods made available to us. By adding a little Cleveland Stadium Mustard, I was able to turn them into a fine delicacy! We also flew commercially available prepackaged food items, such as pudding cups and fruit cups.

All the food that we selected six months earlier ended up packed into small drawers that fit within the middeck lockers. We each had our own drawer, and at mealtime I would float over to my designated locker and rummage through, looking for anything appetizing. There were times when something would tempt my taste buds instantaneously, like a MRE beef in barbeque sauce. At other times, my palette was a tad more discerning; I would search from the front to the back of my tightly packed food drawer and not see anything appealing. *I must have missed the good stuff*, I would think to myself, and begin a second front-to-back search. If nothing jumped out at me during the second pass, I realized that was all I had. So I'd grab something like the spaghetti in meat sauce and a few other items and call it a meal.

Most astronauts lost a few pounds during their flights because it was normal not to have much of an appetite the first couple of days in space and because the food just didn't make you wish you had larger portions.

Food preparation in space was quite simple. Located on the middeck just to the right of the hatch was our food galley that had two main features. The first was a rehydration station, which included a small clamp where freeze-dried packages could be inserted and then pushed into a needle that punctured the package. A small dial allowed us to premeasure the amount of water to be added, and then we pushed either a blue button for cold water or a yellow one for hot water.

Each food pack came with its own instructions for preparation printed on an attached label. The creamed spinach instructions read "2 OZ HOT*2–5MIN" which translated into "add two ounces of hot water and wait two to five minutes before eating." So you dialed '2' into the water amount and pressed the yellow hot water button, and the rehydration station would inject the necessary amount into our freeze-dried food pack. Waiting the recommended time gave the dried food time to absorb the water and soften it up.

For the MREs and other food items that didn't need to have water added to them, we had a small oven located directly below the rehydration station. The oven was about the size of a small breadbox and heated food by blowing hot air that really was excess heat generated on the shuttle. It got a little hot inside the oven and could definitely heat up food packages placed inside, but it had about the same baking qualities as a child's Easy-Bake Oven. You'd never burn your hand on this oven. Typically, it took about five to ten minutes to heat up the food.

For STS-70, I selected the following meals for my second day in space:

Breakfast
Granola with Blueberries   Chocolate Instant Breakfast
Bran Chex   Orange-Mango drink

Lunch
Smoked Turkey   Trail Mix
Tortillas   Candy-Coated Chocolates
Pizza Gold Fish Crackers   Tea with Lemon
Granola Bar

Dinner

Sweet 'n Sour Beef	Granola Bar
Italian Vegetables	Candy-Coated Chocolates
Pizza Goldfish Crackers	Wintergreen Lifesavers
Strawberries	Tropical Punch Drink

That menu featured a lot of comfort foods and was lighter than those later in the mission when I knew I would have more of an appetite. By mid-mission, my lunch and dinner menus were a bit more substantial.

Besides the food drawer that each astronaut had, there also was a small area on the middeck called the pantry. It was located near the ceiling and was about five feet long and about six inches high. Upon opening the pantry door, we could pull out one of the three small drawers that also were loaded with food.

All the food stored in the pantry was considered extra and was available for anyone to help themselves. So if we didn't see anything appetizing in our own food lockers, we were always free to check out what was in the pantry. Also stored there were various condiments, including salt, pepper, mustard, ketchup, and hot sauce.

Because salt and pepper particles would float in space and never settle on our food, normal salt and pepper shakers like we use on Earth were not flown. Instead, we had a small plastic bottle of salt water. Adding a few drops of that heavily salted water to our food accomplish the same thing as a couple of shakes of the salt shaker on Earth. For pepper, we flew a similar plastic bottle that was filled with vegetable oil loaded with pepper particles. Again, a few drops of that liquid pepper could be added to the food without having pepper particles floating into our eyes or inhaling them and causing a big sneeze. Those salt and pepper bottles were simple, practical space-age solutions for tasks that depended on gravity to accomplish on Earth.

We also had a fresh food locker on the middeck. Sometimes it was an entire locker; sometimes only half depending on how tight stowage was. That locker was for special food requests by the crew and for fresh food. Most of the food gets stored on the shuttle days and weeks before launch. The fresh food locker was loaded within twenty-four hours of launch to make sure it was still good to eat once we got to space. So we might fly a few bananas, oranges, and apples. Those would have to be consumed by the astronauts within the first few days of the mission before they began to spoil. Rotting fruit in a closed environment in space was not a good thing. So we tried to minimize what we launched and worked hard to eat whatever was fresh in the first few days.

The problem was that most astronauts didn't eat much the first couple days because they were still adjusting to weightlessness. That's when up to a third of the astronauts experienced different levels of Space Adaptation Syndrome (SAS), also known as Space Motion Sickness. After a few days, that typically passed for most of them and their normal appetites returned.

The fresh food locker also was where we flew astronaut-requested items not on the official food list. Those might include Life Savers, mints, some special chocolate, or food items from our hometowns. That's where we carried the special packs of Stadium Mustard for STS-70. Also included were flour tortillas, small bags of Wheat Thin crackers, and pizza-flavored Goldfish Crackers.

Mealtime Aboard *Discovery*

Mealtime in space was frequently characterized as astronauts enjoying their meals together, much like an idealized family sitting around the dining room table talking about how their day went. Numerous IMAX movies have been made showing a group of astronauts eating together either on the shuttle's middeck or at a table on the Russian *Mir* or the International Space Station. During my four flights, that wasn't exactly how it happened.

In the morning, the middeck was always a busy place. There was so much going on that we never had the luxury or time to prepare our meals and then enjoy them together as a crew. Breakfast on the shuttle was more reminiscent of a modern day busy family with everyone getting themselves up and out the door at different times. If the flight deck wasn't too crowded, I would make my breakfast and float up to one of the overhead windows to watch the world go by as I ate. I tended to eat most of my meals in front of the overhead windows like some people on Earth eat their meals in front of the television. If the flight deck was too crowded or hectic, I'd float back down to the middeck to find some space and a little peace and quiet while I ate.

Lunch was a little different because most of the time all crew members were scheduled for their one-hour lunch break at the same time. Mission Control usually restrained from making any calls to us during our lunch hour unless necessary. Typically, we all gathered on the flight deck to look out the windows while we ate, enjoying the shared experience of watching Planet Earth passing below us. There would be limited conversation about the day's events, and maybe some talk about the next sights we would be seeing or something unusual someone saw. Some of the crew members might use the time to check their email and send a message home, or spend a few minutes organizing their food and clothes for the days ahead.

For dinner, we seldom simultaneously stopped what we were doing. Instead, one by one we would finish some chore—like cleaning air filters, installing a new set of lithium hydroxide canisters to scrub carbon dioxide out of the air, or finishing up an experiment procedure—then we would each select and prepare our dinners and slowly congregate in front of the windows on the flight deck. It could get a bit crowded up there, but we always made room to get all five of our faces in front of the windows to see a sunset or the white beaches breaking waves on the west coast of Africa along the Ivory Coast.

Sometimes I would gaze out the window and wish I could be where I was passing over. *Wouldn't that be great*, I'd think to myself. *I need to visit there someday.* On Earth, I would daydream about being in space and, in space I would daydream about being on Earth. The grass is always greener, especially when it's 165 miles down below you!

On this second day of our mission we also had a great pass directly over something less idyllic—newly developing Tropical Storm Chantal located in the Atlantic Ocean just off the coast of Cuba. It did not have a well-developed eye yet and had not reached hurricane status, but that huge storm was an impressive sight. Tom Jones, our Capcom, told us to get some pictures of it. Nancy replied that all five of us, plus Woody, had our faces crowded around the windows and were getting a good look at the storm.

So with TDRS now successfully positioned in its proper geosynchronous orbit over the Pacific Ocean and with its checkout and testing initiated, and with things set up and experiments underway, we had had a great second day in space. Everyone except Mary Ellen was feeling pretty good and had adapted to being in space. She was still struggling a bit with space motion sickness, and the rest of us hoped that she would be feeling better in a few days.

On my first flight, I had experienced some mild space sickness symptoms myself. I felt like I was a bit hung over for a day or two and didn't want to eat much. It felt like my stomach shut down along with the rest of my digestive system. I never vomited, but my stomach was definitely upset. On STS-70 and my other two flights, I never experienced any of those symptoms again. For many astronauts, the body remembers what it's like being in space and adapts much more quickly on subsequent flights.

Finishing up the second day of the mission, my mood was soaring, and I was totally enjoying the flight. Having successfully deployed TDRS soon into our mission allowed us to kick back, relax, and enjoy life in space without the normal hectic pace associated with typical shuttle missions. We had plenty to keep us busy with nearly twenty experiments needing our attention, but there was also plenty of time to look out the window, talk on the SAREX radio, exercise, and just hang out in space enjoying the experience. It was simply a lot of fun.

Discovery continued to perform flawlessly, and by all measures the mission was going extremely well. Flight Director Kelso told the news media: "I cannot remember another flight where we have tracked zero anomalies [problems] on board the orbiter after twenty-four hours." *Discovery* was an amazing vehicle and was taking very good care of us.

As I got ready for bed, I took a few minutes to write another entry in my journal:

> MET 1 day 6 hours 58 minutes
>
> *Pretty busy day today but this is nothing like STS-65! There is always time to look out the windows. Saw a developing hurricane named Chantal off the coast of Georgia. No eye yet—not very impressive development, but it's big! Maybe we'll see it develop in the next few days.*
>
> *I think I adapted much faster to microgravity this time. After only 1–2 hours I was moving around pretty well. No problems at all—it's such a pleasure.*
>
> *Last night I slept in the airlock and had one of the most terrifying dreams I have ever had. In the dream Simone had had it with our relationship and was leaving me. I couldn't talk her out of it. She was so mad and determined. I was so glad when I woke up and realized it was only a dream.*
>
> *During sleep I just floated in my sleeping bag and it was quite comfortable without using any restraints. I remember it took me 3–4 days on my first flight to get used to sleeping up here.*
>
> *We passed over Cuba again and I got a few shots of Guantanamo Bay for John (my friend stationed down there). This environment is so natural it still amazes me how quickly you can adapt. After only two days my brain no longer uses the conventional up and down orientations inside the shuttle.*
>
> *The launch is still a thrill. It was <u>much</u> more visually impressive than STS-65, but the noise and vibrations didn't surprise me this time and I barely noticed them. I think the visual sensations override all else and being my second launch there were fewer surprises.*

> *I have such a great feeling of respect for the workers who put the shuttle together and the team that gets us safely to orbit!*
>
> *I'm not too thrilled about the food again—maybe it will be better (or seem better) in a few days. I set up the SAREX experiment today and made my first contacts with the ground. As we passed over Houston I talked to Matt Bordelon, the SAREX Team Lead as well as my neighbor and good friend, and asked him to call Simone and tell her I am well. Also talked to John Nickel in Kansas. He helped set up the SAREX contact with Simone last year during STS-65. Also talked with about 20 other folks very quickly. It's amazing how many people are out there wanting to talk!! The airwaves were very active.*
>
> *Landing is one week away. I'll be ready. Need to get some Earth Observation time in at the windows and some dark cabin time to look out at the stars and deep space. I'm happy, doing great, and having a great time!!*
>
> *Saw some amazing rainforest clearings and fires burning in Brazil. It's incredible how much of the rainforest they are cutting down. Also got my first look at KSC off in the distance as we passed over Cuba. I missed seeing it at the end of our first pass yesterday after launch. Things were just too busy getting ready for the deploy.*

At 6:12 p.m. we began our scheduled eight-hour sleep period. While we slept, *Discovery*, under the watchful care of our skilled Mission Control team, continued its silent trek in space, orbiting Earth every ninety minutes and thirty-six seconds in an orbit 195 miles by 165 miles.

Back on Earth, the NASA solid rocket booster recovery ships *Freedom Star* and *Liberty Star* arrived at the Kennedy Space Center with the two burned out SRBs from our launch. After separating two minutes into the flight, they fell into the Atlantic Ocean about one hundred miles off the coast of Florida. Simone went over to Port Canaveral to watch the returning ships pass by. After entering the port they proceeded through a lock before crossing over to the Banana River. Waiting there along the edge of the lock along with a dozen other interested people was my brother Dennis and his daughter, Asja. They both waved to the recovery ships as they passed by and proceeded through the lock. My brother held a large sign he made that read, "THANKS FOR THE GREAT JOB FROM THE THOMAS FAMILY." Just like the crew members, the families also appreciated the thousands upon thousands of workers who helped us get safely to orbit and back to Earth each time.

Chapter Six
Scientific *Discovery*

We were scheduled to sleep until 2:12 a.m. CDT on Flight Day 3, but I woke up early because I was a bit cold. Everyone else was still asleep, so I really couldn't do much but listen to music, read, or write down a few thoughts. I decided to make another entry in my journal:

It's 30 minutes before our official wake up and no one else is awake yet. The cabin is dark so can't even look out the windows. I slept good last night—a good 6 hours with another 1½ hours of lying quietly awake. Woke up a little cold so I moved the 'elephant' trunk hose blowing the cold air out of the airlock. Nancy calls the airlock my little condo. I hope I can spend more time at the windows. Sometimes the flight deck is just too crowded. I'm looking forward to getting up and reading mail from Simone. I hope she and Kai are doing well and sound asleep right now. Two more days until we can talk. Looking forward to that!

I saw 3-4 cosmic ray flashes last night before I fell asleep. Otherwise it was a quiet night. I only heard the WCS [toilet]

> 2–3 times (instead of 10 times the 1st night). Yesterday I wore the Governor of Ohio's watch all day and will pass it on to someone else today. This trip is so relaxed. Wake-up music has just come on! Time to get up.

When our official wake-up time rolled around, most of the crew was already awake. To begin our third day in space, we were greeted with an old song, "Beautiful Ohio," sung by icon Kate Smith. Since most of us were a bit unfamiliar with Smith's repertoire, we all wondered who made the selection. After the song, our Capcom, Marc Garneau, told us Senator Glenn chose it to honor the mission's Ohio astronauts. "Takes us all back home," Tom remarked. It was a great way to start our third day in space.

I decided to wear a rugby shirt that Simone selected. I intended to give it to her after the flight. With long sleeves and stripes of green, purple, and black, it was definitely her style and reminded me of her as I put it on.

Tom and Kevin continued their work with the HERCULES camera but were having some trouble aligning its inertial measurement unit that was critical for pinpointing the exact locations on Earth being photographed. The alignment procedure required them to locate two different stars with the camera in different parts of the sky—very similar to the procedure we used to align the shuttle's navigation units numerous times during the mission.

To complete their task, they needed the flight deck cabin totally dark to better locate and identify the optimal pair of stars. As a result, they closed the shades over the two inter-deck hatches between the middeck and flight deck. That allowed the rest of the crew to continue working on the middeck without shutting off all the lights. Tom and Kevin would spend most of the time during the forty-five-minute night passes performing camera alignments. That made it hard to get much time looking out the windows because with the inter-deck hatch covers closed, we couldn't float up to the windows when we had a few moments of free time. Additionally, they needed full access to the windows for the alignment process and taking pictures of their Earth targets. I felt a bit like I had been banished to the middeck dungeon and tried to grab any window time I could during their breaks for lunch, exercise, or other activities. I also tried to get in some window time early in the morning or at the end of the day when the windows generally were free.

We had another spectacular daylight pass over the developing Tropical Storm Chantal. The maximum winds were reported to be fifty-one miles per hour with the storm moving in a west-northwest direction over the Atlantic Ocean at about five miles per hour. Observing such a large storm in daylight from directly overhead was dazzling. The bright unfiltered sunlight reflecting off the tops of the clouds was so intense that you either needed to squint or put on a pair of sunglasses while looking at it. Even with the bright reflection, we still saw sufficient detail to clearly identify areas with billowing thunderstorms embedded in the tropical storm. It was an awesome sight.

Nancy and I wrote a new message for broadcast via SAREX to help promote the research we were doing aboard *Discovery* during the mission.

> Greetings from the STS-70 crew of Space Shuttle *Discovery*!!
> We are into our third day in orbit and continuing to work on our numerous experiments. We are growing protein crystals to help develop new pharmaceutical drugs on the ground and have a Bioreactor that is growing human colon cancer cells to help understand how cancer cells grow. We are keeping a close watch on a tropical storm over the East Coast of the United States as we orbit the Earth every 90 minutes. We wish you all a good day!!

We updated the message every day to highlight the different experiments and other activities we were performing aboard *Discovery*.

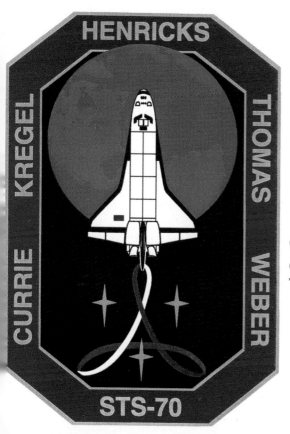

(*Left*) Early design of the STS-70 patch

(*Below*) Artist drawing of operational TDRS fully deployed (NASA)

(*Left*) Proclamation from Governor Voinovich making Kevin an Honorary Ohioan

(*Below*) The All-Ohio Crew after the proclamation was issued. (*Left to right*) Top row: Tom Henricks, Kevin Kregel, Don Thomas; Bottom row: Nancy Currie, Mary Ellen Weber (NASA)

(*Above*) Mary Ellen and I inspect the TDRS at Kennedy Space Center (NASA)
(*Below*) Tom and Mary Ellen inspect our new Block I space shuttle main engine at KSC

Stacking *Discovery* for launch in the Vehicle Assembly Building at KSC (NASA)

(*Above*) First rollout to Launch Pad 39B (NASA)

(*Below*) View of *Discovery* at Pad 39B from T-38 flying into KSC

(*Above*) Crew during practice launch countdown exercise at KSC (NASA)

(*Below*) STS-70 crew with Senator John Glenn during visit to JSC (NASA)

(*Clockwise from top*) The woodpecker attacks our external tank (NASA); Worker at Kennedy Space Center patches holes in the external tank insulation (NASA); Some of the holes made by the woodpecker (NASA)

(*Clockwise from top*) Predator eye balloon attached to the very top of the external tank; Guarding *Discovery* against additional woodpecker attacks (NASA); Worker at Kennedy Space Center installs a plastic owl to scare away woodpeckers (NASA)

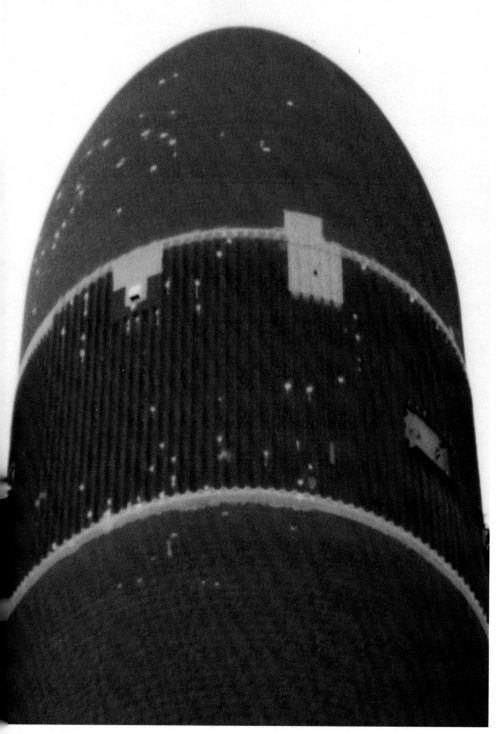

Numerous patches in the external tank from the woodpecker holes

Discovery arrives back at Pad 39B after second rollout, this time with a yellow Predator Eyes balloon attached to the very top of the external tank

(*Above*) Kevin, with plastic woodpecker, stands in front of scaffolding used during repair of the woodpecker holes

(*Below*) Busted quarantine once again! The Orlando Police 'greet' the STS-70 crew upon their T-38 arrival at Orlando International Airport

(*Above*) Crew with spouses in front of *Discovery* the day before launch. (*Left to right*) Jerry Elkind, Mary Ellen, me, Simone, Dave Currie, Nancy, Tom, Becky Henricks, Kevin, and Jeanne Kregel

(*Right*) Governor Voinovich and wife Janet in front of *Discovery* (NASA)

(*Above*) The Thomas Family at the beach house. (*Left to right*) My brother Bill, my mom Irene, my sister Cindy, me, Simone, and my twin brother Dennis

(*Below*) With "The Mother of the Astronaut" two days before launch

(*Above*) The Thomas Family at Night Viewing the evening before launch

(*Below*) Tom, Kevin, Nancy, and Mary Ellen at the base of the mobile launch platform three hours before launch, looking up at *Discovery*

(*Clockwise from top*) Tom leads the crew out to the Astrovan on launch morning (NASA); Nancy took this picture of me just prior to entering the White Room; Closeout crew helping with final suit adjustments in the White Room. *Discovery*'s open hatch is in the background (NASA)

Discovery launches on the STS-70 mission (NASA)

Condensation cloud surrounds *Discovery* shortly after launch (Photo by Peter Gaultieri / *West Kentucky News*, used with permission)

(*Right*) *Discovery* launches on the STS-70 mission (NASA)

(*Below*) Simone and Kai watch STS-70 launch from the top of the Launch Control Center at KSC

Above) The kids of the STS-70 crew on launch morning (NASA)

Below) Artwork drawn by the kids of the STS-70 crew as they waited for launch (NASA)

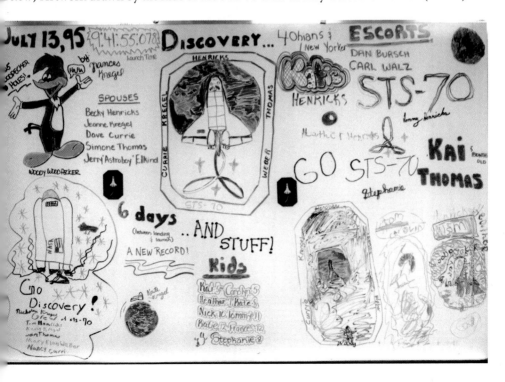

The external tank shortly after separation from *Discovery*. We saw plenty of feathers as we separated but no other visible signs of damage to the tank (NASA)

(*Above*) NASA SRB recovery ship *Freedom Star* tows one of the STS-70 boosters into port the day after launch

(*Below*) My brother Dennis and my niece, Asja, thank the SRB teams for their support (Photo by Peter Gaultieri / *West Kentucky News*, used with permission)

(*Above*) The TDRS satellite tilted up just prior to deployment (NASA)

(*Below*) The TDRS moves away from *Discovery* following deployment. Two pieces of ice can be seen floating near the satellite (NASA)

(*Above*) The STS-70 All-Ohio crew aboard *Discovery* honoring my alma mater, Case Western Reserve University (NASA)

(*Below*) Enjoying some Stadium Mustard on my chicken sandwich (NASA)

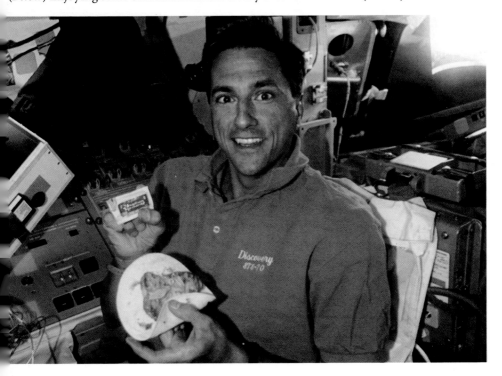

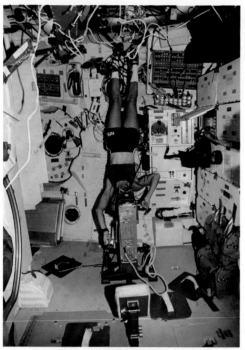

(*Clockwise from top*) Nancy giving her arms a workout during her exercise period (NASA); Tom wrestles with the HERCULES experiment camera to take pictures of the Earth (NASA); Our Woody Woodpecker mascot watching over things for us. The photo is of my niece, Asja, which I carried with me on the flight (NASA); Nancy and Kevin demonstrating there is no up or down in space (NASA)

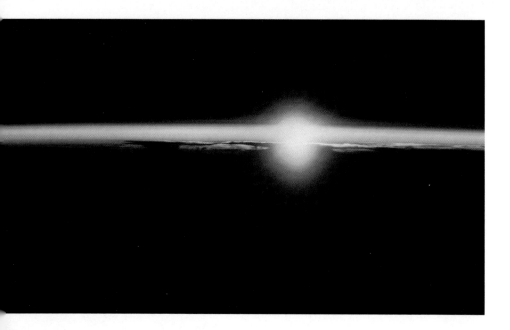

(*Above*) Beautiful view out the aft flight deck windows during one of the 143 sunrises we experienced aboard *Discovery* during the STS-70 mission (NASA)

(*Below*) The Nile River cutting through Egypt. King Tut's tomb is located near the big bend in the river (NASA)

(*Above*) Special message for Kai from the flight deck as we pass over the Pacific Ocean (NASA)

(*Below*) Tom and I just prior to the deorbit burn that would bring us back home

(*Above*) *Discovery* just crosses the threshold of the runway as we prepare to land at KSC. (Photo by Peter Gaultieri / *West Kentucky News*, used with permission)

(*Below*) Kevin deploys the drag chute as Tom lands *Discovery* (NASA)

(*Above*) The STS-70 crew in front of *Discovery* shortly after landing (NASA)

(*Below*) Tom (bent forward) gets a good look at the front tires of *Discovery* while Kevin (left) and I (right) look on (NASA)

(*Above*) The STS-70 crew participated in a Space Day rally at a Rockwell facility in Cape Canaveral. (Photo by Malcolm Denemark / *Florida Today*, used with permission)

(*Below*) The STS-70 crew poses with Woody Woodpecker at the Space Day Rally (Photo by Charles Atkeison IV, used with permission)

(*Above*) Simone, Kai, and myself, along with Governor Voinovich and his wife Janet, at the opening ceremony for the Rock & Roll Hall of Fame in Cleveland, Ohio

(*Below*) Tom, Nancy, and I at the official opening of the Rock & Roll Hall of Fame. Tom holds the Hall of Fame cap we flew on STS-70

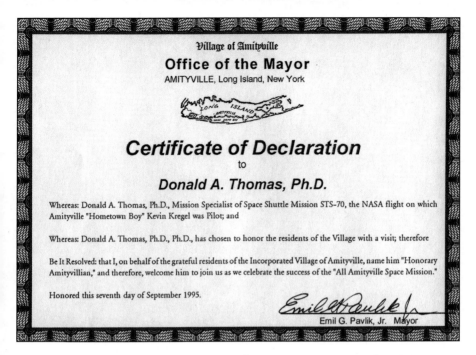

(*Above*) Official proclamation making me an "Honorary Amityvillian" and thus recognizing the "All Amityville Space Mission"

(*Below*) Official crew portrait of the "All Amityville Crew" (NASA)

(*Above*) STS-70 crew at the Ohio Astronaut Reunion hosted by the Glenn Research Center in recognition of NASA's fiftieth anniversary in 2008 (NASA)

(*Below*) Me in front of *Discovery,* now on permanent display at the National Air & Space Museum's Udvar-Hazy Facility (Photo by Kanji Takeno, used with permission)

A Media Darling? No and Yes

It would be fair to say that STS-70 was not an overly exciting mission—just a run-of-the-mill satellite deployment. Been there, done that. Nothing new or unique about deploying a TDRS. It'd been done five times before. So by the time STS-70 rolled around, it was pretty much old hat.

And at about the same time we were scheduled to fly, so was STS-71. That was to be the first docking of a space shuttle to the Russian *Mir* space station. As a result, the JSC public affairs people were of the opinion that ours was a rather boring mission. Nothing interesting. We were pretty much told that it was unlikely there'd be any media interest in our flight, so they weren't going to put much effort into it. It was felt that STS-71 would be one of the more notable shuttle missions and that it would command a lot of media interest and that was where they focused all their attention.

To say the least, we found that attitude a little insulting. We thought STS-70 was an important mission because our satellite would be used for decades and provide for communications between Mission Control, the space shuttle, and the future International Space Station as well as helping to transmit the incredible images taken by the Hubble Space Telescope. We would be risking our lives and, without question, believed it was important for NASA, scientific research, and our country.

I always thought the job of the public affairs people was to make missions like ours seem exciting and interesting and to publicize and promote what we were accomplishing. But that was not the approach used by the NASA public affairs team. What we heard from them was more along the lines of them having more important things to promote.

We knew there would still be a lot of interest in Ohio because we were going to be 'The All-Ohio Space Shuttle Mission.' So as a crew we drummed up media interest on our own, and that led to quite a bit of coverage. Our PAO folks didn't totally let us down. They did throw us a few bones, scheduling several media events for us. My favorite was a conversation Tom had with a World War II veteran in a veterans administration hospital in Wisconsin.

Chatting with a Veteran

To set the stage, Tom would be talking with and answering questions from Harland Claussen, a distinguished World War II veteran who was a patient at the Clement Zablocki Veterans Administration Medical Center in Milwaukee. As an Army infantryman, he escaped from a German prisoner-of-war camp but was recaptured and later liberated by the Allies. Mr. Claussen had quite an impressive background as one of our country's true war heroes.

The conversation would be audio-only, no video. Tom was ready for the event at the scheduled time and waited for their call to start as he floated on the flight deck. Because the flight deck is rather small, the rest of us were on the middeck so we wouldn't disturb or distract Tom during the conversation. Kevin, Nancy, Mary Ellen, and I worked on various experiments as we listened in via the overhead speaker.

The first words we heard from Mr. Claussen were: "Do you hear me? Do you hear me?" Then we heard someone in the background remind him to read the official words provided to him by NASA to start the interview. "Oh. Oh. Okay," Mr. Claussen said, followed by "*Discovery*, this is the VA in Milwaukee. Do you hear me?" He waited a few seconds and, when he didn't get an immediate response, asked, "You on? I don't hear anybody there."

"You're supposed to ask a question and they'll answer it," someone in the room informed him. Mr. Claussen told Tom how proud and honored he was to have been chosen for the call.

"Well, Mr. Claussen, it's our privilege to be speaking with you, and we wish you and all your friends there the best. This is the orbiter *Discovery*. We're about one hundred and sixty miles out over the Pacific Ocean approaching Hawaii," Tom said.

Mr. Claussen wanted to know how fast we were going or whether we were standing still. Tom did a great job providing a brief explanation of orbital mechanics.

After a few more questions, Mr. Claussen joked with Tom about the possible cost of the long-distance call. "I just wanted to tell you something, Commander. They told me I should reverse the charges when I get done talking."

Tom replied, "Sorry, sir. There was some noise in the background. Could you repeat that?" While it can be a bit noisy inside the shuttle, the noise in the background to which Tom was referring was Kevin, Nancy, Mary Ellen, and me laughing at the notion of collect calls to the shuttle.

"I said I'm glad I don't have to pay for this call," Mr. Claussen repeated.

To which Tom said, "I'm glad I'm not paying for it, either. We're very fortunate to live in a country that is free enough to have the ability to conduct operations like this, and we owe a lot of that credit to you and your friends who fought in World War II. Without your efforts, this program wouldn't exist and we wouldn't be making this phone call."

After a short pause, Mr. Claussen sighed and then said, "Let's see. What else do I want to ask?" His next question created another wave of laughter—much more intense this time—from the middeck: "Are the ladies any trouble to you people, to you men up there?"

"No, no. No problem whatsoever," Tom responded. "In fact, this crew has been working together as a team for about a year. Men and women are integrated completely throughout NASA, and this shuttle crew is just an example of how men and women can work together even in a close environment. We respect the fact that we do have some differences, but it's very easy—even in this small environment—to work with that."

Mr. Claussen followed up by wanting to know whether the female astronauts were "the same as [we were]."

Tom said, "Yes, they are. You may be interested to know that one of the women on board is an Army major. She's an active helicopter pilot."

"Oh, for heaven's sake," Mr. Claussen said. How times had changed since his service during World War II! There was about five more minutes of casual conversation, during which Mr. Claussen wanted to know, "Can you be too old to be an astronaut?"

Tom's poignant reply: "In spirit, never."

Then it was over. We congratulated Tom on what a great job he had done during the event, and we laughed for a few more hours about the question whether the women were giving the men any problems.

While probably not the highest profile interview ever conducted while on orbit, the event allowed Tom to eloquently discuss how men and women work together on shuttle missions. We got assigned as a crew and we were all equal members—each with special skills that we brought to the program. The men didn't concentrate on the driving while the women cooked, cleaned, and took care of the men. The early-twentieth century traditional male/female roles were not to be found aboard the shuttle. Men and women were truly equal partners in space.

While at the time none of us fully understood why our PAO folks went to the trouble to set up such an event, we learned later that this VA hospital was celebrating the installation of the first phone in their facility for the free use of patients. In this modern age of cell phone technology, it's hard to imagine the significance of that. But evidently only pay phones were available for the patients before that time. While we were glad to be a part of such a celebration in support of our veterans, I'm afraid it sadly will remain but a simple footnote in the history of telecommunications!

ABC's *Mike and Maty Show*

Mary Ellen, Nancy, and I did an interview via video with an ABC show targeted at moms and kids, the *Mike and Maty Show* featuring Michael Burger and Maty Monfort. We were floating in front of a sleeping bag on *Discovery*'s middeck, and one of the first questions was why our launch was delayed for a month. As I started explaining the reason, either Tom or Kevin off camera floated our stuffed Woody Woodpecker our way, which led to laughter. The Associated Press picked up the image of the three of us and Woody, and it was published in newspapers across the country.

Mary Ellen, a competitive skydiver with over 1,900 jumps, said there was no comparison between her hobby and being on the shuttle in space. "This is pretty awesome. This is pretty remarkable."

Nancy mentioned that the previous day Tom and Kevin oriented the shuttle to fly with its nose pointed straight down at Earth. "Some of us got the sensation that we were falling back to Earth. It was very interesting," she said, recounting the somewhat unusual experience looking out the flight deck's forward windows. It amazed me how small changes in our orientation caused huge differences in how we saw things out the window. Whenever I got a bit tired looking out one window, the simple act of floating over to another window with a slightly different view seemed to change the perspective entirely.

Maty heard that most of us on the crew swapped wedding rings with our spouses and asked us to show her the rings, which we did. She also asked if we had any special messages for our families back on Earth to which Nancy replied, "Tell them not to eat too much pizza until we get back." Mary Ellen told her husband, Jerry, that she loved him and missed him; I told Simone and Kai that I was looking forward to seeing them when we landed in a week. Maty mentioned that Nancy's daughter, Stephanie, wanted to see her dog, Auggie, fly in space. "We would all like to see Auggie in space one day," Nancy said, laughing. It was a fun and lighthearted interview and one that Nancy, Mary Ellen, and I all enjoyed.

Most of the questions were the normal run-of-the-mill ones about life in space and woodpeckers, but one question surprised me. Maty wanted to know "if being an astronaut had made me a babe magnet." I couldn't understand why I was being asked that question. It must have been because I was doing the interview with two female astronauts. I surely had no reputation on Earth for being a babe magnet. Shaking my head while blushing profusely, I stammered out an "Absolutely not," mumbled something about being happily married with one son, and disclaimed any connection with babe magnetism.

About forty-five minutes after talking with Maty, Tom had a fifteen minute interview of his own with a reporter for *The Blade* newspaper in Toledo, which circulated in the areas where Tom grew up—Bryan and Woodville. It was the first media event of the mission with an Ohio connection, but it almost didn't happen. The interview request was approved months before our flight by JSC public affairs, then abruptly cancelled by them. When Tom heard about it, he made one call and helped them change their minds. The interview was reinstated in our flight plan and garnered a front-page story in the Sunday morning edition of *The Blade*.

Not too bad of a day considering JSC public affairs' contention that there wouldn't much media interest in our flight.

Finally, all of us received our first messages from home. During the mission we were each allowed to send one message and receive one message from our families each day. Receiving those messages always made our day and got our spirits soaring.

Mine from Simone told me she was back in Houston and back to work. She and Kai had slight colds, but both slept extremely well the night they got back from Florida. She said watching the launch was an emotionally exhausting experience, and that coming home to an empty house was strange. She congratulated us on the smooth TDRS deploy. She watched it on NASA TV but could only see the back of my head as I was busy looking out the window checking on the satellite. She told me she was looking forward to seeing more of me on TV in the days ahead.

Along with Simone's message I also received a very special one from Kai, who, with some help from his mom on his first-ever letter, wrote:

Hi Daddy!

Hello from Kai! I wanted to let you know that I miss you down here on Earth. Overall Florida was pretty cool, but your launch didn't impress me too much. I liked playing with my cousin Asja even better.

Francis Kregel and Stephanie Sherlock are good buddies of mine by now. They played with me when we had to wait for you to take off. I liked all the attention I got from everybody, but now I'm glad to be home.

Hey, and by the way, I'm able to do something new: I can move a spoon in my hand up to my mouth. Right now there is no food on it yet as Mommy thinks I will make a mess. Maybe she is right.

I miss you, Daddy. I promise to save some really messy diapers for you.
Love you!
Kai

Simone then had Kai hit the keyboard and type in his own special message. It was his first time keyboarding and he wrote me the following:

,mkmk,lx...iokhu mnnbngn nnnnnnnnnnn vr4gf hb4zxcw   hj htb45ty42rr

For a new dad aboard *Discovery* far from home, it brought a huge smile to my face. Sometimes I was asked, "What awaits you after you return from space?" Kai's letter gives a small hint about my answer—changing diapers, taking out the trash, and mowing the yard.

All in all, it had been a great day. Of my forty-four days in space, that one holds the record for the most laughter. We had had a lot of fun as we worked on the various experiments and photographed the Earth. The mood aboard *Discovery* was extremely relaxed. We were scheduled to begin our eight-hour sleep period at 5:12 p.m. CDT. But first, I wrote another entry in my journal:

> **MET 2 days 8 hours 38 minutes**
> *Just getting ready for bed. It's been a busy day! Due to malfunctions, I worked on the Bioreactor experiment one hour later into my presleep period. I did get in a few great Earth Observa-*

tion passes. A few minutes ago we passed over the west coast of South America over Chile where I saw a long chain of snow capped peaks of the Andes Mountains right at sunset. It was beautiful as the last thing to see before we hit 'The Dark Side of the Earth.'

As *Discovery* continued to perform flawlessly, we drifted off to sleep after another busy day.

Flight Day 4 (Sunday, July 16, 1995)

The music for our 1:11 a.m. wake-up was performed by Nancy's eight-year-old daughter, Stephanie, and her Ferguson Elementary School second-grade classmates singing "God Bless the USA." It was the sweetest thing to hear on orbit—a reminder of the loved ones we left behind on Earth. Hearing it made me a bit homesick. After the students finished the song, Nancy radioed: "They did a wonderful job with that, and it's a great way to wake up on orbit listening to a beautiful song like that from a bunch of beautiful kids." It must have been a beautiful moment for Nancy, too, judging by her smile.

Capcom Story Musgrave further brightened the beginning of the day with the good news that the Cleveland Indians were now fifteen games ahead in their division after beating the Oakland Athletics 7-2 in Cleveland the day before.

This day was also special in that it was the twenty-sixth anniversary of the launch of Apollo 11—the first mission to land men on the Moon. The previous year—on the twenty-fifth anniversary—I was aboard space shuttle *Columbia*, and it was a big event for us. But during this flight, it was barely noted. The only mention of it came during our crew press conference.

Such press conferences were scheduled sometime during each shuttle mission so the entire crew would be available to answer questions from the news media. Typically they took place late in the mission—two to three days before a scheduled landing. Waiting until late in the mission gave the crew more to talk about and helped highlight the mission's accomplishments. Because our main mission objective was accomplished six hours into the flight, public affairs scheduled our crew press conference for earlier in the mission, near the mid-point of our flight on our fourth day in space. The event occurred at 7:32 a.m. CDT. Were any reporters up that early on a Sunday morning covering the mission? The PAO strategy must have been to hold it so early in the day to allow reporters plenty of time to file their stories for that evening's broadcasts and for the following morning's newspapers.

The first questions focused on the Apollo anniversary and whether NASA had lost its direction since those heady days of the space program—lost sight of returning to the Moon or going to Mars. Tom said he didn't think so. "NASA's had that goal for at least a decade now—to return to the Moon. But with the fiscal reality of the present economy, we aren't going to be able to do that in the near future," he said. "That's why we're concentrating on endeavors like the International Space Station and cooperating with the Russians to make that an affordable project. But long range, definitely, we will be going back to the Moon and beyond."

We were asked how the early years of our space program, including events such as the Apollo 11 lunar landing missions, influenced and inspired us. I answered about how watching Alan Shepard's launch on May 5, 1961, had gotten me interested in space. Tom said his moment occurred in 1960 watching Echo, an early large inflatable communications satellite, as it passed overhead one evening in northern Ohio. It seemed fitting that he was commanding a shuttle mission whose primary objective was to deploy an advanced communications satellite.

Kevin explained that he grew up watching the Mercury, Gemini, and Apollo programs and that his heroes were Shepard, John Glenn, Neil Armstrong, and Buzz Aldrin. Science fiction books also fueled his imagination and ambitions for flying in space.

Nancy's inspiration came from her dad, Walter Decker, who was a bombardier on a B-29 in World War II. As a small girl, she heard endless stories and went to numerous air shows with her dad. "The love of aviation, the desire to be a pilot, was there from an early age," she explained. "Luckily, my parents, my teachers, none of them ever told me a young girl didn't have a prayer of becoming a military pilot, and by the time I was old enough, the doors were really wide open."

Mary Ellen, the youngest member of the crew, said she caught the fever to become an astronaut much later in life than the rest of us. She was seven years old when Armstrong set foot on the Moon and didn't see a place for herself in the space program when she was young. "There weren't any women astronauts back then. It wasn't that I didn't want to be an astronaut, I just never considered it," she explained. But that all changed in 1983 when Sally Ride broke the gender barrier in the Astronaut Office and became the first female from the United States to fly in space. Mary Ellen was at Purdue University studying chemical engineering at the time. "I think that's where the idea first dawned on me though I don't know that I thought it was achievable," she said. A few years later when she was in graduate school, she visited the National Air and Space Museum in Washington, D.C. where she viewed the IMAX movie *The Dream is Alive*. "There were women flying [in space] at the time. That's when I decided I was going to pursue it," she said.

Nancy said that our mission was special to all of us because we deployed the replacement TDRS for the one lost during the STS-51L *Challenger* accident in January 1986. "I think it speaks very highly of how well NASA has recovered from the *Challenger* accident," she added.

Nancy also described what it was like looking out the window and seeing a world without national boundaries and a cosmos of endless expanse, explaining that for her, it was an incredibly moving and spiritual experience. "You reflect on life in general and your place in the universe," she said. "I am so fortunate to be living what I dreamed about doing"—a sentiment with which most, if not all, astronauts would wholeheartedly agree.

Kevin and Mary Ellen were asked about their impressions as first-time flyers. Mary Ellen said that even though she heard stories about what it was like in space from many other astronauts who had flown, they just didn't compare with the real thing. "Having gone through the launch, I understand why everybody says it's spectacular and indescribable, because it really is. It's just an incredible feeling being on top of this rocket and feeling the sheer power and force of it."

Tom followed up by explaining how the excitement of being in space never wanes. STS-70 was his third mission, and he still found the experience exhilarating and sobering. "The experience of looking out the window is absolutely spectacular. You can't get enough of it," he said. "You look down and realize how fragile our atmosphere is. You realize that everyone on the ground is also traveling on a spaceship—the one we all share called Planet Earth."

The press conference lasted about thirty minutes, with most of the questions being what I would consider friendly. We were all asked about the same number of questions and all got equal time in answering them. Of my four shuttle flights, the STS-70 crew press conference was the most balanced and the most enjoyable because many times previously there was special interest in one or two crew members with the rest of the crew getting pretty much ignored.

STS-70 was a fairly unique mission from all previous shuttle flights because *Discovery* was operating so well. Nothing was breaking and there were no anomalies or unexpected events. Flight Director Randy Stone called it one of the cleanest (trouble-free) vehicles that he could remember. Even the news media was focusing on the trouble-free nature of the flight, perhaps because of our previous woodpecker problems. *The Houston Chronicle* story said, "the toughest problem facing the five astronauts was fixing the vacuum cleaner."

While Kevin and I were using the vacuum cleaner to rid various filters on the middeck and flight deck of dust, lint, and debris, the long electrical cord got pinched as we closed one of the panels where some of the filters were located. It must have short-circuited the cord and caused one of our circuit breakers to pop. While never posing any risk to the shuttle or its crew, that anomaly got mission control's attention. Things were so quiet and problem-free that they had plenty of time to work long and hard on the problem. And sure enough, they developed a set of repair instructions for the pinched cord—splice some wires and wrap the cord with electrical tape. Later that day, it was decided to just stow the vacuum cleaner and not use it for the rest of the mission. Instead, we cleaned the filters using large pieces of duct tape, one of the most frequently used items in space. In official NASA lingo, we were instructed to "use the sticky side of multipurpose gray tape." Because of *Discovery*'s flawless performance, we were able to continue concentrating on our science activities and Earth observation photography.

One of the experiments that kept Mary Ellen and me quite busy was the Bioreactor Development System (BDS), which was about the size of a medium suitcase and was located on the middeck. That experiment was developed at the Johnson Space Center and was investigating the possibility of growing human tissues in space. Starting with individual cells, the objective was to grow larger organized tissues, which theoretically was easier to do in space where normal settling caused by gravity would not take place. That type of bioreactor was thought to have the potential to grow large three-dimension tissues and possibly organs in space one day, which I must admit sounds a bit like science fiction. The bioreactor we flew on STS-70 was one of the pioneering experiments to explore those capabilities.

The cells that were chosen to fly on STS-70 by the researchers were human colon cancer cells. As a former cancer patient myself, it was pretty exciting to think that some of the results from our mission might benefit others on Earth through a better understanding of how those cancer cells grew and developed under the microgravity conditions of space.

The most important component of the bioreactor was its rotating cylinder (about the size of a soup can) in which the cells and tissues were suspended in a special sterile solution (called growth medium), which essentially was the nutrient that allowed the cells and tissues

Scientific *Discovery*

to grow. Almost daily we were required to take out some of the 'old' growth medium and replace it with fresh medium to keep the growing cells and tissues happy and alive. That turned out to be an incredibly time-consuming process.

Contamination was the dreaded enemy of our growing cells, so extreme precautions were taken to prevent that from occurring. It was necessary for us to wear a surgical mask and disposable latex gloves while we operated the experiment. And just about every other step in the procedures had us take an alcohol wipe to clean a needle, the injection port, or something else to kill off any germs that might be lurking around. We went through our share of alcohol wipes on this flight.

Tom created a special video that he downlinked to Mission Control featuring our stuffed Woody Woodpecker working on each of our middeck experiments. While at the Bioreactor experiment, Woody wore a surgical mask to minimize any contamination. He wore a big headset while speaking on the SAREX radio. By showing Woody in front of each of the experiments, Tom hoped to draw some attention to the experiments we were performing.

It had been a busy day. But before turning in, I completed another entry in my journal:

> MET 3 days 7 hours 5 minutes
> *Everything continues to go well. Nancy, Kevin, and Tom are so great to fly with. Tom has been a fantastic commander—I'm really having fun with him on this flight. Still haven't had much time looking out the windows at night. Looking forward to doing so and listening to some Pink Floyd in the next couple of days. Today we had our official press conference. Everyone got asked something and it was all very polite and fun. Tomorrow is my Private Family Conference with Simone. I'm looking forward to that! That's all for now—time for sleeeeeeeep!*

We were nearly halfway through our mission and everything was going extremely well for both the crew and our shuttle. The woodpecker jokes from Mission Control kept coming, and we continued having a great time aboard *Discovery*.

Flight Day 5 (Monday, July 17, 1995)

I got a pretty good night's sleep but woke up a bit early once again. Since everyone else was still asleep, I remained in the airlock floating in my sleeping bag. I wrote a few notes and listened to some music while I waited for the official crew wake-up time.

About ½ hour until wake-up yet—I woke up about 1½ hours ago after 5 good hours of sleep. Just laid here tossing around for a while resting. Saw a few cosmic ray flashes in my eyes, only one medium bright burst. So far on this mission what I miss most are all the new sensations, experiences, and first time views that I experienced during STS-65. <u>Everything</u> was new, different, and memorable on that flight. Last year I tried to write down in a journal all my new experiences each day. So far on this mission there are not many new ones. The one major exception was seeing the launch from the flight deck and viewing it out the overhead windows with my mirror. That was wild and made the launching experience much more visual than on STS-65 where I could only look at lockers down in the middeck. The views of the Earth are still spectacular and I love flying over the rugged terrain. The time for Earth Observation and photography on this mission has been great. The SAREX *has also been a lot of fun for me. During one pass from Mexico to Central America, and then across the southern United States and out over the Atlantic, I spoke with people from Las Cruces (New Mexico), El Paso and San Antonio, (Texas), Louisiana, someone from the American Embassy in the West Indies, as well as someone from London (how did he do that?). It's fun and interesting—passing over the US the radio is just jammed with people trying to make contact—seems like only a small number can actually get through.*

> *Time to rise and shine. As I wake up here I see my picture of Simone and Kai velcroed to the wall. It will be nice to experience a more normal 'family life' once we get back. It was tough the last 4 months before launch. I'm hoping things smooth out ahead.*

Our official wake-up time was at 12:12 a.m. CDT, and the wake-up music was the Cleveland Indians' fight song, "Talkin' Tribe," played for Mary Ellen. Growing up in Cleveland, I didn't even know there was an Indians fight song. It's amazing the things you learn while orbiting the Earth. Mary Ellen enthusiastically responded as the song ended: "Good morning, Houston. How 'bout them Indians?"

Something I was really looking forward to was my scheduled Private Family Conference (PFC) with Simone. Once during the flight Mission Control would schedule each crew member to have a ten- to fifteen-minute somewhat private conversation with our family. Somewhat-private, because even though it would not be broadcast on the normal air-to-ground channels the public could hear, there would still be numerous technicians responsible for communications who just might be listening (only to monitor the quality of the connections for the call, we were assured). With four other crew members always within a few feet of me, the only truly private time on the shuttle was in the WCS, and I was not going to be conducting my family conference from there. During my first flight a year earlier, I was able to talk with Simone once for about five minutes via voice communication only. Things were upgraded for STS-70. We now had the capability to downlink video to our spouses as well. I would have no way of seeing her, but it was exciting to be able to share with her where I was at the moment. I set up a video camera so it showed the airlock I was using as my bedroom.

Because Mary Ellen was still working through some relatively minor space sickness issues, something fairly common on most shuttle missions, my Private Family Conference was delayed a bit and shortened so that she could have a private medical conference with our flight surgeon. Soon after they were done, Mission Control privatized the communications loop for me, and a few seconds later I called down to Simone with an initial voice check. Hearing her voice brought an incredible smile to my face, and I couldn't have been more excited. She filled me in on what she and Kai had been doing. After multiple "I love you" and "I can't wait to see you and give you a big hug after landing" exchanges, our time was soon up, and we said goodbye. I missed hearing Kai's voice—he was at daycare at the moment, but it was so great to talk with Simone. My spirits were soaring afterward!

As the day progressed, Tom and Kevin were working again with the HERCULES experiment. They continued to have problems with the star alignment system but managed to transmit spectacular images of the Earth and completed 95 percent of their planned targets. Nancy kept busy studying the small Medaka fish eggs we were flying as part of the Space Tissue Loss (STL) experiment, investigating the performance and growth of tissues and cells in microgravity. She also continued her work on the WINDEX experiment, investigating the shuttle glow phenomenon. As part of that experiment, Tom and Kevin performed a few thruster firings to assess their effect on shuttle glow. Mary Ellen and I were busy working on the Bioreactor experiment.

Later that day Nancy, Mary Ellen, and I were interviewed by CNN reporter John Holliman. Mary Ellen described the excitement of her first launch: She had dreamed of that moment for years and had worked hard in her career to be where she was. And when the moment of launch finally took place she realized her hard work had finally paid off. Holliman asked Nancy what was going through her mind at launch. She said that coming out of our last countdown hold, she and I gave one another a big thumbs up and we high-fived in anticipation of what was about to happen. She said that since it was her second launch, she was a bit calmer knowing more about what to expect.

Nancy, who had over 3,300 hours of flight time, was asked the difference between being a helicopter pilot and flying aboard the shuttle. "I've covered all the ground that you could cover," she explained. "I've flown at tree-top level in helicopters, flown in the NASA T-38 at forty thousand feet, and orbited the Earth from one hundred and seventy-five miles above. That pretty much covers it."

When asked if there were any changes in vision or other physical changes that occurred in space, Nancy responded, "I always enjoy being in space because you grow taller up here. And being five feet tall on Earth, it's a big deal for me to be five foot two inches for a while when in space." Mary Ellen said her most amazing spaceflight experiences were seeing Earth out the window and witnessing her first orbital sunrise.

Holliman, who referred to me as the zookeeper in space, asked how all our animals were doing. I discussed the rats we were flying and said such research might benefit senior citizens from what we learn about the deterioration of the muscular skeletal system in space. "You're pretty excited about it, aren't you," Holliman asked.

"I sure am," I answered. "I love the research we're doing up here, and you can't get a better laboratory than we have here or a better group of researchers to work with."

I spoke from the heart and from experience when asked if I perceived growing support for what we were doing in space. Explaining that we already had spoken to hundreds of students from five schools around the world using the SAREX amateur radio during the mission, I pointed out that the students were no less enthusiastic about space than I was when John Glenn had launched in 1962. "I see the kids today growing up to be enthusiastic scientists and engineers and carrying our space program forward into the next century." While there is a common misperception that the space program was not as exciting or glamorous as during the glory days of the Apollo Moon landing missions, I said that my experience meeting with and speaking to students around the globe showed their continued strong enthusiasm and support for space exploration.

In one of the more unusual questions I was asked, Holliman wanted to know about some of the weird things that you can or cannot do in space. "Is it true that you can't whistle up there?" Without saying a word, I whistled a short tune for him and put that question to rest. He also asked how bad the food was. I said we had a wide range of food onboard and were able to select our meals for flight. Putting in a plug for my hometown I added, "The same foods I enjoy on Earth I enjoy up here, such as Stadium Mustard from Cleveland, Ohio." Everyone laughed and soon we finished up the interview.

We continued having fun talking to students around the world via the amateur radio. I spoke with students from two different schools, Kevin talked to one in New York, and Tom had a chance to talk with his wife, Becky. In addition, I made about forty other contacts with enthusiastic amateur radio operators around the world. It was amazing how they jammed up the airwaves trying to get through to us as we passed overhead.

A Pacific Ocean tropical depression was starting to form off the southern coast of Baja California that we would be watching as we continued to keep track of the development and movement of Tropical Storm Chantal over the Atlantic Ocean. We never tired of looking out the windows at the beautiful planet below. I wrote in my journal:

> *Had a great day looking at the Earth today! I took hundreds of photos. We had a beautiful pass over Guantanamo Bay, Cuba, and it was spectacularly clear. I sure hope I got some great shots for John! We also had a great pass over South America. The rain forest clearing in Brazil is unbelievable! Everywhere you can see the herring bone patterns of clearings. We also had a number of superb passes over Luxor, Egypt and out over the Red Sea. During STS-65 I just wanted to look out the window and take in the scenery below. This time I want to get some great photos.*

Now past the mid-point of the mission, we were all in a groove aboard *Discovery* and having the time of our lives.

Chapter Seven
Columbus was Right

Since my first flight when I was fourteen years old, I have always enjoyed looking out the window of an airplane and watching the scenery go by. Passing over cities, farmland, mountains, lakes, and rivers, I loved the perspective of looking down below. Whether flying in a small plane at five thousand feet or in one of NASA's T-38 jets at forty thousand feet, the view from above always fascinated me. If I had a map to help me identify landmarks along the way, all the better.

As much as I enjoy looking out the window of a plane, I loved the view from space even more, and studying Earth was always one of my favorite activities when I was on the shuttle. The view was simply incredible from an altitude of two hundred miles. I not only saw cities and states pass by relatively quickly, but also countries, and sometimes entire continents, as they came and went in mere moments. Traveling at 17,500 miles per hour, or 5 miles a second, it only took about eight minutes to fly from Los Angeles to New York City. In that same amount of time—eight minutes—you could also cross the entire continent of Australia.

Even though we traveled at twenty-five times the speed of sound, I never felt that sense of speed looking out the window of the shuttle. Just as it is difficult to sense you are traveling five hundred miles per hour on a commercial jet, you had no real sense of speed on the shuttle because you were not passing any objects in space. Only by looking down at the surface of the Earth did you realize your speed as you saw countries passing by in the same time frame that small cities would pass by looking out the window of a plane.

Before I ever completed my first orbit as an astronaut, I had seen many pictures of Earth taken from space as well as the spectacular IMAX films of our planet. I thought I was pretty well prepared for what I would see. I was not. I gasped as I took my first look out the window. "My God, how beautiful," I exclaimed under my breath. Mine was not a unique experience. During the course of my astronaut career, I flew with five rookie astronauts on my subsequent missions and the reactions were almost universal—first a sudden gasp and then some form of exclamation—"Wow!" or "Oh my God!" or "How beautiful!" Even John Glenn's enthusiastic reaction when he crewed the first All-Ohio mission aboard his *Friendship 7* Mercury capsule on February 20, 1962, was very much the same—"Oh, that view is tremendous!"

It truly was. I made it a point to see as much as I could, knowing what a unique once-in-a-lifetime experience it was—or, in my case, what ultimately would be four times in my life. Minutes after getting my first view of the Earth on my first flight, I had an overwhelming desire to share that view with everyone else. I wanted Simone to see it. My mother, brothers, and sister. I thought everyone on Earth needed to experience that perspective of our planet. I thought it would be great if somehow I could trade places with each of them for only thirty seconds so that they could enjoy, with their own eyes, the majestic beauty of the Earth. I strongly believe it would drastically change how each of them view our planet just like it had changed for me and the five hundred and fifty other astronauts who have ventured to space.

Shortly after my first launch on STS-65, I noticed the thinness of the atmosphere that surrounded our planet and protected us. From Earth, the sky appears to go on forever. But from space, it was clear that our protective atmospheric cocoon was only about twenty miles thick. That was it. Its thinness really jumped out at me and made me appreciate how fragile our planet really was.

Something else that jumped out at me as I took my first glimpse out the window from space was the curvature of the Earth. It was immediately clear that Columbus was indeed right; the Earth is not flat. So to all the members of the Flat Earth Society reading this, I hope that observation does not ruin your day.

Another indelible impression was the blackness of space. In full sunlight, it was a deeper color than any shade of black I ever saw on Earth. I have been spelunking deep within caves and turned off all the lights to experience true darkness, but space appeared much blacker than that. It was almost as if it was glowing—a sort of fluorescent black. It was such a rich, deep-dark blackness that it had an almost velvety appearance.

We astronauts struggle with our descriptions of it because there is, quite frankly, no comparison on Earth. And when I saw the Earth with its thin atmospheric layer glowing bright blue coming up against the velvety black of space, it literally took my breath away.

Being in space always made me feel totally disconnected from Earth. I have passed two hundred miles above Houston, where my home was at the time, but still felt like I was a million miles away—far away from the people, far away from life down there. So it made it all the more enjoyable to see familiar sights from space, even something as mundane and unspectacular as a bridge. I think it was a way for us to connect back to our planet, back to our homes, back to our Earth-bound lives.

One of the most curious things about being in space was the tendency to pick up a pair of binoculars to get a closer look—to look for more detail, to zoom in on whatever was passing below. It wasn't unusual to hear me or one of my colleagues excitedly exclaim something like, "I can see a bridge across that river down there." It was as if we had just made a monumental discovery. We saw bridges over rivers nearly every day on Earth. In those instances, they were remarkably unspectacular. But something happened when we left the planet. We craved those details and delighted in seeing them.

During the sixth day of the STS-70 mission, I spent a significant amount of time looking out the window and saw many more sunrises and sunsets than on any of the other flight days. Because we circled the Earth every ninety minutes, we experienced sixteen sunrises and sixteen sunsets every twenty-four hours. Granted, we probably slept through a third of those, but we still saw those spectacular occurrences many times. And like looking at Earth, sunrises and sunsets were also equally spectacular and profound—no matter how many times I witnessed them.

Guantanamo Bay, Cuba

I kept an eye on our orbital track during the mission so I'd know when we'd be passing over Cuba, particularly Guantanamo Bay, where my mentee and friend John Pickering was serving in the US Army. I had met John five years earlier as a volunteer for a United Way program, Friends of Students, an organization that paired people like me with students. John and I became close friends. Following high school he joined the Army and, after basic training, one of his first assignments was Guantanamo Bay.

Each time we passed over Cuba during daylight, I scanned the island looking for Guantanamo Bay. Once I spotted it, my eyes would be solidly fixed on it and I'd think of him and the good times we had together. I'd also take some pictures for him. A few minutes later, the view of Cuba would be gone as we headed east out over the Atlantic Ocean. One of the personal items I carried with me on the flight was a patch from his military unit. After we returned, I gave him back the patch and some of the pictures of Guantanamo I had taken.

Kennedy Space Center/Cape Canaveral

Nearly all space shuttle astronauts got a kick out of seeing the Kennedy Space Center from two hundred miles above. It was easy to find because Cape Canaveral protrudes into the Atlantic Ocean from central Florida's east coast. Using binoculars, I could see the massive Vehicle Assembly Building, Shuttle Launch Pads 39A and 39B, the Shuttle Landing Facility's three-mile-long runway, and the string of old launch pads at Cape Canaveral Air Force Station from where many of the United States' early rockets were launched.

During one pass just after sunrise, I saw the VAB and the long shadow it cast. That unique lighting condition gave the VAB an almost three-dimensional appearance.

KSC always brought a lot of "oooohs and aaaahs" from my crew mates. When I saw it, I would think, *Man, that's where we started this mission, and look at us now.* On STS-65 as we were completing our first orbit of the Earth I just happened to glance out the window and could see KSC down below. I was absolutely blown away by the thought that I had just traveled completely around the Earth while my friends and family at KSC were probably still fighting traffic waiting to get back to their hotels!

Houston, Texas—Space City, USA

Another locale that always drew great interest was Houston. That's where all of us lived and worked and where our families were. A pass over Houston would have everyone at the windows, noses pressed against the glass, looking for familiar landmarks like Galveston Island and Clear Lake where the Johnson Space Center is located.

During a pass on one of my flights, I pulled out the binoculars and tried to find my neighborhood of El Lago, only two miles from JSC. It was a clear day in April with low humidity, so the air was clear and I was able to make out quite a bit of detail. I saw the Baytown Bridge over the Houston Ship Channel. I followed Highway 146 south from the bridge, and saw where it and NASA Road 1 intersected in Seabrook, about three miles from the Johnson Space Center and half a mile from my house. I could definitely make out that intersection but was unable to see my street—let alone my house. The El Lago neighborhood had too many trees, making it difficult to see any houses from above.

But the fact that I grabbed the binoculars and tried to zoom in to see my neighborhood was, as I said before, an indication of my desire to connect to my planet, my neighborhood, and my home. I found those episodes of trying to see my home exhilarating. Even though they lasted only a few minutes each time, they lifted my spirits and made me feel great.

Rainforests

I continued to be amazed at how much of the Amazon rainforest in South America was being destroyed to make way for development and agricultural use, particularly in Brazil. During my first mission on STS-65 only twelve months earlier, I saw massive fires while passing over South America at night. It was only during daylight that I saw what was burning—the rainforest. During STS-65, nearly the entire continent of South America appeared to be under a smoky pall. Not just isolated pockets or regions of that vast continent, but the whole continent—at least everything east of the Andes Mountains. To see such a global impact from those fires opened my eyes to the impact humans were having on our entire planet.

A year later aboard *Discovery*, I could still see many fires burning at night, and South America still had a lot of smoke over it. But I no longer had the impression that it covered the entire continent. Maybe the cutting and burning was finally slowing down, or maybe they were simply running out of rainforest to burn down.

In some unspoiled areas of rainforest, I saw a single clearing running through the middle of it, apparently to accommodate a small road. On later shuttle missions, that same area had small clearings and new roads coming off the original one. Later yet, those small side roads had their own side roads. Ultimately, large areas would be cleared with virtually no rainforest visible from space. Then gray patches would become visible—small cities forming and growing to support the people moving into the former rainforest.

Seeing that and realizing how much rainforest already had been cleared was another *Wow!* moment for me. It was amazing to see what was being done to the planet. In one sense, it already happened in North America when our early settlers cleared forests to create cities and farmland. So it makes it difficult to criticize the Brazilians for what they were doing, but it didn't make it any easier to look at. It was gut-wrenching.

Our Earth Observation instructors at the Johnson Space Center told us that once the rainforests were cleared away, they would probably never return. We also were told that the soil in the rainforest-cleared areas would support agriculture and be productive for growing crops only for a few years before all nutrients were depleted. And then what? It definitely hurt to see the vast extent of this deforestation from space.

Great Wall of China

At some point during almost every talk I give, I am asked whether the Great Wall of China was visible from space.

The number one, no pun intended, question used to be: "How do you go to the bathroom in space?" But beginning in the mid-1990s, I was asked more and more about the Great Wall.

It had become a sort of urban legend that it was "the only man-made structure visible from the Moon" or "the only man-made structure visible from space." Both probably originate from a 1930s *Ripley's Believe It or Not* cartoon that proclaimed the Great Wall of China was "The mightiest work of man, the only one that would be visible to the human eye from the Moon." In my seven hundred orbits of the Earth, I never was able to see it because it always was well north of my orbital ground track. I never passed directly over it.

But Leroy Chiao, a good friend and crewmate from STS-65, spent six months aboard the International Space Station in 2004–2005. As a Chinese-American, he was greatly interested in seeing the Great Wall and bringing back photographs of it. But during his entire time in space, he too was never able to see it—even with the aid of binoculars. It's too narrow a structure—thirty feet at its widest. And because it was constructed mainly with rocks and dirt from the surrounding area, it tends to blend in with the surrounding landscape.

Leroy did take some great pictures of it in a region that had recently been blanketed with snow. The snow brought out the highlights of the wall much like a long shadow at sunrise or sunset. When the photos were enlarged, you could clearly make out the Great Wall.

But neither Leroy nor any other astronaut, cosmonaut, or Taikonaut (Chinese astronauts) have reported seeing the Great Wall of China from space with their own eyes.

Mount Everest

I was thrilled to see Mount Everest from space on my first mission, but was only able to locate it because one of my crew mates pointed it out to me. Even though it is the highest peak on Earth, towering 29,029 feet above sea level—about five miles high—it doesn't exactly jump out at you from two hundred miles above the Earth. And being in the middle of a huge range of snow-covered peaks of comparable size, it tended to blend in with the others. To see Mount Everest, you need to know exactly where to look.

Before STS-70, I reviewed the maps carefully and carried a few pictures with me in my crew notebook to help me identify landmarks along the approach to Mount Everest. That's because I wanted to be able to find it on my own. One of the more beautiful landmarks high in the Tibetan Plateau—fifteen thousand feet above sea level—was Bowtie Lake, named because of its unique shape. Its bright blue glacial water sitting in the middle of the surrounding snow-capped peaks made it an easy target to let me know we were approaching Mount Everest.

During my four flights, I was able to see Mount Everest about twenty-five times. I considered it the lazy man's way to see its peak because there was no climbing involved. I just looked out the window and sipped a little Tang orange drink as we passed silently overhead.

Great Barrier Reef

I have never been to Australia and, in fact, had never been below the Equator on the surface of Earth until 2007, but I was able to explore much of the Southern Hemisphere from space.

The red soil of the outback in central Australia as seen from the space shuttle was what I imagined the surface of Mars would look like from two hundred miles up. It was so different from the rest of the planet that I almost couldn't believe I was looking at Earth.

What I particularly enjoyed seeing was the Great Barrier Reef. Considered one of the seven wonders of the natural world, it stretches nearly 1,800 miles—longer than the Great Wall of China—and is composed of nearly 2,900 individual reefs and 900 islands.

Reefs in general, and the Great Barrier Reef in particular, were spectacular. The ocean's deep blue water helped highlight the reefs, which typically looked aqua or light blue because we were looking at them through relatively shallow water. The Great Barrier Reef looked like a necklace of turquoise beads. The water appeared crystal clear, which was how I imagine the Earth once looked before humans altered and polluted the environment.

The reef looked vibrant despite the impact human activity was having on the health of that great ecosystem. It was one of those landmarks like Mount Everest that I studied in school when I was young and then one day found myself actually seeing with my own eyes. It was an unforgettable, emotional moment.

Hawaiian Islands

There's no doubt about it. The Pacific is one big ocean. Ask any astronaut who's flown over it, and they will concur without any hesitation.

So how big is it? It's so large that it took nearly twenty-five minutes to transverse it. Keep in mind that's nearly one-third of the ninety minutes it took for the space shuttle to circle the Earth.

I don't think I fully appreciated how big it was until I flew over it a few times. And it's as blue as it is vast. Occasionally I'd see a reef or small island in the middle of nowhere that made me think of

Gilligan's Island. But there's not much else down there, and not much to look at. Typically, if I was looking out one of the windows when we started to cross the Pacific, I would continue looking for a few minutes and then find something else to do until we came upon some land once again.

But that changed when our orbital track took us over the Hawaiian Islands. In the middle of that great expanse of water was a small, beautiful chain of volcanic islands, and they were a magnificent sight to see.

The islands were bright green with white fringes—the crashing waves and exquisite sand beaches that ring the islands. As we'd pass overhead, I'd imagine myself on one of those beaches enjoying a nice vacation. But soon we moved away and were crossing the rest of the Pacific. It was only a few fleeting minutes, but I'm glad I had the chance to enjoy it.

On one of my flights, we passed directly over the Big Island at night. It was pitch black with very few lights. They try to keep light pollution to a minimum to support the large telescope facilities at the top of Mona Kea. Looking down I saw what looked like a river glowing bright red and orange. At first I didn't realize what I was looking at, but it soon dawned on me that it was probably molten lava flowing from Mount Kilauea to the sea. Yet another *Wow!* moment—another incredible sight from a very unique vantage point.

Jet Contrails

Sometimes when you look up at the sky you see straight bright white lines. They form around thirty-five thousand feet when water vapor in the exhaust of jet engines condenses into clouds called contrails.

I had seen them up close on many occasion when flying in formation with another T-38 jet, and loved watching them form, grow, and swirl around before eventually evaporating. One day my pilot decided to try to surf one. With the lead T-38 creating two beautiful contrails, we flew up just above one giving us the impression that we were skipping along and just barely touching it.

Looking down at Earth from space, I could readily see contrails. When conditions were right, I saw many of those thin white streaks crisscrossing the sky. Once I saw the exact point where it was being formed and wondered if I could see the plane creating it. I grabbed a pair of binoculars and studied the view intently. I kept my hands as steady as I could and held my breath to further stabilize myself. But try as I might, I never was able to make out the jet—just the contrail.

Egypt

One of my absolute favorite views from space was looking out the window at Egypt. I had spent nearly two weeks there in 1980 along with my brother Dennis, during which we visited the Sphinx and Great Pyramids, the museums of Cairo, the Nile River, and the Valley of the Kings where King Tut and other pharaohs were buried. Riding camels, venturing to the inner burial chamber deep inside the Great Pyramid, and climbing to the top of one of the smaller neighboring pyramids, it was an incredible experience.

From space the beauty of Egypt was without equal. The bright blue water of the Mediterranean and Red Seas, and the light brown desert with the lush green Nile River passing north-to-south across the country, made for spectacular contrast. And the air appeared to be crystal clear because of the extremely low humidity allowing for unmatched visibility. The detail I could see down below was strikingly spectacular.

To the north I could see the Nile Delta with the ancient city of Alexandria on its west side, the Suez Canal running from the Mediterranean down to the Red Sea on the right. Near the bottom of the delta triangle the city of Cairo was clearly visible as a gray patch down below. With a good pair of binoculars and when long shadows were present during sunrise or sunset, you could sometimes make out the Great Pyramids as small black dots. To see those marvels of ancient engineering while passing overhead in a more modern marvel was such a magnificent contrast. And with as much magnification as I could squeeze out of the binoculars, I always craved more. My eyes strained to see more detail. I wanted to get closer.

Following the Nile River flowing south, the next thing to jump out at me was the big bend in the Nile near the city of Luxor, where only a few miles away the Valley of the Kings and Valley of the Queens are located. Having spent a few days there exploring the Great Temple at Karnak and King Tutankhamun's tomb, I always sought out that area as we passed over Egypt and again tried to see as much detail as my eyes would permit. Further to the south it was easy to make out the Aswan Dam as a thin dark line at the northern end of Lake Nasser with some roads and an airport visible near the city of Aswan.

One of the most striking sights looking down on Egypt was being able to see clearly with the naked eye the border between Egypt and Israel in the Sinai Peninsula. The Sinai is a vast desert area and almost featureless from space on the Egyptian side. Across the border in Israel they had done significant amounts of irrigation with farms and cities easily visible in the Gaza Strip area. That almost day-night contrast across the border was one of only a small handful visible from space. We are used to seeing borders on our maps and globes to differentiate states and countries, but from space our planet was nearly borderless.

I found the entire country of Egypt a most beautiful sight, especially so because of my special connection to it from visiting there once many years earlier. Seeing it from space only made me want to go back to see it all again.

Ship Wakes in Sun Glint

If you have ever been on a boat of any size traveling at almost any speed, you are familiar with seeing the wake behind the boat as it travels through the water. From two hundred miles up, we routinely saw wakes behind larger ships that sometimes extended for hundreds of miles behind the vessel creating them. They could best be viewed during periods of what was referred to as 'sun glint,' when we were looking down at a location on Earth where the sun was reflecting on the water and almost giving it the appearance of being liquid silver or liquid gold. Using those reflections, we could make out incredibly subtle details on the water's surface. Small perturbations greatly affected the reflected light and allowed us to see small differences in the surface of the water.

While passing over the Atlantic or Pacific oceans, Mediterranean Sea, Gulf of Mexico, or other large bodies of water when the sun's reflection was visible, I frequently saw the familiar 'V' pattern of ship wakes. I never saw the ships creating them, but surprisingly I could see the wake for hundreds of miles.

The only times I could see ships themselves were in the Middle East where on many occasions I was able to see some of the large oil tankers in the Persian Gulf. With binoculars I was able to see those massive supertankers (some of which were nearly a quarter mile long!) as small black lines as they passed through the Strait of Hormuz or waited offshore to be loaded with oil.

Earth at Night, Cities and Lightning Included

Once the sun went down, Earth became a totally different planet. Most of the spectacular color was gone. No bright blue oceans, no brown mountains, no green hills and valleys. All that was visible was light—city lights, fires, glowing lava, and flashes of lightning.

In my opinion, the Earth at night was every bit as impressive as the Earth in daylight. I think I enjoyed the nightscapes more. It seemed a bit more serene, a bit more mystical and mysterious. But it was hard to share Earth-at-night views with others because our cameras had a difficult time capturing the low-light-level images the human eye can easily see. They did a good job recording our views in daylight, but once the sun set our cameras could only capture a less-than-perfect representation of what we were seeing. Only recently has NASA started flying new high definition low-light-level cameras aboard the ISS that can capture the full beauty of the Earth at night.

If you've ever flown into airports in cities like New York, Los Angeles, Las Vegas, or Chicago at night, you know the beauty of seeing the metropolis' lights from above.

The view from space was very similar. But because we were at a much higher altitude, we saw a city's broader expanse along with numerous other nearby cities. During STS-70, I saw the lights of Houston, Dallas, New Orleans, Shreveport, Atlanta, Tampa-St. Petersburg, Orlando, and Miami all at the same time as we passed over the Gulf of Mexico just skirting the southern coast of the United States.

Sometimes I found it difficult to get my eyes adapted enough to the darkness so I'd be able to see some of the fainter stars and other phenomenon. With all the cabin lights, lighted instrument panels, and laptop computers, there usually was too much ambient light for my eyes to get truly dark-adapted. Only during brief moments when we would close the laptops and power down the instrument lighting was that possible. It was always a challenge. But once my eyes became dark-adapted, I saw some incredible sights.

Among them were:

Thunderstorms. In the middle of particularly strong ones, I sometimes saw thirty to forty lightning flashes every second from storms four hundred to five hundred miles across. And we saw only the lightning flashes imbedded in the clouds as the cloud tops lit up in various shades of bluish-purple or bright-white flashes. The multiple flashes every second always reminded me of blinking lights on a Christmas tree.

Shooting Stars. My favorite nighttime view was seeing shooting stars or meteors entering Earth's atmosphere. Those momentary thin lines of light were amazing. While most shooting stars seen on Earth are visible when they are twenty to thirty miles above the Earth, aboard *Discovery* I witnessed them about one hundred and forty miles below us. The light trails were much shorter than I was used to seeing on Earth because we were so much farther away from them. It was a bit strange watching them, because on Earth we look up to see shooting stars, but in space we needed to look down, giving it a totally different perspective. Instead of seeing stars in the background, from space you always had city lights, thunderstorms, or fires in the background. During a typical night pass lasting forty-five minutes or so, I would see one to three shooting stars once my eyes fully adapted to the darkness.

Satellites. Down on Earth when you see a small light moving across the night sky, you probably are looking at an airplane. In space, any bright moving light is probably a satellite orbiting the Earth. Occasionally after sunset or before sunrise, I saw such moving stars and knew they were other spacecraft. They were almost as rare as seeing a shooting star. So while I had no idea what specific satellite I was looking at, each time I spotted one it was special. I'd watch it as long as I could until it eventually disappeared after a minute of two. Only once during my STS-83 mission was I able to identify the satellite. Mission Control called up to us that if we looked out one of the windows at a specific time coming up that we would be able to see the Russian *Mir* space station passing by in a different orbit. We passed within thirty-five miles of one another and while it moved fairly quickly and appeared only as a bright star, it was still pretty amazing to watch it zoom by. Fellow American astronaut Mike Foale was aboard *Mir* at the time as he and his two Russian crew mates worked through some major problems related to a recent collision of a Progress cargo spacecraft during a failed docking attempt.

Airglow. This was a faint, hazy, yellowish glow that occurred in the upper atmosphere and was visible on the dark side of Earth as a thin glowing band above what seemed like the edge of Earth. The thickness of the airglow layer was about ten miles or so and occurred about sixty miles above the Earth. The glow was strikingly eerie. It was such a diffuse layer that

I could see bright stars shining through it. After the sun rapidly set and the thin layer of atmosphere around the Earth went through its rapid series of color changes after sunset—from blues to yellows to oranges, then reds, finally to black, the airglow layer remained visible until the shuttle closed in on the next sunrise. It was a strikingly beautiful and graceful glowing band, sort of hovering just above Earth. It was brighter than, but similar to, the Milky Way. It always provided a kind of warm glow around the planet.

Magellanic Clouds. I first saw the Large and Small Magellanic Clouds during STS-65 and enjoyed the unique opportunity to see them again during night passes across the southern hemisphere aboard *Discovery*. Although they look like separate pieces of the Milky Way, they are actually sister galaxies of the Milky Way Galaxy located one hundred and fifty thousand to two hundred thousand light years away. They appeared only as fuzzy patches of light, but realizing that I was looking at two other galaxies orbiting our own made them very special to see. And since they are only visible from the Southern Hemisphere, I enjoyed the opportunity to view them during STS-70 along with the neighboring Southern Cross constellation. Although the smallest of the eighty-eight modern day constellations, the Southern Cross' prominence makes it easily recognizable and is used in the southern hemisphere to mark the southerly direction because there is no south pole star equivalent of the northern hemisphere's star known as Polaris.

One of my favorite things to do during STS-70 and my other missions was listen to Pink Floyd's *Dark Side of the Moon* album as I looked out the window during a night pass—from sunset to sunrise. With the night pass lasting about forty-five minutes and the album forty-two minutes, fifty-nine seconds, it was such a perfect match and made the nighttime Earth-viewing experience even more incredible. It was as if the music added an additional dimension to the experience. For all future space travelers, I highly recommend taking it along for your journeys to Mars or beyond!

One of the STS-70 mission's other unique experiences—one of the more memorable and remarkable moments of my forty-four days in space—occurred because of a meteorite hit on Kevin's side window.

Because of that impact, Mission Control directed us to reorient *Discovery* so we were flying with our payload bay pointed down at the Earth with the tail of the shuttle facing forward. So looking out the larger overhead windows on the aft flight deck, we saw indescribable views of Earth.

At one point, Tom and I were alone on the flight deck. It was perfectly quiet, and there was no radio traffic from Mission Control. I anchored myself into a position sitting on the ceiling so I could look down between my legs that curled around one of the overhead windows and see the Earth below. Tom and I just stared, neither of us saying a word. It was so peaceful. We were crossing the Pacific Ocean looking at the beautiful blue water with occasional white puffy clouds floating by. It was a remarkable perspective. I felt like I was aboard a glass bottom boat.

Soon I could see a beautiful coastline coming into view. It was the west coast of South America in the area of Peru. The snow-covered Andes Mountains were directly below me, and I could see well down the chain of mountains to the south. The terrain was incredibly rugged, something I appreciated even from our altitude. Tucked into the mountaintops were a few beautiful crystal-clear lakes, some deep blue and others more aqua in appearance. I spotted a volcano venting white steam.

The transition from deep blue ocean to rugged mountains was captivating. After a few minutes, we left all that behind as we crossed over the dark bluish-green rainforests of Bolivia, Paraguay, and Brazil. And in another few minutes, we were crossing the east coast of South America in southern Brazil, heading out over the southern Atlantic Ocean on our way to Africa. Spectacular bright blue water once again filled my window. That fifteen-minute ocean-to-ocean pass over South America was an amazing experience. Another one of those *Wow!* moments.

Among all these incredible sights and many others that I experienced while in space, I always tried to see as many of the cities as possible where my mother worked for the State Department during her many years in the Foreign Service. While I was never able to get a good look at New Delhi, India; or Rangoon, Burma, with three of my flights taking place during their monsoon seasons, I was able to see Jakarta, Indonesia; Tunis, Tunisia; and Damascus, Syria, a bit off in the distance. Istanbul, Turkey, where I visited my mom in 1980, was unfortunately just too far north for me to see on any of my flights. Still I followed her adventures around the globe as I enjoyed my own.

One of my disappointments was never passing directly over my hometown of Cleveland during any of my flights. And for all of us All-Ohio crew members, we could barely see the Buckeye State—and honorary Ohioan Kevin's beloved New York State—off in the distance. That's because STS-70's orbital track was a band that extended only about two thousand miles north and about two thousand miles south of the Equator. Our flight path primarily took us over most of Central and South America, Africa, most of India, southern Asia, and Australia. The only portion of the continental United States that we flew over was most of Florida and southern Texas. You could see other areas to the north, but the farther away they were the harder it was to see any detail. That made it nearly impossible to see any recognizable landmarks in Ohio and the rest of the Midwest. I tell people I saw the Terminal Tower, a seven hundred-foot, fifty-two-floor landmark in downtown Cleveland, from space, but that's probably stretching the truth quite a bit (although I would never admit it).

On all my flights I longed to spend more time looking out the windows and studying the Earth. There was just so much that I wanted to see and simply not enough time to see it all. That's why I tried to spend nearly every available moment that I had in front of the windows. I wanted every sight to be permanently imprinted on my brain, making it impossible to forget anything that my eyes had seen. While sadly impossible to accomplish, overall my trips to space had a lasting effect on my life, leaving me with a keen appreciation for how fragile our planet is. I have seen the giant footprints that humans are leaving on its surface, easily visible from two hundred miles above.

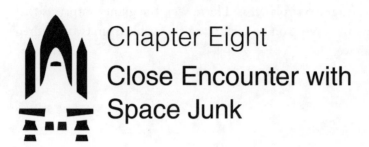

Chapter Eight
Close Encounter with Space Junk

We continued our sleep-shifting, which moved our official wake-up time for Flight Day 6 to 11:12 p.m. CDT Monday. I woke up early once again and floated in my sleeping bag in the airlock while I jotted down a few thoughts:

I woke up about an hour ago—cold, and never went back to sleep. Official wake-up is in about 15 minutes. I think today is Kevin's wake-up music day. Got about 5 hours sleep and feel pretty good this morning.

Tom has been a lot of fun up here. He is really laid back and likes to joke around. Overall, he's a great commander. Only 5 more minutes to wake-up.

I'm ready to start thinking about going home. Only 3 days to go!

After almost a week of Ohio-themed wake-up music, it was time to acknowledge the honorary Ohio astronaut's Empire State roots. Kevin was such a good sport and always kidding about the perks of being an Ohioan: "Butter, cows, Stadium Mustard, and Chief 'Yahoo,'" he summarized. He had been immersed in Ohio culture and lore for most of the past year, but there was still room for some improvement, like getting the name of the Cleveland Indians' mascot, Chief Wahoo, correct. Kevin had morphed into an Ohioan by all appearances, but underneath he was still a very proud New Yorker.

He finally got his recognition and moment in the spotlight with the morning's wake-up music, "The Streets of New York," a classic written in 1906 by Victor Herbert. As soon as the first lines of the song were sung—"In old New York, in old New York"—a huge smile broke out on Kevin's face. The song was for him and only him. He had been trapped inside *Discovery* for the past week talking everything Ohio, so I think he was glad to make a connection to his 'original' home state.

When the song was over, Kevin told Mission Control, "It's good to hear some really good music for a change."

Our Capcom, Story Musgrave, replied, "Well, it certainly is a grand place. It's got everything."

Kevin added, "As old Frankie Baby said, it's one heck of a town." Story asked Kevin whether the Ohio guys were giving him any trouble. He answered in typical New Yorker fashion and with an extra heavy New York accent: "One New Yorker can easily handle four Ohioans." It was classic Kevin. STS-70 was his first flight, but he was performing like a veteran and had earned the respect of all of us on the crew.

When we removed the shades from the windows after waking up, Kevin noticed a small crater in the window just to the right of the pilot's seat. It was caused by a micrometeorite later estimated to be only about the size of a grain of sand. No one heard the impact, so we couldn't be sure exactly when it happened. But it definitely wasn't there the day before.

The impact affected only the outer pane of glass and posed no threat to us because shuttle windows consisted of three separate panes (triple redundancy—two could fail and we still would probably survive). The affected outer-most pane, which was a half-inch thick, provided heat protection during reentry. The inner two panes served as the primary and backup pressure panes for maintaining crew compartment pressure. Each of those panes was three-quarters of an inch thick, making the window's total thickness two inches.

The crater was only about one millimeter across and left a hole no deeper than one millimeter from what we could tell. NASA later estimated the size at $1/16$ of an inch in diameter and $1/32$ of an inch deep. It was a small ding and not very impressive, but it was a sobering reminder that there were many things that could strike us as we orbited the Earth. Even though it was miniscule, it was able to pack a punch because of the high speeds involved. With *Discovery* traveling at 17,500 miles per hour in one direction and the particle potentially moving at the same speed from the opposite direction, the impact speed could have been 35,000 (10 miles per second) or possibly even higher if the particle originated from outside of Earth orbit. Even a tiny speck of dust traveling at such speed can do significant damage to orbiting spacecraft. Needless to say, it got our attention.

There was a lot of discussion onboard about whether we should tell Mission Control about it. Tom decided that was the right thing to do. "It looks like there may be a particle imbedded. And on the exterior of the window, it appears to have a bicycle spoke effect," similar to the ray pattern around impact craters on the Moon, Tom said, describing it as best he could. A short time later, we sent some video of the impact site to Mission Control.

Flight Director Randy Stone later reassured the news media that "There is no hazard to the orbiter. I expect that it meets the criteria for window replacement when we get *Discovery* back to Earth, but that will be the only impact." Total replacement cost was estimated to be nearly fifty-five thousand dollars.

While we were in no danger and never felt threatened or nervous about what happened, it was an eye-opening occurrence and a good reminder that space junk was real and needed to be considered in the design for the International Space Station, which would remain in space for decades.

Capcom Tom Jones told us to keep an eye on our stuffed Woody Woodpecker. "He may have been the one responsible for the ding in the window," he said. We joked that Woody also might have been responsible for the vacuum cleaner problem we had earlier in the flight. "I think we can absolve all other crew members from any blame on this flight," Jones responded. Any and all problems would be blamed on Woody Woodpecker.

Soon we turned our attention to the Microencapsulation in Space (MIS) experiment, which attempted to develop and produce time-release antibiotic medication in weightlessness. The idea was to 'spray' a fine mist of the medication into a small chamber and let it solidify while floating, thus producing a batch of uniform-sized particles that could be put into a time-release capsule, much like the tiny particles of medication packed into the famous time-release Contac brand cold medicine. It was thought that the lack of gravity would allow the encapsulation process to be performed with much greater purity and uniformity than could be achieved on Earth. That experiment was housed in a unit that took up the space of two middeck lockers and was mounted on the forward bulkhead on the middeck.

Without a doubt, it was the simplest experiment I ever operated in space. A detailed and thorough knowledge of rocket science was not needed for that one, far from it. We had a joke in the Astronaut Office and called those types of experiments Pilot Science, meaning that even a shuttle pilot could successfully operate them. It allowed them to be a scientist for a day, but did not require extensive skill or training. Typically Pilot Science experiments had a single on/off switch and maybe an indicator light that illuminated when the experiment was turned on. And that was pretty much all I had to do for the MIS experiment. Still, we made a big deal about it. Nancy was ready with the camera to capture the moment that I moved the experiment's switch from the 'off' to the 'on' position. We may have even conducted a short 10-9-8-7-6-5-4-3-2-1-0 countdown to increase the drama. Switch 'on.' I checked the indicator light. It was green. I was done with that experiment for the day. Automatic timers would run it while I was busy elsewhere, and it would continue operating while we slept that night.

I made contact with a group of students at Fallbrook Union High School in southern California midway between San Diego and Los Angeles as part of the Shuttle Amateur Radio Experiment (SAREX). We began our contact as *Discovery* approached Johannesburg, South Africa, where an amateur radio operator established our connection and relayed the questions from the students. During the session, the students were able to ask ten questions, including why it was important to keep sending humans into space, what it was like during launch, and how long astronauts could remain in space. All was going well until one of the last questions from a former drama student and recent Fallston High School graduate named Nate Golich. He asked what my favorite Shakespearean play was. I then proceeded to break one of the cardinal rules of doing an interview. What I should have said was "I don't know at the moment" or "That's a great question. I like them all." It had been more than two decades since my high school-required reading of Shakespeare, and I sure hadn't read any since. I couldn't think of the names of any Shakespeare plays let alone come up with my favorite. I pressed on with my answer in spite of my lack of Shakespearean knowledge.

I told Nate that my favorite was '*Theodore DeBergerac*,' which got a huge laugh from the students down below. The local TV stations also got a kick out of my answer, which they featured on that night's newscast making light of the fact that a rocket scientist aboard the space shuttle was not very knowledgeable about Shakespeare. After the mission, I received a detailed list of Shakespeare plays from one of the teachers at Fallbrook along with a light-hearted note saying that she hoped the list would help me out in the future. Sometimes the best response is to simply say, "I don't know." Back to media training for me!

We enjoyed contacting students as much as they seemed to like talking to us. On STS-70, we were able to schedule nine schools, one from Argentina and the rest in the United States. One was Hook Elementary in Nancy's hometown of Troy, Ohio. Another was Euclid High School outside of Cleveland, where Mary Ellen's mother was a teacher. For Kevin, students at the Schenectady Museum in New York were able to ask him questions.

As the day progressed, Mary Ellen continued her work on the Bioreactor experiment. She was excited to report observing large clumps of the colon cancer cells coalescing into tissue globules about the size of a pea. After sending some images of the cells to the science team in Mission Control, they enthusiastically reported that those were some of the best cells that had been grown in space to date. That was exciting news for us and illustrated the great promise that future bioreactors might hold for the growth and development of human tissues in space.

As the day wound down, I wrote a quick note to Simone, sharing with her that we had passed directly over Houston about 8:00 a.m. CDT and that it was a beautiful sight. I also told her that I saw a lot of clouds over the Kennedy Space Center and asked her to "see what you can do to have them removed by Friday" for our scheduled landing.

As we got into our presleep period finishing up our sixth day in space, Mission Control played the theme song from the movie *Starman*, composed by Jack Nitzsche. Some have claimed the song reminds them that nothing lasts forever and that everything must come to an end, just like our STS-70 mission and just like the space shuttle program since then. Listen to the song with eyes closed and imagine you are floating above the Earth. It's a moving experience.

Our official sleep period started at 2:42 p.m. CDT, and one by one we were all soon asleep up among the stars.

Flight Day 7 (Wednesday, July 19, 1995)

I started my day with another entry in my journal:

I woke up about an hour early again this morning with a slight headache. Yesterday I felt pretty good, although I was a bit congested. There's getting to be a lot of dust in the air.

I kept myself busy yesterday taking pictures of the Earth and pictures inside the shuttle. I also spent a lot of time on the SAREX amateur radio. I spoke with folks from Hawaii, the mainland of the United States, and South Africa. I even spoke with some folks at Disneyland as well as one fellow whose dad had worked on the Gemini Program. By a stroke of luck I was also able to make contact with Bob Briggs from TRW who built our TDRS satellite. He had tried working with NASA to schedule an official contact with us but was unable to pull it off. I was glad to chat with him for awhile.

While the public affairs folks had quit writing about the wake-up music each day, this day was special indeed. There was one song for each of us on the crew, and this day was mine. Before the mission, I picked a song to honor my hometown of Cleveland, which long considered itself the polka capital of the world and was home to famed 'polka master' Frankie Yankovich who did an outstanding version of the song "Beer Barrel Polka," also known as "Roll Out the Barrel." When I was growing up, that was one of my favorites, so I chose it for my wake-up music.

One of the medical specialists who helped out with physicals when we did our neutral buoyancy underwater training for spacewalks was Gene Hackemack. He was of Czech decent and played in his own polka band on the weekends in the Hill Country outside of Austin, Texas. I asked Gene if he could supply his band's version of "Beer Barrel Polka" to Mission Control, and he gladly complied. It was a selection from their record *Gene and Jason at the Hofbrauhaus*. So after a short five-second alarm tone that acted as a sort of alarm clock for us onboard, Mission Control piped up "Beer Barrel Polka" as performed by Gene Hackemack's band. "Roll out the barrel and we'll have a barrel of fun," it started. I absolutely loved hearing the song, and as we floated in the middeck getting ready for the day, Nancy and I did a bit of a floating polka. It was a great way to start my day.

When the song finished, Story radioed, "Good morning, *Discovery*. Good morning, Don." I pointed out that the song marked another first for the space shuttle program—polka wake-up music in space as five astronauts danced out of their sleeping bags. Story said he hoped we got that on videotape. "I'd love to see a zero-g polka," he said.

After the flight, I received a cassette tape from the Capcom office that included the wake-up music played during the flight. Story attached a note that read: "Good wake-up music." He had been an astronaut for a few decades and probably had seen and heard it all. I think he appreciated my out-of-the-norm wake-up song.

Later in the day I had the chance to reflect on the wake-up music a bit, along with some thoughts about the mission winding down when I wrote the following entry in my journal:

> MET 5 days 19 hours 45 minutes
> *The wake-up music was great today. Gene and Jason at the Hofbrauhaus singing "Beer Barrel Polka." Story Musgrave woke us up with that and we told him we wished he could see us all dancing the polka in the middeck. It was a <u>great</u> way to start today and I had the tune going through my head for the next 1½ hours until I exercised and listened to some other music on my CD player. Tom mentioned to me an hour after wake-up that he still had the tune going through his head! Knowing Gene made it all the more special. I think I made his day as well!*
>
> *Today is one of those unique days in space. I essentially have the entire day free. I've been busying myself cleaning, organizing, photo stuff, etc—whatever I'd like to do. Never had a day like this before and may never again. And this is our last day of experiment work. Tomorrow we start packing.*
>
> *This morning's weather picture showed it clear over Eastern Florida. I hope it remains that way. I have my personal* SAREX

contact with Simone today which will be great! Also made contact with some folks from New Zealand. So far I've got New Zealand, Australia, Mexico, South Africa, British West Indies, and Argentina. Not bad for 28 degree orbit inclination mission.

With the mission winding down, we entered another message from the crew that we would broadcast via SAREX:

Greetings from the STS-70 Crew aboard Space Shuttle *Discovery*!

We are starting our final day of payload operations today and things continue to go extremely well. Tomorrow we will start packing up in preparation for our early morning landing on Friday. We are looking forward to good weather at the Kennedy Space Center. We would like to thank all of you for your interest and support during our mission and we look forward to seeing and talking with you once again when we return.

Of the forty-four days I spent in space, this was the easiest one for me with the least amount of scheduled activities. On STS-65, my days were absolutely jam-packed conducting experiments the entire day. There were times when the ground called up for me to do three things at once and there was barely time to use the WCS. This day must have been some sort of NASA payback for that experience. This day was totally laid-back with only a handful of activities scheduled for me. What a rare occurrence to be on the space shuttle and have lots of free time to simply hang out in space. I savored every minute of it and spent much of the day looking out the windows and taking pictures of the Earth.

One of my main activities was a personal amateur radio contact with Simone. The SAREX team always worked to get each crew member one opportunity during the mission when they could talk to their family or a friend. Many of the astronauts thought the amateur radio was for nerds only, but I found it fun talking with students as well as other amateur operators around the world. And as an added bonus, we each got the chance to "call home" an additional time during the mission.

As the time approached for my contact with Simone, I made sure everything was set up and ready to go. I tuned and retuned the antenna for maximum signal to get the best reception possible. As it turned out, communications were extremely bad, and we were only able to talk for three to four minutes. Most of the time the signal was very intermittent, making it difficult to talk. It was nice to hear Simone's voice, but a bit disappointing we couldn't hear one another very well. We had to keep asking each other to repeat what we had said. Fortunately, the SAREX team came back with a second opportunity for us on the next orbit, and ninety minutes after the first contact, I was able to call Simone as *Discovery* was approaching South Africa. Here I was halfway around the world, 165 miles above its surface and able to speak clearly with her. That was pretty cool! We were able to do that because of a volunteer network of amateur radio operators around the world who helped relay our conversation via telephone lines back to our families in Houston or wherever we were calling to. If you didn't mind having a few other amateur operators listening in on your private conversation, it was a great opportunity.

Simone and I had great communications for nearly six to seven minutes. She told me what was going on at home and passed along greetings from my mom, her parents, my brother Dennis, and our neighbors in Houston, Frank and Carolyn Littleton. We threw in quite a few "I love you" and "can't wait to see you" messages as well. The short time allotted flew by quickly and with one final "I love you" and "Goodbye" we were out of range as *Discovery* headed across the Indian Ocean. That would be our last voice communication with each other until after we landed and were reunited at the Kennedy Space Center. Fortunately, we would still have a few more days of communicating via email. Hearing from Simone and getting news from back home always lifted my spirits.

As we approached the end of our mission, we started finishing up some of the experiments and stowing equipment. With less ceremony than during its activation twenty-four hours earlier, I wrapped up the Microencapsulation in Space experiment. I moved the power switch from 'on' to 'off.' The indicator light went out as expected, and I was done. That experiment couldn't have been any simpler to operate.

After some final data collections, Tom, Kevin, and Nancy completed the HERCULES and WINDEX experiments. I was happy to have both of those finished because the massive cameras took up much of the flight deck. Packing away those experiments provided better opportunities for all of us to look out the windows. During the mission's final days, I spent as much time as I could at the windows taking in all the incredible beauty below.

Some of the time I spent taking pictures, the rest just gazing out at our planet in deep thought. I took the opportunity to listen to Pink Floyd's *Dark Side of the Moon* on my portable CD player. From daylight to a beautiful sunset, to staring out at the infinite universe with the multitude of stars during our night pass, to enjoying yet another sunrise, all while listening to Pink Floyd. For me, it didn't get much better than that.

With the mission winding down, I started thinking more and more about landing. I kept my eye on Florida's weather every pass hoping that everything would cooperate for us to land there in a few days. During one pass, Tom noted that he could clearly see the Shuttle Landing Facility, our primary landing site, but that the rest of Florida was under cloud cover.

At 2:42 p.m. CDT, I began my scheduled sleep period on a natural high. It had been an incredible day for me and made me feel so lucky and privileged to be an astronaut aboard the space shuttle. I slept better than I ever slept in space before or since. And once again, *Discovery* continued to operate flawlessly with Mission Control monitoring all of her systems while we slept.

Flight Day 8 (Thursday, July 20, 1995)

On our next-to-last morning in space Tom finally got his special wake-up music. His wife, Becky, an accomplished professional singer, recorded a tender love song for Tom titled "You are My Destiny," which was played at our designated wake-up time of 10:42 p.m. CDT (Wednesday, July 19). Tom, clearly moved by it, radioed, "That voice is music to my ears. I can't wait to see that singer again soon."

With our landing scheduled for the next day, this one would be spent checking out *Discovery* to make sure all her systems were operating properly, deactivating and stowing most of our scientific experiment hardware, setting up our seats for reentry, and beginning to prepare our suits and other equipment we would need the next day.

The most important and critical activity was the Flight Control System (FCS) checkout conducted by Tom, Kevin, and Nancy. The various systems we would need for landing were powered up and tested one-by-one to make sure everything was still working properly after our seven days in space. After launch, we had reconfigured *Discovery* from a rocket ship into an orbiting laboratory. The following day we would reconfigure her back into a spaceship to fly us home. But before that and before we could commit to landing, we needed to make sure the elevons, rudder, body flap, speed brake, and other flight control systems were functioning properly. Tom and Kevin powered up one of *Discovery*'s three auxiliary power units (APUs) to make certain it was functioning properly. Everything tested A-OK.

Next, Tom and Kevin performed a hot fire test of *Discovery*'s reaction control system jets that we would need to help us steer through our entry profile before we encountered the atmosphere at which point the aero-surfaces would take over control. Once again, all was well. *Discovery* was ready to land. And so was her crew.

We spent a good portion of the day organizing, cleaning, and packing. All loose equipment needed to be stowed in the middeck lockers where it was for launch. We didn't want any extra items crashing around during our reentry and landing or have items on the middeck that might cause clutter and slow us down in case we needed to do an emergency evacuation. It was happy work, and we were all in good spirits because we were all excited to be heading home.

As this was scheduled to be our last day on orbit, we entered a final daily message for transmission with our amateur radio:

> Greetings from the STS-70 crew aboard Space Shuttle *Discovery*!!!
>
> We're sorry to say this is our last full day in space. It's been a lot of fun for us and we are looking forward to talking to you next time we are in orbit. Thanks for your great support!!!!

I really enjoyed talking to people, including many students, around the planet via SAREX. NASA got good publicity throughout the amateur radio community, and that allowed us additional opportunities to educate the public about the different science experiments we were conducting so they could understand and appreciate the work we were doing in space. The vast majority of the amateur radio folks wanted to be able to say they made contact with the space shuttle, so why not educate them as well—which is what we tried to do.

It was amazing how many people would attempt to make contact with us as we passed over the southern United States. During any one of our numerous passes over the United States, from a few dozen to hundreds of people announced their call signs hoping we would hear them and read them back to establish an official contact. The calls would come one on top of the next with most of them unintelligible because only parts of someone's call sign would be heard before someone else stepped on their transmission and started up with their own call sign. More and more people tried to get through as our mission progressed and word got out that we were on the air during many of our passes. It didn't matter what time it was on the ground, day or night, there were plenty of Ham operators trying to make contact. Our experience over the United States was in sharp contrast to the rest of the world. I never heard much over most of Africa, except for an occasional contact from South Africa. The same was true over Australia, where I made only a handful of contacts.

We held one last news media event, a round-robin featuring television stations from across Ohio. One of the best questions was from WJW in Cleveland for Kevin, who it was noted had just spent a considerable amount of time in close quarters with a bunch of Buckeyes, and they wondered whether he ever wanted to see another Ohioan once he landed. Kevin smiled and answered quite diplomatically: "If they are anything like these four up here, I sure would." I wonder what he was really thinking?

One other notable question was why so many astronauts came from Ohio. Tom noted the many pennants and banners from Ohio colleges and universities on the middeck wall behind us which included ones from Ohio State, Ohio, Bowling Green, Case Western Reserve, Defiance, and Toledo. Tom credited the state's outstanding educational system for preparing us so well for our mission.

Later in the day, we had the unique opportunity to speak with another crew on another space shuttle. The STS-69 crew was strapped inside space shuttle *Endeavour* participating in their Terminal Countdown Demonstration Test at Kennedy Space Center's Pad 39A. Mission Control configured our communications so that Tom could speak with *Endeavour*'s commander, Dave Walker. Tom mentioned he had heard their crew watched our launch with our spouses a week earlier, adding with a smile, "We hope to do the same with your spouses and girlfriends here soon."

We also did a special event to thank our STS-70 flight surgeon, Dr. Larry Pepper, who had been a NASA flight surgeon for many years and supported many shuttle crews and their families. He would be leaving NASA after our flight for, as Tom put it, "a higher calling"— to become a missionary doctor in Uganda. We told him we would be looking for him every time we passed over Africa on our future flights.

I received a final message from Simone that she and Kai would be leaving for Florida in a few hours, signaling an end to our mission. I could tell she was exhausted and couldn't wait for me to get home. I also received a note from Kai who told me that he learned to roll over from his back to his stomach all by himself. Being only five months old, I had missed nearly ten percent of his life during the mission. Simone ended her note:

> Tomorrow we will be looking out for a little white dot in the sky and the noisy sonic booms that will announce your homecoming. Love you in a big way.

As we finished our work for the day, we set up the seats for landing and did as much preparation work as we could so that we would be ahead of the timeline for the always-busy landing day the following morning. I tried to spend as much time as I could getting in some final glimpses of the Earth. There might be a few moments to do so the next morning, but I knew what a busy day it was going to be. A few more sunsets. A few more sunrises. Who knew if and when I might get another opportunity to be up there.

I went to bed exhausted, both physically and emotionally. We had been preparing hard for this mission for the past twelve months. We successfully deployed the TDRS satellite more than a week earlier, and we completed all the other experiments. We achieved all the mission's objectives. It was a great feeling of accomplishment.

> *MET 7 days*
> *We just finished dinner and before that there was a massive photo frenzy—almost 3 hours straight—crew photos and taking photos with some of the special personal items we had brought with us. It's only 2:00 p.m. Thursday, but I'm dead tired. We need to set up the middeck seats yet tonight, but otherwise we're in good shape for entry. The 1st weather call this afternoon was very promising for our 1st and 2nd attempts into KSC. Later on at our Private Medical Conference, Dr. Phil Stepaniak told us there may be ground fog on the 1st attempt. We'll have to wait and see. I'm ready for landing and excited about getting home. Eight days is enough in these small quarters (plus seven more in quarantine). We made a morning pass over Houston and it looked great. I was looking for Simone. She and Kai should be on their way to Florida by now. And very soon we will follow!*

Curt Brown, our entry Capcom, told us the weather looked promising for the first landing attempt. We all anticipated eating dinner on Earth the next evening. I started thinking about what I was going to eat for my first meal when I got home. There was only one choice really—a large double-pepperoni pizza.

As we slept, Mission Control in Houston once again transitioned from the new Mission Control Room to the old one in preparation for our landing. Everything worked extremely well with the new facility over the past week, but NASA wanted to evaluate it a bit further before committing to use it for launches and landings. I so looked forward to being home and sleeping in my own bed. It had been a great flight, but we were all ready to go home.

Chapter Nine
Fireball to Earth

Landing Day wake-up time was 10:42 p.m. CDT Thursday, July 20, and everyone was excited that morning. There was a sense of great anticipation. Everything was looking good for us. Even my horoscope was optimistic: "The time, energy, or money you have invested in a special venture pays off today" (Jillson 1995). We all felt great about the mission and looked forward to getting back home and seeing our families. No tears were shed because it was our last day in space. We couldn't wait to get back on Earth.

For our final morning in space, we had one of the flight's stranger wake-up music selections. We received a list of the mission's wake-up calls on the first day of the flight to give us a heads up as to what was coming. That way if a special song was being played for a crew member one morning, they would be prepared to recognize it and respond back to Mission Control after the song was played. For that final day, the official wake-up music was listed as "Aggie War Hymn," the Texas A&M University fight song. Located about two hours from Houston, it was the alma mater of our lead flight director, Rob Kelso, and approximately half the people working in Mission Control. The other half attended University of Texas. I guess Mission Control had had enough of the Ohio-themed wake-up music and decided to honor Texas that last morning.

When the wake-up music was finally played, it turned out that it wasn't the fight song from Texas A&M, but rather that of the University of Texas, "Eyes of Texas." You have to understand the intense rivalry that exists between Texas A&M and University of Texas to appreciate what was going on. That rivalry all but dominates Texas culture, like cowboy boots and rodeos. Half the people think A&M is best, the other half think UT rules. And just like red and blue states in politics, there is no middle ground. So why wasn't the A&M song played as advertised? It seemed that another of our flight directors, Randy Stone, pulled the old switch-a-roo, and changed the wake-up music to the UT song. The rivalry continued. All this hardly mattered to us onboard. We were just glad that the song played honored half the folks working in Mission Control that morning. Maybe next time they would honor the other half. We thanked our Capcom for the song and got on with our work getting ready to come home.

With our mission quickly winding down I took the opportunity to write a few final thoughts in my journal:

> MET 7 days 18 hours zero minutes
> LANDING DAY MORNING!
> It's about 3:50 a.m. at KSC where Simone is (and Kai!). Hope they are sleeping well. Less than 4 hrs until landing and I'm all wired up for DSO 603C experiment (Blood Pressure Monitoring). A few quiet minutes before we begin suiting everybody. I woke up about 3 hours early today (excitement about landing, I guess). I went back to sleep and had my second nightmare of the mission. I dreamed I came back from a long trip and Simone had made plans to go to some party instead of being home with me. She left without me and when I showed up, she totally

ignored me and was mad and told me to go home. It was terrible and I was so glad when I woke up to find myself in space and not living that experience. Well, a few more looks out the window. I feel the need to soak in all I can, while I can! Only 3 more orbits to go.

We had done most of the prep work getting things ready for landing the day before, yet we spent much of the morning doing our final packing and stowing of equipment and cleaning up a bit. We also took the opportunity to take in a few more views of the Earth during those final couple of orbits, and I jotted down a few final impressions.

MET 7 days 19 hours zero minutes

We just made a spectacular night pass over Texas and the Gulf Coast. From San Antonio to Houston to New Orleans to Tampa over KSC and Miami. All beautifully lit up. Over Miami I could still see Houston on the horizon! And as we passed over the East Coast I could see the first light (glowing on the limb) of sunrise. What a spectacular pass—all in about 5–10 minutes. Over KSC there were some patchy clouds and I could see a thunderstorm over the Gulf just west of Tampa. I could see [shuttle] pads 39A & B lit up and the SLF! Maybe our 1st landing is go!

It's starting to get pretty cold in here!

Only 2 more orbits and we'll be on the ground. The day before yesterday Tom Jones called up as we started our 100th orbit. I was glad to hear that news. Made me say "Wow!" What a trip!

Suiting Up

We started getting into our bulky, one-piece bright orange Launch and Entry Suits (LES) a few hours before our deorbit burn. On most flights, the commander goes first, followed by the pilot, and then Mission Specialist 2, in this case Nancy, who acted as the flight engineer. I led the suiting up process on STS-65 a year earlier and, together with Mary Ellen, was handling things for this mission. It was imperative that everything be done properly because the suit represented our last line of defense in the event of an emergency—cabin depressurization or a bail-out scenario. If the proper connections were not made or if something was not configured correctly, it could mean the difference between life and death. The most important lesson I learned from STS-65 was that the process must be done in a well-organized manner. Doing it exactly the same way each time and using the same techniques the suit technicians at KSC employed to get us ready for launch helped ensure that nothing got missed and that everything was done right. It also helped speed things up.

The first step was to don an adult diaper and then put on our dark blue long underwear (top and bottom) along with a pair of thick wool socks. Because we were prechilling the orbiter as a defense against the heat of reentry, it got quite cold in the crew cabin. So getting dressed helped warm us up. In a bit of irony that wasn't lost on any of us, we then put on our liquid cooling ventilation garments (LCVG). That consisted of small plastic tubes filled with chilled water that helped keep us cool once we were wearing our spacesuits. While we initially didn't need the cooling from them, we surely would later on during reentry and after landing.

Next, we slipped into a g-suit which looked like a set of cowboy chaps. During reentry, we manually adjusted that suit by dialing in different levels of inflation causing its various bladders to compress the thighs, calves, and abdomen. That helped restrict blood flow to the lower extremities and allowed more blood to get to the head so we wouldn't get dizzy or faint as we returned to gravity.

Then it was time to get into the LES itself, more affectionately called 'pumpkin suits' because of their characteristic color. It was usually a little more difficult getting into the LES for the return to Earth than it was on launch morning. Part of the problem was that most astronauts grew one to three inches while in space. Without the pull of gravity, the spine expanded a bit making us taller. The suits we were supplied were sized to our normal Earth height. In addition, since we were floating as we were putting it on instead of sitting in a chair as on Earth, it was a bit more difficult because there was less to grab hold of to steady ourselves when suiting up.

Putting on the LES required quite a bit of flexibility and some contortions. The hardest part was always getting your head through the rubber neck dam without straining your neck. I've seen crew members nearly effortlessly pop right through, and others struggle for minutes with their heads buried inside the suit working up quite a sweat as they struggled to get the suit over their shoulders and their head through the neck ring. While one of us wrestled to don the suit, other crew members would help hold us down to provide a bit of stabilization and made sure we didn't hurt ourselves by bumping into something during the struggle. As the head emerged through the neck dam, I always imagined the sound of a champagne cork popping—and frequently other crew members helping out would make such a sound which was always followed by a round of applause.

With the LES and boots on, we got into our one-piece parachute harnesses. Then red chemical light sticks were activated and placed in small clear pockets along the top of each shoulder to aid in any search and rescue operation should that be necessary in the unlikely (we hoped!) event of an emergency evacuation or bail-out.

After Mary Ellen and I finished with each crew member, I took a moment to make certain that everything was donned correctly, that all connections were made, and that no straps were twisted that might cause problems later. With the once-over completed, I gave them their bag of personal items—wrist watches, pens, pencils, etc.—as well as their gloves and helmets—and sent them on their way up to the flight deck.

So with an understanding of the suiting-up process, I now can explain one of the funniest things that happened during the flight.

Tom grabbed his long underwear, wool socks, and LCVG, and headed to the WCS where he changed. When done, he floated up to the flight deck and started getting ready for the deorbit burn. Next came Kevin. He donned his long underwear, socks, and LCVG, and then joined Tom on the flight deck. Once there, we all heard Kevin say, "I can't believe how much I've grown up here. The sleeves on my LCVG are soooo short now!" Everyone was busy getting ready for reentry, so no one paid particular attention to Kevin's apparently much-elongated arms. Then Nancy changed into her long underwear and LCVG, and made her way to the flight deck.

When Mary Ellen and I finished getting the LESs ready, it was her turn to change into her long underwear and LCVG. But, Houston, we had a problem. Mary Ellen couldn't find her LCVG. So a search ensued. It wasn't in her reentry clothes kit where it should have been and we didn't see it anywhere on the middeck. Then I found a plastic bag that contained the top portion of a LCVG. *Eureka*, I thought, *we found it*. However, upon closer inspection, the bag had 'PLT,' NASA shorthand for 'pilot,' written on it. It wasn't Mary Ellen's, but was Kevin's! My mind immediately flashed back to Kevin's comments about how much he had grown in space and the snugness of his LCVG shirt. It didn't take a rocket scientist to figure that one out. A big smile came across my face accompanied by some hearty laughter when I realized that Kevin, who was well over six feet tall, somehow acquired the LCVG top issued to five-foot-something Mary Ellen.

I took Kevin's LCVG top and floated up to the flight deck and gave it to him. With the rest of the crew howling in laughter, Kevin floated back down to the middeck and changed.

"This one fits much, much better," he said. Being his first flight, he had assumed he experienced space growth—something he had heard about from many previous flyers.

I still laugh when I think of him floating back down to the middeck wearing Mary Ellen's top that looked like someone had shrunk it a few sizes in the dryer. It was a classic moment in the annuls of human spaceflight and one of my favorite memories of the mission.

Years later, for a book about personal stories of shuttle astronauts, Tom wrote that the STS-70 crew discovered another side effect of being in space. "It turns you into a cross-dresser," Tom wrote in reference to Kevin wearing Mary Ellen's cooling garment shirt.

After regaining our composure, we went back to work getting the middeck ready for landing and began the process of getting my crew mates into their Launch and Entry Suits. I made a mental note not to put Kevin into Mary Ellen's. The suiting-up order was Tom, Kevin, Nancy, and Mary Ellen, with me last. Soon Tom, Kevin, and Nancy were all set in their seats on the flight deck.

Mary Ellen was running behind schedule, and I was uncomfortable with how long it took before she got into her flight deck seat. I had to make sure she was strapped in properly, and then return to the middeck to deactivate the WCS, shutdown the galley, and stow all sorts of remaining equipment that we had un-stowed during the mission before I could strap myself into my middeck seat. I would have preferred to be ahead of schedule, not behind. That last hour before our scheduled deorbit burn was pretty hectic for me since I was working alone with everyone else up on the flight deck.

Something else that took a lot of my time before reentry was the donning of special medical hardware for one of our experiments. Under my suit I was wearing a blood pressure cuff, three to four electrodes on my chest, and a data recorder box for an experiment that was looking into something called orthostatic intolerance—lightheadedness or dizziness that astronauts experience on landing. Putting all that hardware on and testing it to make sure everything was working properly always took quite a bit of time.

Another medical experiment I was to perform during reentry involved the evaluation of the effectiveness of a medication called Florinef for minimizing the effects of orthostatic intolerance. Usually thirty to forty-five minutes before the deorbit burn, crew members took a number of salt tablets and drank a lot of water in order to load their bodies with fluid. Fluid loading helped minimize dizziness at landing associated with fluids shifting from your head back to your lower body, the exact reverse of the fluid shifting astronauts experienced when they arrived on orbit. The Florinef tablet was taken instead of the salt tablets with the thought that it would provide a better fluid loading for the body. The good part was it needed to be taken five hours before landing so I didn't need to take the salt tablets and chug all that water right before the deorbit burn. The bad part was that I felt water-logged and bloated for a longer period of time before landing.

I was told before the mission that I was the last test subject needed to complete the Florinef study. The previous results indicated that it was not very effective compared to salt tablets. The researchers were anxious to get the data from me so that they could prove once and for all that Florinef was not effective and should be discontinued. That irritated me because we weren't looking to prove something worked, but rather were attempting to verify that something didn't work. But I always agreed to participate on the experiments I was being asked to perform, no matter how I felt about some of them. I believed I had a responsibility to be a test subject and that it was an important part of being an astronaut. With the good comes some bad. So instead of being bloated and feeling water logged for the last hour of the flight, I would experience that sensation for the last five hours of the flight. Lucky me.

We waited to get the 'Go' from Mission Control to perform the deorbit burn. Tom, Kevin, and Nancy were monitoring systems and watching the countdown clock. The weather forecast from the previous day was rather hopeful. Partly cloudy skies for our planned landing time of 7:54 a.m. with conditions forecast to improve for the second attempt at 9:30 a.m.

Another Day in Space

Unfortunately, as it turned out, the weather forecast was a bit off. As we approached our scheduled burn time, we got a call from Mission Control waving off the burn and delaying the landing attempt until the next orbit. Low clouds and fog enveloped KSC. Fellow astronaut Steve Oswald, flying weather reconnaissance at KSC in the Shuttle Training Aircraft to evaluate the conditions that *Discovery* could expect for landing, reported that the area around the runway was 'clobbered,' meaning that the fog and low clouds were so thick that there was no chance of attempting a landing on the first opportunity. In fact, only the bottom half of the Vehicle Assembly Building was visible with the upper half obscured by the clouds. It was definitely 'No Go' weather for our first landing attempt that day.

There was hope that as the sun came up the fog and clouds might burn off enough to permit a safe landing at KSC. The delay meant another trip around Earth—ninety more minutes, another twenty-five thousand miles.

I floated up to the flight deck, took a few pictures of the crew all strapped in, and spent a few minutes catching my final glimpses out the window. It was a pretty quiet time with not much to do except wait for the occasional call from Mission Control. Periodically they relayed weather updates. As we approached our second deorbit burn opportunity, we were extremely hopeful that the weather would clear and that we would be allowed to land. I floated back down to the middeck and got into my seat to be ready in case the burn took place. We sat and waited. At 8:10 a.m. EDT, as we approached the Indian Ocean with the burn only a few minutes away we got the word. Instead of clearing, the conditions had actually gotten worse.

"Unfortunately, the weather didn't cooperate, and we are going to call a 'No Go,'" Capcom Curt Brown radioed.

"Okay, that doesn't upset us a bit," Tom replied.

Tom, Kevin, and Nancy immediately got into the Deorbit Burn Backout Procedure and started to reconfigure the shuttle for another day in space. While they did that and reopened the payload bay doors, Mary Ellen and I got out of our suits and reactivated the WCS and the galley. Then one by one Nancy, Kevin, and Tom floated down to the middeck and we helped them out of their suits.

Things were a bit cramped with the five of us and our five suits all floating about. We temporarily stowed the suits best we could to get them out of the way and had nearly four hours before our scheduled bedtime without much to do. We got a few calls from the ground, but there weren't many actions required from us. We just spent the time up on the flight deck taking advantage of the extra time to enjoy the view. A few more sunrises. A few more sunsets. I savored each one. I hoped to return to space for another mission, but you never knew what was in the cards for you. I tried to soak in as much of the experience as I could—each and every second of it.

We had a fantastic pass over South Africa with a great view of Johannesburg. I could see many of the mining areas for which that region is famous. The sky was clear and the terrain rugged, which made for a spectacular view.

Then we had another great pass over South America. Brazil continued to fascinate me with all the burning of rainforest. Giant areas of that country were being deforested. As we made one of our final passes over the area, I shook my head seeing how much smoke was rising from the region where they were burning the felled trees.

A few orbits later we made a night pass over the South Pacific where I got a final glimpse of the Large and Small Magellanic Clouds as well as the Southern Cross. The Milky Way appeared brighter than normal to me with incredibly dark patches in it from interstellar clouds of gas and dust. I was able to witness two more shooting stars as yet more material from space entered Earth's atmosphere and burned up in a final blaze of glory. As sunrise approached, I saw Venus and Mars together in the morning sky. Just the day before, they had appeared close enough together in the sky that I could cover them both with the tip of my pinky finger held at arm's length. Now, just twenty-four hours later, Venus had moved noticeably further from Mars. As the predawn sky turned from blue to red then orange, I was struck once again by the fragility of the planet. The atmosphere appeared only as an infinitesimally thin band.

One unwelcome call from the ground was confirmation they wanted me to take a second dose of Florinef and repeat the same fluid loading procedure for our landing attempt the next day. I was already urinating prolifically as the fluid loading properties of the first dose of Florinef were wearing off. I didn't think they would have the nerve to ask me to repeat that rather unpleasant protocol. But they did. They also recommended that I eat some dried banana chips to build up my vitamin B a bit because one side effect of Florinef is to reduce vitamin B in the body. The ground team in Mission Control was really doing their best to look out for me!

I continued urinating right up to bedtime. We kept the airlock closed after our wave-off so my normal sleeping area was not available. I wedged myself into a corner of the middeck and went to sleep. One thing was certain: good weather or not at KSC tomorrow, we would be landing. The Edwards Air Force Base landing site would be activated for us. We would have two chances to land at KSC. If the weather didn't cooperate, the plan was for us to come down in California. I had mixed emotions about an Edwards landing. I was anxious to see Simone and Kai as soon as possible and landing in California meant they wouldn't be there. But I always thought it would be cool to land at Edwards, where years earlier historic rocket-propelled aircraft, including the X-1, X-15, and the first space shuttle, *Columbia*, pioneered the way and landed on the huge dried lake bed there.

NASA preferred that we land at KSC, but there was also a bit of a rush to get us back on the ground and not wait another day for the weather to clear there. That was because *Discovery* was scheduled to undergo a series of inspections and modifications after the STS-70 mission. The weather forecast for KSC once again called for the possibility of fog and low clouds that could prohibit landing there. The forecast for a California landing called for excellent weather with only high, scattered clouds and light westerly winds. This would be our last night in space on this mission.

Before falling to sleep, I made an entry in my journal:

MET 8 days 2 hours 30 minutes
O.K., so I was wrong!
　First landing attempt we waved off due to low clouds. For the second attempt I suited everyone up and was just about suited up myself when they waved us off for the day. I was a little disappointed because of all the work we had gone through

in preparation for landing and now we would be getting home a day late. The positive side was more free time for window viewing.

We are definitely landing tomorrow either at Edwards or KSC. I have the feeling it will be Edwards. The forecast doesn't sound fantastic for KSC, but we will try anyway! I think we will be in our suits a long time tomorrow, fluid loading all day long (Florinef). I hope we land in Florida, but I'll be happy to land anywhere tomorrow. But you just can't beat this window time. During the last 3 hours we listened to "Dark Side of the Moon" and Beethoven's Ninth.

This Time It's For Real

We were awakened by Mission Control at 10:42 p.m. EDT Friday, July 21. No wake-up music for us this day. It was all business. The deorbit burn for our first landing opportunity at KSC was scheduled for 5:26 a.m. EDT. That meant we had almost seven hours to get ourselves up, into our spacesuits, and complete final preparations for landing.

Because we left as much as possible configured for landing from the day before, our workload was much lighter. Still, there was plenty to do. But having completed much of the work the day before and having just run through everything we would need to do to get ready for landing, things went very smoothly. It was definitely a more relaxed day, not as hectic and rushed as the day before. We started the day by changing into our long underwear and thermal socks because we would be suiting up once again in a very few hours.

And once again about five hours before landing, I took my Florinef pill and started performing the prescribed fluid loading protocol. Got water?

During our last orbits of Earth, we took the opportunity to catch more final glimpses of our home planet.

Time passed quickly as we got ready for landing. Kevin and Nancy worked through the procedure to close the payload bay doors. I watched silently over their shoulders as the doors closed and latched. Everything went flawlessly; there would be no need for an emergency spacewalk to secure the doors. I kept my upper lip stiff, but sighed slightly. That was my last EVA possibility for the STS-70 mission. I knew that doing such a contingency EVA was a long shot. In the history of the program, there had never been a need to manually close or latch the doors for reentry.

Next, everyone got suited up and strapped into their seats. I told Mary Ellen I would take care of the middeck and not to worry about helping me down there in an effort to get her settled and into her seat earlier than the day before. That helped quite a bit. With Tom, Kevin, Nancy, and Mary Ellen all settled on the flight deck, I busied myself on the middeck deactivating the galley and WCS and stowing last minute equipment that we had left out. I floated up to the flight deck with one of the cameras and took a few final pictures of everyone in their seats—all dressed up and ready to go.

As we approached the appointed time for the first deorbit burn attempt, Mission Control told us the weather wasn't exactly cooperating. Even though weather looked good enough for the first opportunity, flight controllers opted to pass on it to allow weather conditions to improve a bit. We were waved off for the first attempt and would orbit the Earth yet one more time. Another ninety minutes, another twenty-five thousand miles.

I had a few minutes during that final orbit to make one last entry in my journal:

> *Well, they waved us off for a rev while they check out weather. We passed over KSC a little bit ago and it looks like it's improving. We're all just hanging out looking out the windows. We had a great pass over Africa and Madagascar. Lots of burning going on down there. It's still so amazing to look out the window. Also had another great pass over the Gulf of Mexico including passes over Houston and New Orleans. Clear and beautiful—I don't say it often, but I can't wait to get back to Houston and especially can't wait to get home.*

The lights of Houston shined bright on that unusually clear July night. It reminded me a bit of the view we got from forty thousand feet while aboard the NASA T-38 as we'd be flying back to Ellington from a trip. Houston's a sprawling city and its bright lights would be visible from hundreds of miles away. It wasn't my favorite location on the planet, but it was my home and I couldn't wait to get back there.

We also passed a little to the south of the Kennedy Space Center on what we hoped would be our last orbit. I floated at one of the overhead windows and had an excellent view of central Florida and the Cape Canaveral area. I could see some patchy clouds, but I could also clearly see the lights around KSC. I reported to Tom that the weather was looking good and that I thought we were likely to land on the next orbit. With the lights of the Cape fading in the distance, *Discovery* continued it silent cruise 167 miles above the Earth at 17,500 miles per hour as we headed out over the Atlantic Ocean. I watched one final sunrise that would be my last for the STS-70 mission. *Discovery* would experience yet one more sunrise just before our scheduled landing, but down in the windowless middeck I would have to miss that one.

Deorbit Burn

I remained on the flight deck for a time getting drinks and snacks for the rest of the crew during our final orbit. But as we approached the Indian Ocean for our deorbit burn, I was strapped into my seat. We finally heard what we had been waiting for—we were, according to Capcom Curt Brown, "Go for the deorbit burn." We cheered. I was glad we would be landing at KSC. I would see Simone and Kai in just a few hours.

Tom, Kevin, and Nancy worked methodically through the deorbit burn checklist. One of the steps called for Tom and Kevin to reorient *Discovery* so our tail would be leading the way into our direction of travel. With our main engines and the Orbital Maneuvering System (OMS) engines facing forward, firing the OMS engines would slow us down. We needed to reduce our speed by several hundred miles per hour so the shuttle could begin its fall from orbit. Gravity would then do the rest.

It was pretty quiet in the cockpit as we approached the deorbit burn time. Shortly before, we had a five-minute period without communications with Mission Control as we passed through what was eerily called the Zone of Exclusion (ZOE), during which we had no communications with Houston. Upon exiting the ZOE, Capcom Brown informed us that we had one minute to go before the burn.

It was critically important that the burn begin precisely on time so we could land on our designated runway at the Kennedy Space Center. When traveling at five miles a second, it doesn't take much of a delay in the burn to overshoot the targeted landing site. All eyes on the flight deck were on the gauges and computer displays checking and rechecking the status of our engines, fuel quantity, tank pressures, and other critical items, including the countdown itself. I sat quietly on the middeck, listening intently to the conversations describing our progress. After all, there wasn't anything else for me to do but follow along.

With about ten seconds to go before the burn, Tom said, "Okay. Last chance to change your mind. Everybody ready to go to Earth?" Even though there was no time for a reply, we were all very ready. And right on time—midway between Africa and Australia over the Indian Ocean—the two OMS engines fired. It sounded like the muffled blast from a cannon and I instantly felt the jolt letting me know ignition had taken place. After weightlessly floating in space for the previous nine days, the acceleration felt quite strong even though it was only one-tenth of the force that we normally experience on Earth. It had been a great mission. As much as I treasured every minute I was in space, I was tired and wanted to get home.

I had watched course-correction OMS burns out the aft flight deck windows on a few occasions during my first flight on STS-65. There was a visible flash from the engine nozzles at ignition. Even though it lasted for only a fraction of a second, a few astronauts got incredible pictures of the event. However, for deorbit burns with the payload bay doors closed and locked, we no longer had a view of the OMS pods at the rear of the shuttle. While we couldn't directly see the engines firing, there was no doubt they were both burning and slowing us down.

"No stopping us now," Tom called out one minute into the burn. We had slowed down and lowered our orbit enough that we were irrevocably committed to returning home. The burn lasted 2 minutes and 55 seconds and decreased our speed by only 343 feet per second—a mere 230 miles per hour, but that was sufficient for us to begin our free-fall back to Earth.

After the burn, all was quiet again inside *Discovery*. Everyone on the flight deck was checking the gauges and displays to make sure everything was as it should have been.

"*Discovery*, Houston. We saw a good burn," Capcom Brown radioed.

"We concur," Tom replied. He and Kevin then reoriented *Discovery* into a nose-first attitude for our reentry. We were less than an hour from our scheduled touchdown at 8:02 a.m. EDT.

As soon as they received word that we successfully performed our deorbit burn, the Orbiter Recovery Team's approximate twenty-five vehicles moved from their staging point near mid-field of the Shuttle Landing Facility to the south end of the runway. The first thing some of the team's members would do after we landed was make certain that there were no threats to our safety while exiting the vehicle and make sure that all was safe for the numerous ground personnel that would be helping out—no aft thrusters leaking any toxic fuels and no toxic fumes from the shuttle's ammonia spray boilers, auxiliary power units, or other propulsion systems.

Meanwhile, astronaut Oswald once again was airborne evaluating weather conditions. He started flying approaches in one of our T-38s about 3:00 a.m., and two hours later switched to the Grumman Gulfstream Shuttle Training Aircraft (STA), which had been configured so it could as precisely as possible mimic the shuttle's landing characteristics. He was flying landing approaches similar to the one

Tom would make with *Discovery* to evaluate the visibility, winds, and other weather conditions we would encounter. In a nutshell, the weather was looking very good for our KSC landing.

After the burn, Tom put on his helmet because we now were 100 percent committed to landing. Kevin, Nancy, and Mary Ellen did the same, and soon I had my helmet on too. Tom and Kevin continued working through the reentry checklist with Nancy, our flight engineer, closely watching over them. Kevin was still enjoying himself and was humming away as he got ready to start our auxiliary power units (APUs) that provided hydraulic power for our landing systems. "Dum-dum-dum dum dum," Kevin hummed aloud. He was relaxed and loving everything that was happening. I knew that Kevin, as a rookie pilot on his first space shuttle mission, was more than a little excited about the landing.

Our flight path took us just off the northern coast of Australia. A few minutes later, we crossed the Equator for the last time as we passed over Indonesia. We were forty-eight minutes from landing as we headed out over the Pacific Ocean toward Baja, California. *Discovery*'s on-board computers commanded the orbiter's nose to pitch upward for our reentry attitude.

As I sat all alone strapped in my seat on the middeck, my blood pressure cuff would periodically automatically inflate, squeezing the heck out of my right arm while taking measurements for the vestibular system experiment being performed on me. As Tom and Kevin called out events and milestones such as speed, altitude, and gravity level, I repeated them into a small microphone I had attached to the neck-ring of my suit. It was supposed to record comments I made about how I was feeling during the descent and after landing as part of the experiment. However, I decided to take advantage of the recording to narrate my experiences as well—what I could hear and feel on the middeck—and repeat some of the observations made by my crew mates on the flight deck.

Everything was quiet as we crossed the Pacific Ocean. The flight control team was busy monitoring *Discovery*'s systems and keeping a close eye on our speed, altitude, and ground track. We were right on course and all of *Discovery*'s systems were operating flawlessly. As we continued our silent plunge, we all started to inflate our g-suits. Some astronauts didn't like the squeezing of their abdomen during the landing, especially if they were experiencing vertigo or motion sickness symptoms. The g-suit isn't a thing of comfort, but it does its job well if you inflate it properly. I always tried to inflate it to the recommended levels on each of my flights.

As we descended into the upper layers of the atmosphere, Tom noted the orange-pinkish glow that was enveloping the shuttle. The thickening atmosphere would continue slowing the shuttle and the frictional heat created a plasma layer around the shuttle as we descended. During the Mercury, Gemini, and Apollo flights, the plasma layer would cause a condition known as radio blackout when there was no communication—voice and telemetry—between the spacecraft and ground control. The shuttle had multiple antennas distributed through the vehicle. Because we were using a TDRS for communication and relying on the antennas on the top side of the shuttle where the plasma was least strong, we were able to communicate all through reentry.

Tom called out as we were passing 84.5 miles above the Earth noting that our speed had started to drop as we were now traveling at Mach 24.

"Let the light show begin," Nancy remarked enthusiastically.

"Sorry, Don," Tom quickly consoled. I was facing a wall of middeck lockers and had no view outside whatsoever. There was a five-inch window on the hatch on the middeck, but that was well behind my seat over my left shoulder, and besides, it was covered for launch and landing.

Tom and Kevin periodically exercised their legs during the descent to get their leg muscles working again after nine days of floating in space. They would need full control of their legs for landing. At one point, Kevin, with his best exercise instructor voice, called out, "One and two and one and two. Everybody now." Kevin was still quite relaxed, joking around and loving every minute of the experience.

Just north of the Hawaiian Islands, we passed through four hundred thousand feet (about seventy-five miles), a point that marks the shuttle's entry into the upper thin layers of the atmosphere. We called it 'entry interface' or EI, and it also meant that we were four thousand, three hundred miles from touchdown and descending at about five hundred feet per second (over three hundred miles per hour). We would be back on Earth in another thirty-four minutes. From here our ground track took us across the Pacific to Baja California. *Discovery*'s computers commanded a few roll reversals during which we slowly banked from left to right in a big 'S' pattern to control our descent.

"The weather's clearing and starting to look really nice," Capcom Brown said of things at KSC. I had noted the clearing during our last pass over KSC a little more than an hour earlier, and it was good to hear things had improved even more.

Everyone on the flight deck did a great job keeping me informed about what was happening, including descriptions of the view out the windows. Kevin told me there was a pretty good light show out the front. Nancy said there was a great one out the back as well. Her seat was behind Tom's and directly underneath the two large overhead windows in the rear portion of the flight deck. With the large helmets getting heavier by the minute, it was impossible for Nancy to tilt her head up and see directly out the window. Most astronauts strapped a small mirror to their leg during reentry (as I had done for our launch) so they could use it like a rear-view mirror to see the incredible light show out the overhead windows. Every now and then there would be a bright flash as the plasma discharged almost as if a camera flash was going off. It would really light up the flight deck for an instant each time it flashed.

We continued our plummet toward Earth and soon crossed the magical fifty mile altitude mark that defines the edge of space. "We are no longer in space—below three hundred and fifty thousand [feet]," Kevin reported.

Kevin and Nancy continued their descriptions of the light show going on outside: "Starting to get a pink-orange glow out the front." "Yellow and orange out the aft windows."

Kevin summed it up best: "I'd say it's hot out there."

As we passed through forty miles, the light show continued—a bright pinkish glow out the front and a more yellowish glow out the side windows with bright flashes continued out the overhead windows. We were strapped in our seats comfortably ensconced inside *Discovery* while only a few feet away temperatures were nearing two thousand degrees Fahrenheit.

"Pretty smooth ride," Kevin observed. Overall, the descent was surprisingly smooth, especially when compared to the wild shaking and vibrations during the first two minutes of launch. There were a few moments when we experienced light turbulence, but much milder than anything you might experience while landing in a commercial aircraft during a rainstorm. It also was quieter. I heard an occasional thruster firing but these gradually tapered off as we got further down in the atmosphere and our aero-surfaces started to become effective.

As the gs started to gradually build, Kevin blurted out in his best exaggerated New York accent, "No more floatin'." We were experiencing less than half the normal gravity on Earth, but it felt like two to three times gravity because we had spent the last nine days in weightlessness. Moments later, Kevin got excited when he was treated to a most unusual sight—a small crescent moon rising above the Earth that was visible through the pink-orange glow. He jokingly asked me to bring him a camera so he could take a picture, but we were all securely strapped in our seats and experiencing the strong pull of gravity.

A few seconds later, Tom saw Venus rising and ten seconds later Mars rose as well. We had been watching those planets preceding every sunrise while we were on orbit, and Tom knew that another sunrise—the last for our mission—was coming up soon.

We were still racing at Mach 21 and the gs had built to $0.6\,g$. There was a small thump as something near Mary Ellen fell to the floor. "I just heard something drop," she reported. "It's been a while since I last heard something hit the floor." The sound when something falls and hits the floor is such a familiar one on Earth. But in space, you never hear it because of weightlessness. An object hitting the floor is definitely an Earth-based sound.

We were passing through two hundred thousand feet (forty miles) as we approached Baja California—twenty-six hundred miles from KSC, only twenty-four minutes until touchdown. From there, we crossed over northern Mexico as we headed east.

Tom told Kevin to look out the window to "see how fast the clouds were whistling by." "That's bookin'," Kevin responded.

"And that's from two hundred thousand feet," Tom noted.

"X-15 territory," Kevin said. Even at that altitude, everything below seemed so close compared to what we had seen the previous nine days while one hundred and seventy miles above the Earth. Tom saw some land below us and thought it might be Los Angeles because, as he noted, there was a lot of green down there.

Soon we flew directly over Houston where some of the workers in Mission Control went outside to watch us pass. It was still dark and they were treated to an incredible light show of their own, we were later told, as a long plasma trail followed us across the sky. As we flew out over the Gulf of Mexico, the plasma trail behind us appeared from nearly horizon to horizon for the folks on the ground during our five-minute pass over Houston. It reportedly looked like a streaking star. About five minutes later, those still outside were treated to a sonic boom as our shock wave made its way through the atmosphere. But by that time, we were already passing east of New Orleans. Tom saw the Mississippi River Delta out his left window, while all Kevin could see to his right was water. We were passing over the Gulf of Mexico just south of the southern coast of the United States. With six hundred miles to go and traveling at Mach 13.5, *Discovery*'s computers commanded another series of roll reversals.

Gravity was still building and I constantly felt like I was sinking lower and lower into my seat. It definitely felt like someone was pushing me down. My arms started to get a little heavy and the helmet also became noticeably heavy. Tom asked Kevin how many *g*s he thought we were experiencing. Kevin estimated about 3.5 *g*s.

It was only 1.4 g. We had become acclimated to zero-gravity and now everything was feeling heavy. We were still racing at Mach 15 with only about fifteen minutes until landing.

We passed just north of Tampa Bay at 142,000 feet (27 miles) descending at 400 feet per second (270 miles per hour). We were two hundred miles from KSC with only eleven more minutes until landing.

Touchdown!

Besides feeling the pull of gravity, there were other indications we were returning to life on Earth. Kevin saw the contrail of an airliner as we passed one hundred and thirty thousand feet having slowed down to Mach 6.2. Tom saw Florida's east coast. We were getting close. As we passed through one hundred and twenty thousand feet (twenty-three miles) still at Mach 4, the pace was picking up for Tom, Kevin, and Nancy. Tom asked everyone to stop any extraneous talking so we could concentrate on our landing eight minutes away.

At 75,000 feet (14 miles), NASA's long-range TV cameras were able to pick us up and broadcast the images on the space agency's cable channels and shared them with news media outlets, including numerous TV stations in Ohio. We were 40 miles from KSC and our descent had slowed to 260 feet per second (175 mph). Only 5 and a half minutes to go.

Soon we were directly over the Kennedy Space Center with the classic double sonic booms announcing that we were back in town. To those on the ground, it sounded like cannon shots. But we didn't hear them inside *Discovery*. In fact, it was relatively quiet on the middeck. I pulled the earpiece from one side of my communications cap away from my ear far enough to get an unfiltered listen to the sound of *Discovery* gliding to a landing. I could hear the rush of air outside, but it was not very loud or impressive. More of a subtle 'whooshing' sound, not much different from the sound you hear inside a commercial jet while landing. We were minutes away from touchdown.

Tom took manual control and was now flying *Discovery*. As we completed the 237-degree turn around the Heading Alignment Circle, NASA TV broadcast an incredible view of yellow-golden light from the rising sun reflecting off the bottom of *Discovery*. The black tiles that covered the shuttle's underside looked as though they were made of gold. Two beautiful condensation trails flowed off the trailing edges of *Discovery*'s wings.

As we continued our slow turn to line up with the Shuttle Landing Facility's runway, a familiar sight, the Skid Strip runway at the Cape Canaveral Air Force Base, was visible off to the right. There was a little early-morning haze from humidity, but visibility was excellent. Kevin reported that he could see the five hundred foot tall Vehicle Assembly Building, where *Discovery* had been stacked for launch before being rolled out to the pad.

"Runway in sight, Houston," Tom reported, with ninety seconds remaining until touchdown. We were passing ten thousand feet and dropping like a ton of bricks.

Kevin lowered the landing gear at three hundred feet as he continued to apprise Tom of our airspeed and altitude. Before he was chosen as an astronaut, Kevin was a NASA Shuttle Training Aircraft instructor pilot who taught Tom how to land the shuttle in the modified Gulfstream STA. "Looking good," Kevin told Tom.

We crossed the threshold of Runway 33 nearly twenty-three feet above the ground. "Twenty feet. 10...5...4...3...Touchdown!" *Discovery*'s main landing gear made contact with the SLF's concrete runway. Kevin then deployed the drag parachute and, seconds later, exclaimed once again—"Touchdown!"—as the nose gear slammed onto the runway at 8:02:11 a.m. EDT.

It was very evident to me when we landed even though I was sitting in the windowless middeck. The main gear provided a good solid bump similar to a pretty heavy or rough landing on a commercial airliner. But what really got my attention was when the nose gear hit. Bam! It was much more abrupt and felt like the nose of the shuttle slammed down hard. When that occurred, I was thrown forward in my seat and immediately felt my shoulder harness straps lock to hold me firmly in place. Anything loose flew forward and then dropped to the floor. We were back on Earth where gravity rules.

As Tom applied the brakes and steered to keep *Discovery* on the runway's centerline, Kevin read off our airspeed until, at 8:02:57 a.m. EDT, Tom radioed Mission Control: "Wheels stop, Houston."

Their reply: "Copy. Wheels stop. Welcome home Tom and crew."

When our wheels stopped rolling, the STS-70 mission officially ended. Total time since liftoff was 8 days 22 hours, 21 minutes, and 2 seconds during which we traveled nearly 3.7 million miles. There was nothing quite like the adrenaline rush you got returning safely to Earth after a successful mission. It was a fantastic feeling and we were all on top of the world.

Kevin, ever the instructor pilot, congratulated his commander. "Beautiful landing there, Tom."

Nancy added, "Nice one, Tom." Nancy certainly would know because she previously was a Shuttle Training Aircraft Flight Simulation Engineer.

The Orbiter Recovery Convoy immediately headed toward *Discovery*, with the convoy commander welcoming us back to KSC. "It's good to be back," Tom replied. "It looks like a nice morning here."

Tom, Kevin, and Nancy immediately started working through the postlanding checklist just like we had trained hundreds of times in the simulators in Houston. Our motto in the astronaut office was "Train like you fly and fly like you train." It was familiar territory for us.

Down in *Discovery*'s middeck, as soon as we came to a stop I initiated a final set of blood pressure measurements for the DSO 603 experiment, the blood pressure cuff squeezing my arm painfully tight. *I won't be sad to finish up this experiment*, I thought to myself.

I inserted two safing pins into position inside a small white box that covered two T-handles just to the left of my feet. Pulling one of those handles would have depressurized the crew compartment. Pulling the other would have caused explosive charges to blow open the side hatch in case we needed to do an emergency evacuation. Since we were now safely on the ground, we no longer needed those capabilities. The pins were reinserted into the handles to prevent the charges from accidentally being fired. The last thing the Orbiter Recovery Team needed was the side hatch detonating and flying out toward them. With the pins installed we were safe.

I disconnected my parachute harness from my parachute, unfastened my seat belt, and took off my gloves and helmet. I sat there in my seat for a moment and stretched my legs to exercise my muscles a bit before attempting to stand up. When I felt ready to give standing a try, I grabbed onto a strap I installed on the middeck lockers directly in front of me and slowly pulled myself to my feet. I felt heavy, but was surprised by how great I felt. I was feeling much better than I had a year earlier after my fifteen-day STS-65 mission. I was absolutely ecstatic.

We were safely back on Earth and I was feeling tremendous. Adrenaline surged through my body. I could have screamed at the top of my lungs, "We made it!" It sure felt good to be home.

I unzipped and removed my parachute harness and put on my lightweight headset so I could still communicate with my crew mates and listen to what was going on. I was feeling a little hot in my suit as I moved around without the cooling attached. I wiped my perspiration-soaked forehead with a towel.

Tom asked how we were feeling and the entire crew said they were doing great. I was walking around a bit on the middeck working on getting my strength and balance back.

On my first flight, I literally felt like I weighed two thousand pounds after landing. As I sat in my seat after STS-65, I tried to lift my right foot off the floor but could only manage an inch or two. I tried with all my might but I could barely lift it. I wondered whether my foot was stuck on something so I looked down to investigate. Nope, everything was clear. I was having difficulty lifting my foot for two reasons. First, my muscles had deteriorated and were much weaker than they were fifteen days earlier when we left Earth. It's similar to the muscle loss people experience after being in the hospital for a week or two. It would take a few days to regain the muscle strength lost during the mission. The second reason was because my brain had forgotten how much force it took to lift my foot on Earth. We had been floating around weightless. Suddenly we had the tremendous pull of gravity acting on us. It took my brain a bit of time to recalibrate and relearn the normal force and exertion level required to lift my foot.

But since this was my second flight and my second return to Earth, my body remembered how much force it took to lift my feet. It wasn't a surprise. I had to lift hard but there was no problem getting up and lifting my foot. The body and brain remembered. I felt heavy but had no problems moving around. And having the middeck all to myself, I had plenty of room to move around. On my first flight, there were three of us on the middeck during landing so there was little or no room to try walking around before leaving the vehicle. This time, my goal was to readapt as fast as I could. I was amazed I wasn't feeling very dizzy, only slightly so. After about five minutes I was feeling better than I had ten hours after landing on STS-65.

Tom called down to the middeck and asked how I was doing. "I'm up and walking around already," I told him.

"Strong crew," Tom said.

To which Kevin added, "You got a beer for me?" Tom joked that I would be setting up our exercise bike while I waited on the middeck, and Nancy said she thought she heard me riding it during reentry. Everyone was very relaxed and there was plenty of casual joking as we worked through the postlanding checklists. It was just like we were working through them in one of the simulators back in Houston!

As we shut down systems and powered off equipment, the Orbiter Recovery Team vehicles gradually came into sight. "We got a parade waiting for us," Tom joked. He then reported to the team's commander that the side hatch had been safed. When the lead vehicle was one hundred feet away, it stopped and two technicians wearing white emergency response garments got out. They looked like aliens from another world and were carrying equipment to sample the air around *Discovery*.

They were looking for any fuel leaks such as the super-toxic hydrazine used to power our auxiliary power units (APUs) that provide hydraulic power to move the aero-surfaces on *Discovery* for our landing. The saying went 'if you ever smell hydrazine you're already dead.' The stuff was that toxic. So before they opened the hatch, they carefully checked that there were no leaks and made sure the air was of good quality. After satisfying themselves that everything was all right, the rest of the team approached the shuttle and hooked up power and cooling lines. They also set up a giant mobile wind machine called the Vapor Dispersal Unit near *Discovery*'s tail to blow away any fumes that might have vented after landing. A lot of precautions were taken because safety was the number one concern. While I was anxious to disembark, I was totally pleased that everyone was taking their time to make sure we were safe.

Tom reported that he was going off comm for a minute, meaning he wouldn't be able to hear or talk while he took off his helmet and put on a lightweight headset.

Kevin proudly informed the rest of us, "Second in command is in control," to which Nancy jokingly moaned, "Oh, no."

At the Johnson Space Center, Brewster Shaw, a former shuttle astronaut who at the time was space shuttle program director, and Rich Jackson, the STS-70 entry flight director, held the postlanding news conference shortly after touchdown.

Shaw summarized the flight, highlighting the flawless performance of the new Block I main engine and noting that our TDRS satellite was on station in geosynchronous orbit where it was being checked out and that, so far, everything looked good.

Jackson praised Tom's landing as "picture perfect." He noted there were no major and only eight minor in-flight problems during STS-70. He praised the ground processing teams at KSC who readied *Discovery* for the flight.

The reporters wanted to know why we had been waved off on our first landing attempt when the weather conditions had been forecast to be favorable. They seemed to raise some concern that NASA had wasted a good landing opportunity in favor of waiting for a better one. Jackson told them that while the first landing opportunity was indeed forecast to be favorable, they had seen fog develop quickly about the same time the day earlier; it was decided to wait for the second opportunity after sunrise in the hopes that it would ensure better conditions at that time.

Overall it was a pretty routine postlanding news conference. It lasted less than ten minutes.

Time to Say Goodbye

Tom saw the Crew Transfer Vehicle approaching as the Orbiter Recovery Team members were getting ready to open *Discovery*'s side hatch. "I'll meet them at the door," I told Tom.

Before the Crew Transfer Vehicle, NASA used a set of mobile stairs that allowed shuttle crews to disembark. But when missions started lasting two weeks instead of three or four days as they did early in the program, NASA recognized the need for a safer way to get potentially weakened and/or dizzy astronauts off the vehicles. NASA adopted the Crew Transfer Vehicle concept which was modeled after the 'people movers' used at Dulles International Airport outside Washington, D.C. The people movers are much like buses or maybe more correctly, mobile lounges, and are circa 1960s. Passengers boarded the people movers at the designated gate and were driven to a waiting airplane. The passengers walked directly from the people mover onto the plane. NASA purchased a few of those vehicles and then modified them for use at shuttle landings. One was kept at the Kennedy Space Center and a second was stationed at NASA's Dryden Flight Research Center at Edwards Air Force Base in California. Besides making it easier for the astronauts to get off the shuttle, the CTV also allowed the astronauts to do so away from public view.

I could hear some activity at the hatch as they worked to get it open. I watched intently as the hatch handle began to rotate, but after a single revolution it stopped. No more rotation? Were they taking a break? Hey, what's the problem? I half felt like doing the job myself, but just sat back and let them do their work. Everything came to a stop and all was quiet for about five minutes. It seemed like an eternity.

Finally, I could hear the latches moving once again and saw the hatch handle rotating. I later learned that the delay occurred because the team had not gotten official approval from the convoy commander to open the hatch. They had to stop and wait for that.

The latches finally opened and seconds later the hatch was being lowered outward.

"Welcome home," I heard and I, in turn, welcomed them aboard *Discovery*. I backed up a few steps and made way for our astronaut support personnel, Joe Tanner and Steve Smith. They immediately went to the flight deck and started removing our helmets and other equipment. I felt kind of funny having to ask them, "Hey, can I get out of here?" They wondered whether I was ready. Was I ready? Were they crazy? "Yeah, I'm more than ready," I said and then carefully made my way to the hatch and crawled out with no problems. Our flight surgeon, Dr. Larry Pepper, was there to assist me. Afterward, he boarded *Discovery* and climbed the ladder to the flight deck to check on the rest of the crew.

As I crawled out, I definitely was going against the flow. There was a line of people waiting to get inside to start carrying equipment off. Besides our helmets and gloves and personal effects, there were rats and other experiment samples that needed to be unloaded as soon as possible. "Look out! Coming through!" It felt like the opening of an ant hill where there is constant movement and commotion.

Once outside the hatch, I surprisingly didn't notice the fresh air. We had been stuffed inside this vehicle for nine days without any fresh air. The toilet was full and old garbage filled the storage compartments underneath the floor of the middeck. The rat experiment was full of rat feces and urine. It had to have smelled inside, but we must have just gotten used to it.

I always wondered what it was like for the ground crew who opened the hatch, but it wasn't until just after my final shuttle flight that I finally found out. STS-94 landed in Florida, and NASA decided the crew would spend the night there before flying us back to Houston the next morning. That evening we were allowed to leave KSC, so my wife, son, and I drove to Frankie's restaurant in Cape Canaveral which was noted for its Buffalo-style chicken wings. After nine days of eating freeze-dried space food, I was craving something hot, spicy, and—most importantly—not-dehydrated. After placing our order, I looked across the restaurant and saw Bill Todd and a few other suit technicians. They were the folks who suited us up on launch morning and helped strap us into our seat and who had been there to help us off *Columbia* after landing. We had trained over and over with them and I knew Bill pretty well. So I walked over and thanked them for their support.

Then I asked THE question—Does it smell when you open the hatch after landing? Bill blinked a time or two, quickly glanced down at his food, took a swig of beer, and finally looked back up. "No, no. Not at all," he replied. I knew he was lying, but why? Perhaps it just wasn't possible because astronauts are national heroes and national heroes don't stink. But I smelled something rotten, so I pressed him a bit.

"Come on, Bill. Tell me the truth. Does it stink or not?"

He slowly smiled, looked from side-to-side as if to make sure nobody was listening to our conversation, and then looked me straight in the eye and said, "It stinks like hell."

After that revelation, I walked back to our table. "It does stink when we land," I told Simone. She didn't seem too surprised. I think she suspected that all along. The spouses knew that astronauts were not super-humans. Sometimes even national heroes could stink.

Overall it was probably a good thing that our senses slowly adapt to changing conditions and that we couldn't notice the smell of rotting garbage and a full toilet. While it's sad to think we'd been living happily under those conditions being totally clueless, given the circumstances, I'd rather that it went on around me unknowingly!

So I never had the pleasure of enjoying that first full deep breath of fresh air upon exiting the shuttle, but it did feel great to get out. I slowly walked ten or fifteen feet inside the Crew Transfer Vehicle and took a seat in one of the large comfortable overstuffed chairs. It felt great sitting down. Then they handed me an ice-cold bottle of water and suggested I drink as much as I could. They were encouraging me to rehydrate because my spacesuit was quite hot and I was perspiring quite a bit in it. And even though I had done a lot of 'fluid loading' with what seemed like endless amounts of water before landing, I knew it was good to be fully hydrated to prevent dizziness. The water was ice cold and tasted great. I sucked down the first bottle within seconds and started working on the next.

I was as happy as I could be. I was elated. I was flying high. Our mission was a 100 percent success. We accomplished everything we set out to do. We set sail the final shuttle-deployed TDRS and all the experiments had gone well. I had done my part. And the rest of the crew had done theirs. The mission would go down as a success and I was extremely gratified as I sat in that chair.

Emotionally I was soaring and physically I felt wonderful. My legs were strong, my balance was 90 percent, and I couldn't imagine that I could ever feel any better than that after coming back from space. The adrenaline rush continued. I was ecstatic!

Soon Nancy made her way into the CTV, then Mary Ellen, Kevin, and finally Tom. The commander was usually the last to leave the ship. We all had huge smiles on our faces.

The Walk-Around

About five minutes after Tom joined us in the crew transfer vehicle, a quick assessment was done to see who was feeling well enough to do a vehicle walk-around. It was a tradition that started with the very first shuttle landing—*Columbia* at Edwards Air Force Base in California on April 14, 1981. Astronauts John Young and Bob Crippen exuberantly bolted down the stairs and walked around *Columbia* to give it a quick inspection following its first flight. Since that time, most crews did it, too, depending on how everyone was feeling and whether there were any medical tests that needed to be completed immediately after landing.

The STS-70 crew was all in for the walk around. Tom got out of the CTV first, followed by Mary Ellen, Nancy, me, and then Kevin. From the CTV's door to the runway, it was a relatively simple walk down eight or nine steps. As simple as walking down nine steps sounds, it was quite a challenge when your vestibular system was not working properly and you're experiencing various degrees of dizziness. There were handrails on each side of the steps and smart astronauts used both of them to steady themselves.

With great care and intense concentration, we all made it down the stairs without incident. We were once again standing on Mother Earth. We were greeted by Steve Hawley, a former astronaut who was deputy director of the Flight Crew Operations Directorate at JSC; Launch Director Jim Harrington; and Mike McCully, another former astronaut who was director of Shuttle Operations for Lockheed Martin at the Kennedy Space Center. We shook their hands and heard the accolades—"Great Job," "Nice Landing," and "Welcome Home."

Since we were right at *Discovery*'s nose, we posed for a few pictures, including the traditional one with all giving the thumbs-up sign as the shuttle rested majestically in the background. Mission accomplished!

Then we walked over to the orbiter and Tom took a close look at the two nose wheels sitting virtually dead-center on the runway. He had brought *Discovery* to a stop less than six inches from the centerline. That was an impressive accomplishment. We then took a look at the main landing gear. I put my hand up near some of the black tiles on *Discovery*'s underside and was amazed that I could still feel the heat from reentry. Less than two hours earlier those tiles were experiencing temperatures of one to two thousand degrees Fahrenheit and I could still feel the heat as they slowly cooled down.

As we continued our walk around *Discovery*, my legs felt heavy but strong. I felt only minimal dizziness when I turned my head, and it was no problem for me to step over the large twelve-inch diameter yellow hose spread across the runway that was used to blow cold air into *Discovery* to help cool her down. When I had approached that hose during the STS-65 walk-around, it appeared in front of me like an insurmountable obstacle. I prayed that I would be able to step over it without tripping and falling. I somehow managed to successfully scale it. This time, I carefully concentrated as I stepped over it, but it was a piece of cake. As Kevin came up to it, he said, "Okay, time for the high hurdles." We all made it across the dreaded giant hose without a single misstep or tumble.

It would have been nice to hang around *Discovery* for a while to take it all in. But it was time to get in the Astrovan for the short drive to crew quarters where our families were waiting. I took one last look at *Discovery* on the runway. I had just safely returned to Earth aboard her. She protected me during the fiery reentry through the atmosphere. She was my home for the previous nine days shielding me from the vacuum of space, the temperature extremes, and the radiation. It was humbling to stand there and gaze at her. I had a feeling of awe (Did we really do this?), a feeling of great pride (What an amazing space vehicle this is), and a sense of relief (So glad to be safely back on Earth). I took a deep breath and shook my head in disbelief as I exhaled with a "Whew!" It was hard to take it all in. I focused my eyes on the American flag on the side of the shuttle and then on the name *Discovery* in big bold letters right below and aft of Kevin's side window. "Thank you, *Discovery*." I quietly said to myself.

With that I turned and walked to the Astrovan and carefully climbed the three or four steps. Here was another opportunity to trip or stumble, so it was with great concentration and deliberation that I made it inside the Astrovan and took a seat. With all of us aboard, we headed to our family rendezvous five miles away. I recall a few jokes, but otherwise the ride was quiet. Even though the adrenaline was still running through us, we were all tired and emotionally wrung out. I was still thirsty—half from sweating in the hot Launch and Entry Suit and half possibly from the Florinef I took prior to landing. I chugged down another quart of water while in the Astrovan.

Reunited

I took a seat near the front of the Astrovan because I was anxious to see Simone and Kai as soon as possible and wanted to be the first off. When we got to the Operations and Checkout Building, I was hoping our families would be outside to greet us as we got out of the vehicle. But nobody was there. They were waiting upstairs in the more comfortable crew quarters because it was July in Florida and although it was only mid-morning it was already starting to get hot.

We got on the elevator and rode it to the third floor. When the doors opened, I looked straight ahead for Simone. Not there. I looked to the right and then to the left. Nothing. I saw all the other family members, but no Simone and no Kai. I was disappointed and moved off to the side and watched my crew mates greet their spouses and children. Maybe ten seconds or so passed—it seemed more like an eternity—and Simone finally appeared. I gave her a big hug while still searching for Kai. Where was he, I asked. Simone had given him to Trudy Davis, our official family escort, while she had made a quick trip to the bathroom. I soon spotted Trudy and took Kai from her. Thankfully, he was only five months old and not too heavy for a weakened astronaut. Simone, Kai, and I did a three-way hug for a few moments. After my first flight, Simone had to hold me upright as we hugged. This time I was much steadier. The hugs felt great, but all too soon it was time to head for medical check-ups, debriefings, and a few postlanding experiment data collections.

After we were helped out of our suits, we went back to our rooms, got out of our long underwear, and put on our more comfortable blue flight suits. Then we rode the elevator to the second floor where I was examined and debriefed by our flight surgeon and from there I participated in a few other medical examinations. One of my personal favorites was the orthostatic intolerance stand test. It began simply enough. We laid down on a stretcher for a few minutes while blood pressure and heart rate data were collected. So far, so good. After that, six to eight people surrounded me and slipped their arms underneath me. In a well-rehearsed and well-choreographed maneuver, there was a 3...2...1 countdown, then the command "lift" was given, and instantly I was stood up on my feet while they continued to monitor my blood pressure and heartbeat. "Let us know when you feel dizzy or are about to faint," they told me. Once upright, I felt the blood rushing from my head as gravity pulled the blood down to my lower body. The sensation is hundreds of times more intense than when you jump up quickly after lying in bed and you get a momentary feeling of dizziness. I felt weak and dizzy and the team helped keep me steady so I wouldn't fall over. Within seconds I was feeling pretty woozy as if about to pass out.

"I'm about to faint," I said and they gently laid me back down. That experiment was looking at how well the body can maintain blood up at the brain after standing quickly after spaceflight. I was about to pass out all in the name of science.

After a few more tests and the donation of seemingly endless test tubes of blood, I met privately with our crew surgeon where I was asked how I was feeling, how I had slept, and what medications I had taken while on orbit. I had taken the usual sleeping pills, vitamins, a few aspirin for mild headaches, and some antihistamines for decongestion. I then was given a brief physical examination.

Afterward, I was asked if I wanted anything to drink and I requested an ice-cold soda. I was craving something carbonated after drinking powdered drinks for the past nine days and I wanted it to be ice-cold. Someone brought me a soda and I put it right next to my face to let a little of it spray in my face as I opened it. The cool spray felt great. Man, was I glad to be back on Earth. After a few sips, I was told I could head back to crew quarters to take a shower and visit with my family.

There was just something fantastic about that first hot shower after returning from space. And after spending the previous five or six hours inside my suit, I was soaked in perspiration and in great need of a shower. After washing with the rinseless soap and cleaning my hair with rinseless shampoo, showering with real soap and real shampoo seemed like an incredible luxury. And best of all, there was nobody in line waiting to use the shower. I took my time and let the hot water spray down on my head as I moaned and groaned and made other strange noises enjoying the experience. The showers in crew quarters all had a small built-in bench so we could sit down if feeling weak or dizzy. So I sat down and just let the hot water rain down on me. NASA recommended that we take warm—not hot—showers because they were concerned that the hot water might cause vestibular problems or facilitate a fainting spell. I took my chances and went with the hot shower. I was fine.

Afterward, it was time for lunch—hopefully nothing freeze-dried. In fact, returning shuttle crews traditionally were served lasagna, salad, and fresh rolls prepared by the crew quarters' staff. They were well known for their hospitality and delicious meals. I grabbed another ice-cold soda and filled my plate with a lot of lasagna and a little salad. The lasagna was simply excellent, made all the more so because it was the first fresh food any of us had eaten in the past nine days.

While we were eating lunch, we got a call from Ohio Governor George Voinovich, who congratulated us on our successful mission and told us how much he and his wife, Janet, enjoyed being at KSC to see us off at the launch nine days earlier. He invited the entire crew to be his special guests in two weeks at the opening day of the Ohio State Fair in Columbus. Tom thanked the governor and without knowing what our schedules would be for the next couple of weeks, accepted his invitation.

Besides the usual ready supply of chocolate chip cookies in the crew quarters cookie jars, we also had the official STS-70 launch-day cake for dessert. The cake, decorated with our mission patch, was used primarily as a table decoration on launch morning as we ate our last Earth meal for a while a mere five hours before liftoff. The cake was kept in a freezer until we landed.

After lunch we were driven over to the KSC press site for our traditional postlanding press conference. Tom made an opening statement thanking the KSC team for their excellent work getting *Discovery* ready for launch and for "recovering from the woodpecker damage and getting us off on time."

The first question from the handful of reporters in attendance was about the STS-71 SRB O-ring problem. They wanted to know if we were told about the problem while we were on orbit, and if we thought NASA needed to wait longer between flights to have more time to resolve any problems detected after a mission. Tom said we weren't told about the SRB problem because, at that point, our SRBs were long gone and there was nothing we could have done about it anyway. Tom said he didn't think we were rushing the flights but simply needed to keep the flights going in order to demonstrate the kind of launch schedules we would need to meet once the space station became a reality.

Phil Chen, a regular at KSC-JSC press conferences, asked Kevin if an 'All-New York space shuttle mission' might be in the works for Kevin, Jim Wetherbee, Ellen Baker, Mario Runco, and Dan Bursch: all New Yorkers.

"There is a good chance that someday there will be an All-New York crew because even though my STS-70 crew members are quick to say that there are more astronauts from Ohio than any other state, New York is number two and we're catching up," Kevin said. While unfortunately the space shuttle program came to an end before an All-New York mission ever materialized, New York did go on to eventually surpass Ohio in having the greatest number of astronauts.

We were asked what we thought our mission meant to the people of Ohio. I replied that we had just spoken with Governor Voinovich, who told us that just as John Glenn inspired our generation of astronauts, the All-Ohio crew had inspired a new generation of future astronauts.

We also announced that the governor invited us to the opening of the Ohio State Fair in two weeks and that we were looking forward to being there with the governor and giving him back his watch which flew with us on our mission.

Tom was asked if there were any future missions he was interested in flying. In true astronaut fashion, he replied, "The next one available."

Tom's landing of *Discovery* was lauded, but Tom immediately shifted the praise to Kevin and Nancy. But Kevin apparently felt he had to set the record straight. "Tom is too modest. On a four point scale, his landing was a five. It was perfect. Right on speed. Right on touchdown." Then he jokingly added, "Don was sitting down on the middeck and didn't even know we had landed—it was that smooth." Tom was a great commander and an excellent pilot.

Mary Ellen said her most memorable moment of the flight was her first view of Earth against the blackness of space. A reporter noted that Mary Ellen was the first astronaut born after John Glenn's 1962 *Friendship 7* mission. "I'm just proud to be serving in the footsteps of Senator Glenn," she replied.

At the end of the press conference, Tom reached into one of the pockets of his flight suit and produced a buckeye. "We look forward to returning these and our other Ohio items when we get back there in the next few weeks," he said. And with that, our just over ten-minute press conference was over. It was pretty painless as far as press conferences go.

Right after that, we were driven back to crew quarters and we went to a large first-floor conference room where some of our extended family members were waiting. Normally only immediate family members—spouses and children—were allowed at crew quarters for postlanding visits. Other family members—parents or brothers and sisters—were not allowed. So NASA scheduled a thirty minute visit with our extended family members who had come to see *Discovery* land.

There was always a flood of family and friends on hand for launch. For the landing, which I think is an incredible experience in its own right but is considerably less dramatic and a bit anti-climatic, almost nobody showed up. For the STS-70 landing, my mom was my only extended family member there and it was great to be able to see her so soon after the flight. She had stayed in central Florida since the launch with a friend from her days working for the Department of State. My mom had a huge smile on her face which probably mirrored the smile on my own. I gave her a big hug and told her I loved her and she just replied, "I'm so proud of you! I'm so proud of you!" She was wearing her infamous T-shirt that said 'As a matter of fact I am…' on the front and 'the mother of the astronaut' on the back. My mom and I chatted for a few minutes about what she had been doing in Florida and what I had been doing in space.

Before the mission, I had given her a gold necklace and a gold pendant with the STS-70 mission logo on it which she wore nearly every day. I had taken an identical one with me on STS-70 and asked her if I could swap the two. I gave her the flown version and she returned the unflown one. I did the same thing for each of my four flights.

Our NASA minders encouraged us to finish our visits because they wanted to get us back to Houston that afternoon. So I said a few quick hellos to some of the other parents and crew members' guests. As I said goodbye to my mom, I promised her I would visit her soon in Indiana and tell her more about the flight. After one more hug during which I told her, "Thanks again for everything you have done for me which helped make this moment possible," we said goodbye and I was whisked away.

Headin' Back to Houston

Soon Simone, Kai, and I were on a bus that took us to the Skid Strip, a runway at the Cape Canaveral Air Force Station just a few miles away where two NASA Gulfstream jets were waiting for us. Shortly before 3:00 p.m. we were taking off for Houston. It was about a three-hour nonstop flight to Ellington Field, during which I talked with Simone, played with Kai, slept a bit, ate some of the snacks provided, and drank anything nonpowdered that I could find, mainly sodas and bottles of ice-cold water. It all tasted so good.

There was a long-standing shuttle program tradition that allowed the families flying home to enjoy a selection of baked goods made by fellow astronaut Marsha Ivins. Each of our planes had its own box of goodies, and we enjoyed treats such as chocolate brownies, chocolate cake, cheesecake, and other desserts that do not exist in space. They were absolutely delicious.

I wanted to reacquaint myself with Kai who I had only seen for seven days during the previous five weeks. He was only 157 days old and I had already missed one-fifth of his life. I felt like I had some catching up to do. As I was playing with him, he was fidgety and crying a bit. Someone jokingly commented that "Kai was so good the whole time till you got here." That didn't make me feel very good, but that was one of the tough transitions astronauts made after they returned from space—to reintegrate themselves with their families.

We arrived at Ellington Field shortly after 5:00 p.m. CDT. It was a beautiful sunny afternoon and as hot and humid as always in Houston during the summer months. Our Gulfstream jets taxied to a hangar at the north end of the airfield where a small welcome home ceremony was planned. Some drew large crowds, with even presidents attending a few of them. But for STS-70 on a late Saturday afternoon in the middle of summer, our crowd consisted of only the most dedicated and faithful—friends, neighbors, our training team, and others who worked some facet of the mission including those in Mission Control, public supporters, and a few extreme space fans. On the side of the hangar was a big banner that read "WELCOME HOME CREW" that was put up by the Manned Flight Awareness Office at the Johnson Space Center.

A red carpet was rolled out that led from the door of our plane to a small red-carpeted stand. Such amazing treatment for astronauts returning home from space. The stand had six seats on it, one for each crew member and one for the JSC Director, Dr. Carolyn Huntoon. Behind the seats were the US and NASA flags; in front of us a podium bearing the STS-70 mission logo. There was a strong breeze so the flags were fully waving in the wind. As we got off the plane, we were greeted by Dr. Huntoon. Then, we headed down the red carpet. We took our seats on the small stage, while our families sat in seats set up in front of the stage just for them. To climb up on the stage, there were two small steps. We took extra care walking up them to avoid any possible mishap. It took a day or more for us to get our balance back to normal after returning from space, so occasional trips and missteps were common during the hours after landing. But thankfully none at the event.

Dr. Huntoon thanked everyone for coming and then introduced Tom. We each got to speak three to four minutes and each of us thanked different groups of people who helped get us ready for the flight or who supported us in Mission Control. I, proudly wearing a bright yellow baseball cap with Woody Woodpecker embroidered on the front, thanked my neighbors, Frank and Carolyn Littleton, who helped Simone and me so much in the past year. Then I thanked the TDRS training team—Allen Burge, Robert Graubard, Ginny Young, and Bill Preston—who prepared me so well that TDRS's deployment was flawless. After our brief remarks, the ceremony came to an end. We spent a few moments shaking hands, signing autographs, and talking with those who were there to welcome us home—including our Capcom Tom Jones who held high a home-made sign that read 'USED TO BE A BABE MAGNET.' It was a reference to a question I had been asked during an on-orbit interview about whether or not being an astronaut made me a 'babe magnet.' I laughed at the sight of Tom holding that sign in public. He wasn't about to let me forget that interview! Astronaut Ron Sega, a good friend and fellow Clevelander, was also there to welcome me back.

Because of the dizziness most astronauts experienced after returning from space, NASA didn't want us driving, so they provided us with a ride home. We headed to vans that were waiting to take each of us and our families to our residences. The Ellington crew already loaded our suitcases and the vans were ready to roll. Simone, Kai, and I were joined by Tom and Becky. It was always an impressive sight to see the astronauts departing Ellington and was done with great fanfare and police escorts. The Harris County Sheriff's Office provided police car escorts complete with blaring sirens and flashing emergency lights. It was all so impressive for the first two or three miles, but then suddenly our police escorts fled the scene and we were once again on our own. I'm sure the Apollo crews returning from the Moon got police escorts all the way home. Times had changed and astronauts were no longer the big stars they once were. Shuttle crews only warranted escorts part of the way home!

Suddenly it was quiet in the van which I found refreshing. I had had enough of the attention and the noise. I was looking forward to getting home to some peace and quiet. We first stopped at Tom and Becky's townhouse which was located just outside the Johnson Space Center's back gate. We said goodbye as they unloaded their luggage and soon we were rolling toward our home, a ten minute drive away. As we turned onto Shadow Creek Drive in El Lago, I saw four large American flags flying in front of my house. It really made me feel good. I moved to El Lago seven years earlier specifically because it was the home of many of the Gemini and Apollo astronauts. Neil Armstrong lived in El Lago. So had Frank Borman and Ed White. I wanted to live there, too. As an added convenience, it was only two miles to work. Only ten minutes door to door.

El Lago was a small town with some big traditions. Since the early days of the NASA Manned Spacecraft Center—now known as the Johnson Space Center—the City of El Lago placed American flags in the front yards of returning astronauts who lived in their city to welcome them home. I didn't realize that until my first flight when we turned onto my street and I saw the flags. I have to admit it was a beautiful sight. It made me feel like a part of history, and I was proud to live in El Lago. It was an older community. The houses were mostly twenty-five to thirty years old (ancient structures by Houston standards) and very few astronauts chose to live in El Lago at that time. A few veterans were still there, like Hank Hartsfield, Don Peterson, and Apollo 16 moonwalker John Young. But not many other astronauts lived there anymore. They seemed to gravitate to the newer communities with the spacious new homes. But I always liked El Lago—great neighbors and a small town feel to it. Plus all the history and tradition. I liked calling it home and never once considered living anywhere else when I was working at NASA.

So with the four large American flags flying in our yard, we unloaded our bags and, with the help of our NASA driver, carried them into the house. I said thanks and goodbye to our driver and then closed our front door. The silence was magnificent! We were home once again. It was quiet once again. And best of all, there was privacy once again. It felt so good to close that door and leave the rest of the world outside for a while. After living in close quarters with my crew mates during the mission, it was great to be alone with my family once again. Home. I made it back home.

Greeting me just inside were eight balloons and a small sign:

"Welcome home, Dad!
Welcome back to the real world
of diaper changing and sleepless nights!"

It was a clear reminder of what awaited me. I was no longer in space. I was at my home, with all the usual issues. But somehow it still felt good to be home.

There was a single message on our telephone answering machine from my high school friend, Tony Cuda, welcoming me back. It brought a smile to my face and I couldn't wait to call him back in a day or two to tell him about the mission.

We had a large futon in front of our TV and I just crashed right there, too lazy to walk up the stairs to change out of my blue flight suit. Simone ordered a double-pepperoni pizza for me that had never tasted so good. We then went next door to see our neighbors, Frank and Carolyn Littleton. Carolyn was a high school teacher; Frank was an engineer and on the management team at the Johnson Space Center. There were no better neighbors in the world. Frank mowed my lawn the day we landed to make sure it looked good when I returned. We took along a bottle of champagne to celebrate with them and after a half hour or so we headed back home. I went to bed about 10:00 p.m. after being up for the previous twenty-four hours. I was exhausted, but it felt so good to be back on Earth and to be home again.

I am frequently asked where I prefer to live—in space or on Earth? The answer is quite simple. While I loved visiting space, I strongly prefer to live where my family, friends, and favorite foods are, down on Earth.

References
Jillson, Joyce. 1995. "Joyce Jillson's horoscope, Creator Syndicate." *The Houston Chronicle*. July 21. Pg. D-8.

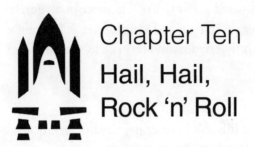

Chapter Ten
Hail, Hail, Rock 'n' Roll

After a good night's sleep, I woke up about 6:00 a.m. Sunday and got out of bed a half-hour later. My calves were sore and my legs felt tired and a little weak like I had been hiking in the mountains for ten hours the day before. But overall, I felt pretty good. After a few wobbly steps, I got my legs working properly and went downstairs and had a few minutes to myself.

What I wanted most upon my return to Earth was privacy, peace and quiet, and the opportunity to relax a bit. I wanted to spend some time reflecting on what I had just experienced. I wanted things to be a bit unstructured, with more freedom, instead of following those infamously rigid NASA schedules. Most importantly, I wanted to be with Simone and Kai as much as possible.

But unfortunately none of that was possible to any significant degree because we moved quickly into the third phase of our mission—what NASA called postflight. It was when we tied up all the loose ends associated with STS-70 to finish up the mission. There was a lot of work that needed to be done—writing reports, preparing presentations, and debriefings with our trainers and managers about lessons learned. We also needed to thank the many people who worked so hard for so long to keep us safe and help us return home. As a result, less than twenty-four hours after arriving back in Houston, we were back at work. It would be as busy and as hectic as our preflight training and our flight itself, but generally was much more relaxed and involved a bit more fun.

If the weather had cooperated, we would have landed on Friday and had Saturday off—just what a weary group of astronauts needed. But that didn't happen because we landed on Saturday. And for better or worse, we scheduled our postlanding crew party for Sunday. It was a decision we made before launch and decided to keep because it would have been too difficult to reschedule. The event was our way of thanking our training team and many others who prepared us for the mission and supported us while we were in space. Tom chose Carrabba's Italian Grill, a restaurant midway between the Johnson Space Center and Ellington Field. The party, scheduled for 1:00 p.m., would be our first semi-official postflight event.

After enjoying nearly thirty minutes of quiet time that Sunday morning, Kai woke up at 7:00 and was soon downstairs with me. After getting him dressed and fed, he and I drove to the JSC Astronaut Crew Quarters to pick up some things I left there before our flight. I also went to my office to look through the mail that had piled up while I was gone. The first-floor hallway in the building where the STS-70 crew office was located had been decorated by Tim Terry and the rest of our training team with big 'Welcome Home' signs plus numerous references to woodpeckers. I took a few minutes and slowly walked down the hallway to look at the pictures, notes, cartoons, and other decorations. The display was another long-standing NASA tradition that allowed references to and depictions of many of the private jokes between the training team and the crew, along with anything else noteworthy and humorous that might have taken place during the flight.

After about fifteen minutes in the office, Kai and I drove to the Seabrook Classic Cafe, a diner not far from our house. There I enjoyed an unbelievable breakfast of French toast and sausage (neither of which had been freeze-dried) and a big glass of fresh (not powdered) orange juice. I even treated myself to an order of beignets generously covered with powdered sugar. I savored every bit of breakfast, especially being able to share it with my son. I'm sure I walked out of there with a modest amount of powdered sugar on the front of my shirt, which could be attributed to living back on Earth with its relentless pull of gravity.

We then drove home and I played with Kai for a while until it was time to get ready for the 1:00 p.m. party. Simone, meanwhile, got to enjoy sleeping in for the first time in quite some time.

At Carrabba's, Tom thanked everyone for their hard work the past year, with special kudos to our training team led by Tim Terry. They had prepared us well for the mission. Then, each of the other crew members took a few minutes to thank and acknowledge others associated with STS-70. It was a great event and I enjoyed that opportunity to treat the folks who had been working so hard on our behalf for the past year.

Afterward, Tom invited the crew back to his house where we all hung out for a while and took a collective sigh of relief. We were all tired but glad to have completed our postflight party. Scratch one item off the 'to do' list of things we needed to take care of.

From Tom and Becky's, Simone, Kai, and I drove home and we relaxed a bit that afternoon. I made a few calls to family members and friends and we later took a walk around the block. It felt really good to get out and walk around the neighborhood. I enjoyed the freedom of being home.

Medical Check-ups and JSC Debriefings

Monday morning came quickly, and our first scheduled activity of the day was a postflight medical examination. I gave blood and saliva samples. They checked my balance and blood pressure. They collected some data as part of the medical experiments we conducted during the flight. Overall the physical was uneventful. I was cleared to return to normal activities like driving and flying, but was cautioned to take it easy for a few days. That was about it.

Oh, but wait. I forgot the urine sample saga.

As we were leaving Ellington after our return from KSC on Saturday, each crew member was given a small shoulder bag containing empty plastic bottles and ice packs. As part of our postlanding physical, we were supposed to collect our urine for the next thirty-six hours. We were given plenty of those urine collection bottles, but not quite enough for me. Remember, I took a large dose of that fluid retention drug, Florinef, not once but two days in a row. Hours after I returned home, the drug began to wear off and soon I began to urinate. I urinated and urinated and urinated. And then I urinated some more. I had so much fluid in my body I must have sloshed around the room as I walked. By the end of the evening—just a few hours after I arrived home, I had filled up all of my urine collection bottles. With more than thirty hours left before my physical, I would need more bottles. Lots more bottles. The urine was still flowing.

Being out of the official urine collection bottles, I turned to plastic gallon milk jugs. I had the first filled by noon the next day and was working on my second one by the time Monday morning rolled around. When I dropped off my samples, the technician looked at me in disbelief. "What's with all this urine?"

"I guess it was the Florinef," I told her. "Maybe I should have been issued more of the official urine collection bottles, eh?" You think so?

I produced more urine during those thirty-six hours than the rest of the crew combined. Nancy thought my urine sample record was a hoot. I, on the other hand, didn't think it was all that funny. After all, I did it for NASA, my country, and science. I was just glad the experiment finally was—hum, let's see—water under the bridge.

After the medical stuff, we met at JSC crew quarters and began going through material for our postflight crew report and various debriefings. We also had a chance to review the pictures we took during our flight. We were still using film cameras back then and the film was one of the first things unloaded from *Discovery* after we landed and sent immediately to Houston for processing. That expedited film processing was done so that NASA could release pictures from the flight while there was still media interest, before our completed mission became old news.

Most of that first week back in the office was spent on debriefings. We talked to the JSC upper management team and Tom summarized the flight and pointed out a few things that could have gone smoother, such as our quarantine. He also acknowledged a few people who did an outstanding job for us. We continued our debriefings with various groups at JSC to answer their questions and provide additional details on how things went for our mission—flight crew equipment, middeck payload operations, the TDRS deploy, and our EVA training. We identified things that worked well and things that needed attention to improve the processes, training, and equipment for future missions. In reality, many of the suggestions simply got ignored. But for big safety of flight issues, NASA did a pretty good job addressing those and fixing them as they were identified.

Opening of the Ohio State Fair

When we spoke with Governor George Voinovich shortly after we landed, he invited us to attend the opening of the Ohio State Fair on August 4. Obviously, we were thrilled at the prospect and very much looking forward to it. A few days later, we were pleasantly surprised to learn Governor Voinovich had invited us to spend the night at the Governor's Residence.

We flew to Columbus the day before the opening of the fair and went directly to the residence where we were warmly greeted by the governor and the first lady. They welcomed us like we were family visiting for the holidays and did everything possible to make our stay as pleasant as could be. After a brief tour of the residence, we were shown to our rooms on the second floor where we dropped off our bags, freshened up a bit, and then met downstairs for the trip to the state capitol to tour the governor's office. Then it was back to the residence for dinner.

We ate in the formal dining room which featured oak paneling, a carved stone fireplace, and a bay window along one wall. A huge gold and crystal chandelier hung above the table. With a bright white linen tablecloth, official State of Ohio china, and shining silverware, it was an impressive sight—and nothing like what I was used to.

We sat down at our assigned places with Tom sitting next to the governor and me sitting across from Janet, the governor's wife. She was such an incredible hostess and the warmth and informality of the dinner made it a most special occasion.

We started with a salad and then were served the main course of grilled salmon. The serving staff all wore formal attire—black pants and short white jackets. Once my plate was in front of me, I noticed Janet attract the attention of one of the servers. She gave him a nod, indicating to me that something was up. Seconds later, one of the servers walked up to me wearing white gloves and carrying a silver platter. On it rested a single jar of Stadium Mustard and a small silver knife. In the most official of voices and with as straight a face as possible, he asked, "Would you like any Stadium Mustard with your salmon this evening, sir?" What a tremendous surprise. I couldn't help but laugh. I glanced over at Janet and she had a huge smile on her face as well.

When everyone else saw what was going on, they joined in the laughter. Kevin seemed a bit envious that I got first crack at the mustard. He had become hooked on it during our mission, frequently asking if he could "borrow a pack of Stadium Mustard." Once the look of surprise and delight left my face, I admitted to Janet that I had never eaten salmon with Stadium Mustard before. But noting that everything tasted better with Stadium Mustard, I enthusiastically put a liberal amount on my plate before it was offered to the rest of the crew, the governor, and, of course, Janet. What an incredibly thoughtful gesture by the Voinoviches. I was flattered that they would even think of such a thing, and it turned out to be a great ice-breaker that set the tone for a fun and informal dinner.

We had a most enjoyable evening with the Voinoviches before we headed to our rooms. We were up early the next morning and served a nice breakfast before we got ready to head to the fairgrounds. I believe I even saw a small jar of Stadium Mustard on the breakfast table that morning.

We joined the governor and Janet, both wearing T-shirts that said 'Ohio Proud' and bearing the STS-70 mission logo, at noon to cut the ribbon officially opening the fair. The event took place at the George V. Voinovich Livestock and Trade Center. Also in attendance were former Governor James Rhodes and Ohio State University President Gordon Gee, who promised Nancy that she would get to dot the 'i' during the performance of "Script Ohio" at one of her alma mater's future home football games. Tom told the crowd that he thought it was appropriate that he, an Ohio farm boy, was talking about the STS-70 All-Ohio Space Shuttle Mission in the state fair's livestock building.

Afterward, we ate a lunch of fair-appropriate food in one of the fairground's dining halls. Then we boarded some wagons pulled by a tractor and set off on a tour of the fairgrounds during which we saw many typical fair attractions, such as cows, pigs, cookies, cakes, and home-made pies. As we walked through a section that showcased Ohio companies and Ohio products, the governor proudly pointed out a special exhibit he knew I'd particularly enjoy featuring Stadium Mustard and its parent company, Davis Foods, of Cleveland. We also visited a booth featuring Sandusky County, where Tom had grown up, which displayed numerous STS-70 items, including a huge depiction of our mission logo.

From there we walked over to a milk exhibit where they asked us to pose for some pictures drinking milk. They wanted to get a photo of us with milk mustaches, an advertising promotion for drinking milk still popular today. Unfortunately, real milk doesn't leave that much of a visible mustache and nothing like seen in the advertisements. But we drank the milk anyway and they took pictures of our less-than-impressive milk mustaches. I don't think much ever came of those pictures.

We also got to see a cow sculptured in butter—a state fair tradition; visited a NASA exhibit; were serenaded by the Ohio State University marching band—a tip of the tuba to OSU alum Nancy; and went fishing with the governor and a bunch of children at the youth fishing pond stocked with bluegills and catfish that was part of the fairground's Natural Resources Park.

It was at the park, much to the consternation of an Ohio State Highway Patrol trooper, that Kevin actually got to see how buckeyes came to be. The buckeye trees were bearing their fruit, a two to three-inch green seed pod with a smooth but bumpy exterior. A man walking with Kevin reached up and pulled one of the pods from one of the trees, which officially became the state tree in 1953. The trooper, apparently determined to protect the state's heritage, approached and began chastising the man for picking the pod. The governor quickly intervened, explaining that Kevin and the man were with him. The trooper backed off, and the man peeled away the pod so Kevin could see the beginnings of the more-familiar-to-him mature chestnut brown buckeye seed with its light tan circular eye that Native Americans thought looked very similar to the eye of a buck deer. Hence the name. Also, according to folklore, the buckeye is considered a good luck charm. Apparently Florida woodpeckers weren't aware of that.

After a couple of hours touring the fair, we thanked the governor and first lady for their hospitality and then boarded our T-38s for the flight back to Houston.

During my seventeen years as an astronaut, not once did I have the opportunity to visit the White House to meet the president, as frequently happened after many shuttle missions. While I am sure such a visit would have been an amazing experience, I can't imagine it could ever compete with the warmth and hospitality the Voinoviches extended to my crew and me after STS-70.

Postflight Presentation and Space Flight Medals

The postflight presentation to the news media and JSC employees is a long-held tradition during which returning crews narrated a fifteen-minute movie about their flight, showed some pictures of Earth, and answered a few questions. It was usually held within three weeks of landing, and that gave us a little time first to review all our film and images. Ours was held August 11.

After a brief welcome and introduction, George Abbey, the deputy director of JSC, along with Dr. Carolyn Huntoon, the JSC director, pinned the Space Flight Medals on each of us. We were called up one at a time in order of our position on the flight. Tom was first followed by Kevin, me, Nancy, and finally Mary Ellen. Suits, ties, jackets, and medals! What an impressive group we were that day looking so unlike the normal astronaut walking around the space center in polo shirts and jeans or khakis.

After the medal presentations, we went right into our crew presentation during which we showed our STS-70 movie with highlights from the flight and then a series of slides showing some spectacular images of the Earth, including scenes of an erupting volcano in Argentina, dust storms over the Red Sea, the blue waters of the Bahamas, and the colorful Tiffernine Dunes in northern Africa.

Next, Tom, in a bit of a surprise, presented Kevin and Mary Ellen with their gold astronaut pins which traditionally are presented to rookie astronauts after their first flight. The pins flew with us on *Discovery* and normally would have been presented at the crew postflight party. Since our party was the day after we landed, Tom wasn't able to get the pins in time. As Tom placed the pin on Kevin's lapel, he told him, "I hope we get to do this again someday." Whether Tom knew it at the time, he and Kevin were soon assigned to the same flight—STS-78, the Spacelab Life Sciences mission.

After answering a few questions from the audience, we signed a few autographs for NASA employees and visited with some friends and co-workers. Then we headed to our office to get back to work on our other debriefings, which we completed over the next two weeks. After that, we went on the road for a number of postflight appearances.

Space Day Rally in Cape Canaveral

The STS-70 crew flew from Houston to the Kennedy Space Center on August 22 to visit a NASA contractor, Rockwell Aerospace, in

Cape Canaveral for that day as part of the NASA Manned Flight Awareness Program and spend time at KSC the next day to thank everyone there for their help getting us to orbit and back home safely.

Landing at the Shuttle Landing Facility in our T-38s was, for me, a poignant moment as I recalled that just one month earlier we touched down on the same runway in space shuttle *Discovery*. Only this time, unlike the last on the shuttle's windowless middeck, I could actually see the runway as we landed. I liked the view much better this time around.

NASA sent astronauts across the country to visit companies that manufactured space shuttle parts or provided services to the space agency that supported the shuttle program. These visits allowed us to commend the employees for their good work, thank them for watching out for our safety, shake their hands, sign a few autographs, and present a few Silver Snoopy Awards.

The Silver Snoopy is the astronauts' personal award presented to NASA employees and contractors for outstanding achievements related to manned spaceflight safety or mission success. The award, which dates back to the late 1960s, depicts Snoopy, a character from the *Peanuts* comic strip created by Charles Schulz, wearing a spacesuit and helmet.

Rockwell Aerospace, whose parent company, Rockwell International, built the space shuttle orbiters, was in charge of shuttle logistics, making sure we had all the spare parts needed to keep the shuttles flying. After we arrived in our T-38s, we put our bags in our rooms at KSC crew quarters, had a quick lunch there, and then headed to Rockwell's facility in Cape Canaveral for the visit.

It was a pretty festive atmosphere for our visit at the event billed as a Space Day Rally. We split up and spoke with various groups of employees, acknowledging their critical role as part of the space shuttle team. I always thoroughly enjoyed meeting as many people as possible who worked on the program. Every astronaut knew who the real heroes of the program were—those folks and the thousands like them, both civil servants and contractors.

When we went outside for the rally itself, we were greeted by a Woody Woodpecker character that had flown over from Universal Studios in Orlando. I immediately knew this was going to be a fun event. On the side of a palm tree decked out to look like an external tank was a sign reading *'Endangered Space Hardware'* and Woody Woodpecker pretended to hit his beak on the tree trunk as if to make more holes in one of the tanks. Nearby an employee, in apparent protest, held a sign that read, *'Save the External Tank.'* Everyone enjoyed the woodpecker humor. We were each presented with a stuffed animal of Woody wearing a special T-shirt that bore an image of Woody Woodpecker in a spacesuit floating in space outside the shuttle. Woody seemed to be almost everywhere we went after the flight. For the employees at the rally, there was a Woody Woodpecker sound-alike contest during which each contestant had to imitate the cackle laugh for which Woody was so famous. I was asked to represent the crew in this contest which I enthusiastically did. Unfortunately my cackle only warranted second place with one of the Rockwell employees winning the grand prize of two tickets to Universal Studios in Orlando.

We spent the night at crew quarters and were greeted at breakfast the next morning with a headline in *Florida Today*, the local newspaper that read, "Woody Woos Space Program," and coverage of the previous day's event.

After breakfast, we headed over to the KSC Administration Building for our official postflight presentation. The auditorium was filled with KSC employees and a few children on hand for the occasion. We were introduced by the KSC Director Jay Honeycutt and then Tom thanked the entire KSC team for doing such a great job getting *Discovery* ready to fly, making special mention to the team responsible for patching up the woodpecker holes in our external tank. Tom also acknowledged the teams that had prepared the TDRS satellite, the Inertial Upper Stage, and the rest of our middeck payloads. All of those had worked perfectly during the mission which was a tribute to the hard work and dedication of the thousands of KSC workers who had gotten all the hardware ready for us. It took every single worker focusing on every imaginable detail to make each mission successful, and the KSC team had done an outstanding job for STS-70.

We then showed our postflight highlights film, and got extra laughs when a clip of a Woody Woodpecker stuffed animal floating through the cabin of *Discovery* appeared. The KSC team had gotten the stuffed Woody from Universal Studios in Orlando and stowed it onboard as part of our Special Flight Data File. After signing autographs for the workers and the children, we split up to present a few more Silver Snoopy Awards.

Then we headed back to Houston. After takeoff from KSC, we were treated to a great view of the various launch pads along the Space Coast. What a great place to visit. There was so much history that took place on that few-miles-long stretch of land. I thought back to the views of KSC we had aboard *Discovery* as we passed overhead on many of our orbits. As we climbed to altitude on our departure, I wondered when I would be coming back to KSC. With two flights under my belt, I already was looking forward to the possibility of flight number three. But there were no promises in the spaceflight business. It was always wait and see. Two hours later, we landed at Ellington Field and everyone drove home to spend some much needed time with our families.

At that point, most of the official postflight activities were done and the STS-70 crew started to head in their own directions. We each had some postflight appearances to do on our own, such as returning items that we flew for various organizations and doing hometown visits.

Back to Ohio

Tom, Nancy, and I would spend the 1995 Labor Day Weekend at various events in northern Ohio, starting off with a visit to Put-in-Bay on South Bass Island. There are a handful of wonderful Lake Erie islands, most of them just a few miles off the coast. Put-in-Bay features a beautiful Victorian-era village and we spent that night at a cozy bed and breakfast. The next day we went to Perry's Victory and International Peace Memorial that honored the Battle of Lake Erie during the War of 1812. There we were honored with a parade, each of us riding in the back of an open convertible.

From there, we went to the grand opening of the Rock and Roll Hall of Fame and Museum in Cleveland. There were some events to which we were invited and—while we had no idea of why or how we got the invite—we were thankful for the opportunity. This was one of those events. As it turned out, Tom had arranged to fly a Rock and Roll Hall of Fame baseball cap on our mission, which probably prompted the invitation to the September 1 event.

After years of debate and competition between cities—mainly New York, Philadelphia, and Cleveland—the decision was made to build the Hall of Fame downtown on the Lake Erie shoreline. The glass pyramids designed by famed architect I. M. Pei are a beautiful sight.

The official ribbon-cutting ceremony in front of the Hall of Fame drew countless rock 'n' rollers, politicians, rock fans, and media from around the world. Also in attendance were Tom, Nancy, and I representing the STS-70 crew. There was a big stage in front of the Hall of Fame and forty to fifty dignitaries sat there for the speeches that preceded the ribbon cutting. Among them were Little Richard, Yoko Ono, Paul Shaffer, I. M. Pei, Governor Voinovich, Cleveland Mayor Michael White, and Tom, Nancy, and me. Again, why and how we were included up on that stage with those rock legends was beyond me. I credit Tom for the incredible opportunity.

Most of those on the stage representing the rock and roll industry were dressed in colorful, somewhat flashy attire, typical in the entertainment industry. There were lots of sunglasses. The many politicians were clad in their trade clothing—suits and ties. And then there were us astronauts wearing our matching crew polo shirts with the STS-70 mission logo embroidered on them, khaki pants, and loafers. Maybe people thought that because we were sitting together wearing our matching outfits that we were members of a band called 'STS-70.' I felt completely out of place, but didn't much care because it was such an amazing opportunity to be there.

After the opening, it was fun to visit with Governor Voinovich and Janet once again where I was able to introduce them to Simone and Kai.

Tom also arranged for us to spend the next day at the Cleveland Air Show at Burke Lakefront Airport, just a few hundred yards from the Rock and Roll Hall of Fame. There Tom, Nancy, and I were given the honor of serving as the show's grand marshals.

While in Cleveland, I took the opportunity to present the single and only surviving pack of Stadium Mustard flown on STS-70 back to David Dwoskin, president and owner of the Davis Food Co. that sells and distributes the mustard. I flew ten packs on the flight, but had set one aside (at great personal sacrifice and under heavy protest by Kevin) to present back to David after the flight. Having traveled 3.7 million miles around the Earth during STS-70 and a few thousand extra miles by aircraft from the Kennedy Space Center to Houston and then to Cleveland, that pack held the world record for the most miles traveled by Stadium Mustard until years later when some of the mustard was flown aboard the International Space Station. Along with the pack of mustard, I gave David a few pictures I took of the mustard pack floating aboard *Discovery*.

The highlight of the press conference was when David presented me with one of the most prestigious, if not most cherished, awards I have ever received. It is the little-known 'Certificate of Mustard,' which authorized me to receive a lifetime supply of Stadium Mustard. The certificate is one of my favorites and still hangs prominently above my desk at home. True to his word, David sends me a dozen bottles of Stadium Mustard each year, one for each month.

On To Amityville

In order to get the crew to New York to do some events in Kevin's part of the country, Tom arranged a visit through the Manned Flight

Awareness Program to Monitor Aerospace, a manufacturing plant on Long Island which produced parts for the space shuttle and, conveniently enough, was located near his hometown of Amityville.

For that first evening, Kevin thoughtfully arranged for us to attend a party in our honor not at some posh county club, fancy restaurant, or Knights of Columbus hall. Instead it was at 112 Ocean Dr., a large Dutch colonial house made famous by a book, *The Amityville Horror*, and used in the 1979 movie by the same name, both based on a supposedly true story about that house being haunted after six people were murdered there. What a great place for a party. I had to wonder whether any of the ghosts would be attending.

Kevin knew the current owners of the house (Kevin knew everybody in Amityville) and thought it would be a great place to have the event. Just as Kevin endured listening to endless stories and legends about Ohio for the past year, Tom, Nancy, Mary Ellen, and I endured Kevin's stories about Amityville as well. It was a pleasure to finally experience Amityville, including—but hopefully not too much—the horror house itself.

It was a nice party in the backyard with plenty of good food and cold beverages, and it culminated with a tour of the house and the upstairs bedrooms where the murders occurred and the paranormal activity was said to have taken place. It was a fascinating tour, but no ghosts were seen and there were no strange occurrences noted.

The highlight of the evening was meeting the mayor of Amityville, the Honorable Emil G. Pavlik Jr. During an informal presentation, Kevin thanked everyone for hosting us and presented the mayor with the Amityville flag that flew with us on *Discovery*. And then to our great surprise the mayor presented Tom, Nancy, Mary Ellen, and I each with an official 'Certificate of Declaration,' which made each of us Honorary Amityvillians prompting the mayor to proclaim STS-70 to be 'The All-Amityville Space Mission.'

So there you have it, STS-70 was in reality a two-fer—All-Ohio and All-Amityville.

Go Browns!

There's nothing quite like a football home opener. All the better when it's the Cleveland Browns and we, the STS-70 crew, were invited to attend. It was a beautiful sunny day and the Goodyear blimp floated overhead. Tom, Nancy, and I were given tickets to sit in a special loge in Cleveland Municipal Stadium. At half-time, Tom and I went onto the field with a Browns jersey with the number 70 that I had flown aboard *Discovery* to present to the NFL franchise. The crowd of 61,083 roared their approval. We were on the field for only a brief minute with our recognition lasting less than thirty seconds. But for a life-long die-hard Browns fan, it was a memorable and exciting moment. The day became even more special when the Browns ended up beating the Tampa Bay Buccaneers, 22–6.

STS-70 Winds Down

There was never any definitive end of a shuttle mission for the astronauts, quite unlike how the mission started the moment we were told we'd been assigned to a flight. It just sort of faded away. There were, from time to time, a few postflight appearances that popped up after all the major ones were finished. But as time moved on, there were less and less of those. STS-70 references became few and far between.

I did a few postflight appearances by myself in the months that followed. One of them involved a trip to New Ulm, Minnesota to be the grand marshal for their annual polka festival parade. Since the 1940s, New Ulm has proclaimed itself as the 'Polka Capital of the Nation.' That must make Cleveland the 'Polka Capital of the World.' The organizing committee heard about my request for polka wake-up music and, seizing the moment, invited me to their festival. Simone and I rode in the back of an aqua-blue Thunderbird convertible along the parade route through town, while my mom watched the parade along the street with Kai. After the parade, we ate some great bratwurst and learned some local Minnesota phrases like, "Yah sure you betcha!"

Tom was nothing short of a hero in Toledo. Astronauts from Cleveland were a dime a dozen, but Tom was the Toledo area's first and only space traveler. He was genuinely famous in northwest Ohio and everyone in the area knew his name. As a result, many requests came in to the astronaut office from schools and community groups for Tom to make appearances, most of which he worked hard to accommodate. There was such a flood following our STS-70 mission that Tom couldn't honor all of them, so Nancy, Mary Ellen, or I would pitch in and do some of those appearances.

I volunteered to do one at Glenwood Elementary School in Rossford, a Toledo suburb. After a full day of presentations at the school, my host, Linda Cutler, a dedicated third grade teacher at Glenwood, took me to a place called Tony Packo's for dinner. Packo's is famous for their Hungarian hot dogs and a spicy chili-like hot dog sauce among other things. The restaurant became famous nationally in the 1970s when Toledo native Jamie Farr's character of Maxwell Q. Klinger on the TV series *M*A*S*H* talked about Tony Packo's in some of the episodes. Along with his classic Toledo Mud Hens baseball cap, he single-handedly assisted in bringing Toledo some national notoriety. For years and years, celebrities and famous politicians have visited Packo's to sample its cuisine. And beginning with actor Burt Reynolds in 1972, they would be asked to sign a plastic replica of a hot dog bun that the restaurant displayed on its walls. I signed one for them that evening, and it's still on display in a prime location—right by the front door.

I fell in love with Packo's that night and have enjoyed their hot dogs and other delicacies ever since. On my next two shuttle flights, I flew a few cans of the Packo's hot dog sauce which I enjoyed tremendously. As famous as Astronaut Tom Henricks is in and around the Toledo area, I don't believe he has a signed hot dog bun on the walls of Tony Packo's yet. Sorry, Tom! Tony Packo's and the City of Toledo have adopted me and that great city has become a second hometown for me.

Discovery's Postlanding Inspections

Within a few hours of landing, ground personnel made their initial assessments of *Discovery*, powered it down, and readied it to be towed back to one of the Shuttle Orbiter Processing Facility (OPF) hangars located next to the Vehicle Assembly Building (VAB). The trip back was a little quicker than the one mile per hour rollout to the launch pad. They started towing *Discovery* from Runway 33 at 11:45 a.m. and had it safely inside OPF bay No. 1 a little more than an hour later. Technicians spent that weekend off-loading the cryogenic reactants, such as liquid oxygen and liquid hydrogen as well as the hypergolic propellants used to fuel our Orbital Maneuvering System engines and the Reaction Control System thrusters. A few days later, the payload bay doors were opened and technicians began removing the IUS support equipment used to launch TDRS into space. Over the next few weeks, technicians continued to unload hardware, removed the engines, and prepared *Discovery* for its major modification trip to a Rockwell facility in Palmdale, California. After every couple of flights, the shuttles were routinely sent to Palmdale where they were thoroughly inspected and any upgrades were performed. Minor modifications could be done at KSC, but NASA preferred to do the major ones in California which helped keep many shuttle technicians there employed for many years after the orbiters were built.

Careful inspection of the shuttle's protective heat-resistant tile system showed that *Discovery* sustained a total of 127 hits, nine of which were larger than one inch across. The more sensitive bottom side covered with the black tiles sustained 81 hits, five of which were greater than one inch across. The damage was caused by debris coming off during ascent (ice from the external tank or some of the ET's foam being shed and impacting the tiles), small debris on the runway getting kicked up by the wheels during landing and rollout, and other miscellaneous reasons.

As alarming as those numbers sounded, they were well within the normal limits of what was typically experienced in the shuttle program. Overall *Discovery* had slightly less than the average number of hits (131) and significantly fewer hits larger than one inch than normally experienced (21). Some previous shuttle missions experienced up to 290 total hits with one flight having fifty-five impacts greater than one inch in diameter.

It was also discovered that the nose landing gear tire had been damaged during landing. There was a small slit or tear visible in it and four pieces of the tire were found on Runway 33 after we landed, nearly a mile from where we touched down. The damage was traced to one of the runway's centerline light covers which was protruding a quarter inch above the surface. There were 298 of the twelve-inch wide centerline lights which normally were flush with the concrete runway. But that one light either was installed improperly or somehow had lifted up a fraction of an inch. That small protuberance caused the tire damage as the 195,800-pound *Discovery* slowing down from its initial landing speed of about 220 miles per hour made its way down the runway. The light problem was corrected before the next shuttle flight, STS-69.

All in all, *Discovery*'s damage was relatively minor and much less than experienced on previous shuttle missions. NASA officials proclaimed her a "relatively clean bird," meaning we had taken pretty good care of her.

SRB Problems

Just after we landed, reports started to surface regarding some problems with the solid rocket boosters during the STS-71 launch, the one just before ours. "Shuttle down; red flags up," a headline in the *Florida Today* newspaper stated the day after we landed. A small leak in one of STS-71's two SRBs was found during postflight inspections and analysis. That immediately raised concerns within NASA because it was a leak in one of the SRBs that led to the *Challenger* accident in 1986 that killed seven of our colleagues and destroyed the orbiter. Leaking SRBs got everybody's attention.

The STS-71 SRB problem had been discovered on the fourth day of our mission, but no one mentioned it to us because we already were safely in orbit at the time. First word of the problem was released to the news media just as *Discovery* was touching down and we weren't informed about the problem until a few hours after we had landed, right before our postlanding press conference. Questions were being raised once again about the safety of the shuttle program.

Three days after we landed, inspectors found evidence of similar leaks on one of our SRBs. Engineers reported that hot gas had snaked through an insulating rubber material that served as a barrier for the O-ring seals in a nozzle joint of the rocket. The leak caused four tiny dot-like marks on the O-ring itself, similar to, but less extensive than those noticed on the STS-71 booster.

With two consecutive missions experiencing that potentially serious problem, NASA knew it could not launch the next flight, space shuttle *Endeavour* on the STS-69 mission scheduled for August 5, until the problem was fully understood and the issue resolved. NASA proceeded cautiously, but the general mood was that it was not a major engineering problem. It was felt that a solution could be found without a major impact on the shuttle program schedule.

By mid-August, after thoroughly examining the STS-70 boosters and running numerous tests, NASA officials believed they understood the problem. *Endeavour* was repaired on the launch pad, and STS-69's SRBs were cleared for flight. That mission successfully launched on September 7, 1995, without incident.

Discovery Moves On

While we each finished up our postflight appearances and slowly got our lives back to normal, *Discovery*'s inspections and upgrades in Palmdale would take nearly nine months. Besides routine maintenance, *Discovery*'s airlock was being moved from inside the middeck to the outside where a new docking adaptor would be added in preparation for future flights to the International Space Station.

That would almost double the size of the middeck allowing NASA to carry more equipment and supplies to the orbital space outpost.

They would also be adding a fifth set of liquid oxygen and liquid hydrogen tanks that would allow *Discovery* to stay in space several more days beyond the ten or so day capability we had on STS-70.

On September 27, *Discovery* was flown on the back of a 747 to Palmdale to begin its major modification period.

Chapter Eleven
Onward and Upward

Astronauts assigned to shuttle missions were molded into close-knit teams, performed the missions for which they were trained, and once the flights were over, returned to the pool of their colleagues to wait their turn for assignment to another mission. New crews were formed, new close-knit teams were developed, new missions were flown, and what seemed like an endless cycle continued. That was the normal progression until the end of the shuttle program in 2011.

When a mission was over and the last postflight appearance was completed, it was a little sad seeing the crew disperse. But it also was an exciting time. While knowing that our names were probably at the very end of the very long astronaut mission assignment list, there was a bit of anticipation about when our next spaceflight assignment would come and what that mission would entail. Plus, there was the prospect of the new job we'd be doing within the Astronaut Office.

But that didn't necessarily mean that the personal relationships and friendships developed during our previous assignments came to an end. After training so hard and spending more than a year together preparing for a flight, our friendships were quite unique. We experienced something very few human beings had or ever would. In some cases, we very much became an extended family. Some of the friendships I formed with my crew mates were lifelong, others less so. That's just human nature.

Where Are They Now?

So what happened to everybody after the All-Ohio Space Shuttle Mission?

Well, as our postflight appearances started to wind down, I was assigned to the shuttle program's Payloads Branch where I supervised the development and testing of various experiments that would fly on future shuttle missions. With my extensive payload experience from STS-65 and STS-70, that assignment was a good match for me and one that I thoroughly enjoyed.

My home life returned to normal as best it could. Kai, at the time about six months old, kept Simone and me quite busy. We pretty much fell into the normal routine of parenthood with a newborn baby. Except for the fact that somehow, I seemed to be the only one who ever heard him crying in the middle of the night, necessitating frequent trips downstairs to get him a bottle to help him fall back to sleep.

After my rapid reassignment to STS-70 only a month after STS-65 landed, I was looking forward to a more normal lifestyle, without the tension and frenetic pace associated with being assigned to a flight. Don't get me wrong. I definitely wanted to fly on the shuttle again, but I also welcomed the break between flights. It was a time to catch my breath. But the sabbatical didn't last very long.

On January 16, 1996, less than six months (178 days to be exact) after landing aboard *Discovery* on STS-70, I was told that Ken Cockrell, the chief of the Astronaut Office, wanted to speak with me. Once again my initial reaction was, "What have I done wrong now?" I walked down the hall to find out. I was soon informed that I was being assigned to the crew of STS-83. Like my first flight, it would be another Spacelab science mission aboard space shuttle *Columbia*. The launch was tentatively scheduled for the spring of 1997 with no specific date yet announced.

I was elated by the news (Who doesn't like getting assigned to a space shuttle mission?), but at the same time I thought, *Here we go again*. Training would begin in a month or two and would involve quite a bit of time at NASA's Marshall Space Flight Center in Huntsville, Alabama. I waited until I got home that evening to tell Simone. She was happy for me, but couldn't believe it was happening again so soon. She had some of the same thoughts as when I was assigned to the STS-70 mission only one month after STS-65. *Couldn't NASA give us a little break?*

Commanding STS-83 would be Jim Halsell, with whom I had flown on STS-65. Our pilot would be Susan Still, a Navy aviator making her first spaceflight. Susan was the second woman assigned to pilot the shuttle. Eileen Collins, one of my astronaut classmates, was the first in 1995. Our payload commander was Janice Voss and the other Mission Specialist was Mike Gernhardt, who was born and raised in Mansfield, Ohio. Rounding out the crew were two payload specialists, Roger Crouch and Greg Linteris. I soon found myself with a busy training schedule that sent me to Huntsville just about every other week. So much for trying to lead a normal family life. I guessed that would have to wait until after STS-83. Soon we were assigned a target launch date—March 27, 1997—a little over one year away. Life became hectic once again for the Thomas family.

While always busy, the training flow was quite familiar and very similar to what I experienced for STS-65. The only real difference was that STS-83 involved far fewer international experiments which meant no overseas training trips. Many of the experiments were developed and managed by the then-Lewis Research Center in Cleveland, so we had a number of trips scheduled to visit there for training which in no way, shape, or form disappointed me.

Flying back and forth between Houston and Huntsville with the occasional trip to Cleveland and over to KSC, training went pretty much as planned. Everything was going well until January 29, 1997, a little more than two months before our scheduled launch. During some routine shuttle evacuation training, I twisted my ankle and heard an ominous crack as I was walking down some stairs. I immediately knew I broke my ankle and some extremely off-color language floated through my mind. I was taken to the JSC medical clinic for an x-ray. The doctors weren't sure if my ankle was broken, so I was driven to an orthopedic specialist in downtown Houston for a look. He, unfortunately, was quite certain about my situation. My ankle was indeed broken. I was devastated.

The good news was that it was a clean break. My bones were still perfectly aligned and I wouldn't require surgery—no pins, no plates, no screws. Even so, I wasn't too elated. My overriding concern was whether NASA would replace me on the flight. Fortunately, my orthopedic surgeon, Dr. Jay Oates, was sympathetic to my cause, and gave NASA a very positive prognosis for my recovery. He saw no reason why the bone wouldn't be healed in time for the flight. He fully supported my continued training. NASA, on the other hand, was a bit more cautious. While they didn't pull me from the flight, they decided to assign a backup in case my recovery didn't go as expected. Cady Coleman was assigned to begin training in case they needed to replace me.

I continued with my normal training schedule as best I could, and was excused from anything involving heavy activity or being in the water. No T-38 flying with a broken ankle, so I traveled to and from Huntsville via commercial airliners. It was a bit humiliating being pushed in a wheelchair while transferring from one gate to another at the Nashville airport so close to launch.

I wore a boot cast on my foot and got around on a pair of crutches for the first five weeks. At my next appointment with Dr. Oates, he told me the bone was nearly healed and said I should start walking on it. I began one of the most intense physical training programs of my life, doing just about everything imaginable to expedite the healing and recovery. My physical trainer was Beth Shepherd, wife of astronaut Bill 'Shep' Shepherd, and she was one of the best. She worked with me six to seven hours a day, seven days a week. I started walking in a swimming pool and then graduated to walking down long halls. Back and forth. Back and forth. Beth had me jumping over little plastic cups she would line up on the floor. Her number one motto was "Good pain is good." I received heat treatments, ice treatments, ultrasound treatments, electric stimulation treatments, and ankle massages. We tried them all. I needed to get healed and we didn't have much time. If good pain was good, things couldn't have been going any better for me.

Our practice countdown test was scheduled at KSC one month before launch. I joined the rest of my crew at KSC, but Cady Coleman would be filling in for me during the exercise. Even in Florida, Beth and I continued the relentless physical therapy sessions. By the time we arrived back in Houston a few days later, my ankle was doing much better. I was no longer wearing the boot and could walk on my own, albeit with a slight limp. With an extremely positive report from Dr. Oates, NASA made the decision less than one month before STS-83's scheduled launch to keep me on the mission. The only people disappointed by the decision were Cady Coleman and Cady Coleman's mom.

As we were about to launch April 4, 1997, a woman sitting next to my mom in the family viewing area at KSC said to her, "My Cady is so disappointed that she isn't flying on this mission." It was Cady Coleman's mom. My mom replied with just the smallest hint of sympathy, "My Don is so glad that your Cady is so disappointed."

One of my STS-83 highlights was a meeting with Neil Armstrong the day before launch. I had invited Armstrong to attend the launch and was thrilled when I received word from him that he would be there. The day before launch, as Simone and I were relaxing at the beach house, I received a call from NASA management that Armstrong and his wife, Carol, wanted to meet me. Simone and I immediately drove back to crew quarters where we spent nearly an hour with them. Ron Woods, who helped suit up Armstrong before his

Apollo 11 launch, joined us as we showed them around our suit-up room and the crew quarters area. The rooms had been upgraded over the years, but much of crew quarters was pretty much the same as it had been on July 16, 1969. Neil and Carol couldn't have been nicer. When it was time for them to leave, Neil wished me a great flight. I asked him how long he and Carol were planning to be in Florida. He looked me straight in the eye and, with a slight grin, replied, "How long are you going to be in Florida?" I was so flattered that he was planning to stay as long as it took to get us off the ground. Fortunately, we launched the next day right on schedule.

Launches never got boring and that one—from where I sat on the middeck—was no different. After a quick ride to orbit, we soon found ourselves settling into normal operations for our planned sixteen-day mission and I was able to see one of the greatest sights I ever saw while in space. Comet Hale-Bopp was easily visible with the naked eye reaching its peak brightness the day we launched. I couldn't wait to get a look at it from space. During one of our first orbits as the sun set and my eyes became dark-adapted, I saw it. It was an impressive sight. It was brighter than it was when I saw it back on Earth. Slowly, the comet sank lower and lower in the sky. After watching it for only fifteen seconds or so, it suddenly disappeared as it passed behind the Earth. I continued staring at the blackness of space for a few more seconds until it dawned on me. I had just watched a comet-set. During my life, I had seen countless sunsets and moonsets, but now I saw my first comet-set. It was one of those *Wow!* moments. And one of the coolest things was that I would be able to see it all happen again in another ninety minutes—once every orbit. I tried to catch as many of those comet-sets as I could during the flight.

On our second day in space, we received word from Mission Control that one of *Columbia*'s three fuel cells used to generate our electricity was acting up. I thought nothing of it at the time and was certain our Mission Control team would figure out a way to keep us safe and in space. I was more than a little surprised to learn the next morning that NASA decided to bring us home early. They explained that the fuel cell might catch fire or explode, so they instructed us to shut it down and begin preparing to return to Earth. It wouldn't be an emergency landing, but rather an expedited return with the plan calling for us to land on day four of the mission. That was a huge disappointment for the entire crew, especially the three rookies. We had trained so hard and long for the flight and would now have to return home without accomplishing the science we set out to do.

By the time landing day arrived, we were given some hints that NASA was going to refly the mission in a few months and that we would all be given the opportunity to try it again. I thought it was great news. The entire crew was excited about the possibility, but I knew Simone wouldn't be too thrilled. Launches were tough on the families; to put them through two launches within a few months was something none of them looked forward to.

After an uneventful landing, I found myself in a NASA bus being driven home from Ellington Field with Simone and Kai once again. As we drove past JSC, I felt like I had been in the simulator for a few days practicing for the mission. Actually, I had spent the past four days in space, but it just didn't seem real.

Within a few days, NASA announced that our new launch date would be July 1, 1997. But we had a new flight number—STS-94— which had been a place holder on the shuttle manifest with no mission assigned to it yet. After a week off for some vacation, we resumed our training and launched on time only eighty-eight days after the STS-83 landing. It is one of the only records I hold in space and one that I share with the rest of the crew: the shortest turnaround between flights.

After conducting numerous microgravity experiments looking into the effects of zero-gravity on how crystals grow and how fires burn, we landed safely at KSC on July 17 after successfully completing our mission. It was great to be back home and once again I found myself looking forward to the more familiar routine life on Earth.

For the next two years, I worked as chief of the Payloads Branch within the Astronaut Office and had astronauts K. C. Chawla, Dave

Brown, Laurel Clark, and Israeli payload specialist Illan Ramon working with me in my branch. A year later, they were assigned to fly on the ill-fated STS-107 mission aboard *Columbia*. After STS-94 I had been asked if I was interested in flying on STS-107. Many of the experiments scheduled for that flight had flown on my previous missions—STS-65, STS-83, and STS-94—and I had worked on many of them. But I said no and let NASA officials know that I was interested in flying on one of the upcoming International Space Station missions. After flying four times on the space shuttle, I was looking for a new experience and knew an ISS flight definitely would provide that.

NASA asked me in 1999 if I would consider being the director of operations in Russia and heading up the NASA office at the Gagarin Cosmonaut Training Center in Star City. That was the Russian equivalent of JSC. After talking it over with Simone, I accepted what was to be a six-month assignment after which I was to return to JSC. Simone and Kai joined me and, while life was a bit tougher than in Houston, we embraced the incredible Russian culture and traveled as much as time permitted. As my six-month assignment was coming to an end, I received no response from JSC when I asked about who my replacement would be. After a series of emails and unreturned phone calls, I finally got word that they had no replacement lined up. I was asked to stay another six months.

NASA had been sending astronauts to manage things in Star City for a number of years, but nobody had been asked to stay more than six months. I was the first civilian that NASA assigned to the position, with all previous managers coming from the military. NASA apparently thought only pilots could fly the desks in Star City. But when they ran out of interested candidates, they settled for a civilian. It turned out to be one of the most interesting years of my life and luckily Simone and Kai got to share it with me. Besides my management duties, I took Russian language classes and signed up for many other training classes involving the Russian portions of the ISS.

In the fall of 2000, I was assigned to be a member of the Expedition 6 prime crew aboard the space station. For the next two years, the training and travel were intense and nothing like I experienced before. We spent one month training in Houston, then one month training in Star City. Then back to Houston for a month followed by another month in Star City. Simone and Kai, who were living back in Houston, would just get used to me being gone and I then would return. After a few weeks of getting used to me being home, I would suddenly depart again. Life was full of adjustments all the time. I had one life in Houston with my family, and quite a different one nine time zones away in Star City. Life couldn't have been crazier as I missed countless birthdays, anniversaries, and holidays back home.

As our scheduled STS-113 launch aboard space shuttle *Atlantis* approached, the training became even more intense. Ever so slowly in the months before the mission, I gradually separated from my family. Physically I was gone from home half the time. Soon the emotional bonds with home were being broken as we all prepared for me to be absent for the next five months.

That all came to a screeching halt one day while I was in Star City with only two months to go before our scheduled launch. One of the doctors at NASA Headquarters in Washington, D.C. was having issues with me flying to the ISS. Because of my thyroid cancer twelve years earlier and the radiation treatments I received following the surgery, the doctor decided to pull me from the flight because of his concerns that the doses of radiation I would experience aboard the ISS could lead to further cancers in the future. It was an outlandish concern, one not shared by the rest of the NASA flight doctors and medical community who had been taking care of me at JSC for the previous twelve years. I was unceremoniously pulled from the mission. I felt like I had been traveling one hundred miles per hour and suddenly hit a brick wall. I found myself with absolutely nothing to do the next day. Nor the day after that. Nor the day after that.

I pretty much took a year off from NASA and spent as much time as I could with Kai. I took him to school in the morning and picked him up in the afternoon. Once or twice a week, I would go to his school and eat lunch with him. After all the traveling I had done for

NASA the previous seven years of his life, I took advantage of the windfall of free time and spent as much of it as I could with him. We went to afternoon businessman's baseball games during the summer to see the Houston Astros play. We played miniature golf. We went jet-skiing. We hung out on the beaches of Galveston. We had a great time together.

I was offered the position to be the program scientist for the International Space Station in 2003, a position I held for nearly three years. I never made it to the ISS myself, but I was able to contribute to the program and help plan and coordinate its science activities.

I left NASA in 2007 and accepted a position at Towson University outside Baltimore as the director of the Hackerman Academy of Mathematics and Science. I am actively involved in outreach programs to get more young students interested in careers in math, science, and engineering. Most days you will find me in an elementary, middle, or high school somewhere in Maryland working to inspire the students to follow their dreams just as I had done in my life. From time to time, I also do programs at the Kennedy Space Center Visitor Complex and at the US Space and Rocket Center in Huntsville, again hoping to inspire the next generation of scientists, engineers, and explorers.

My passion for space exploration has never waned and is every bit as strong today as it was on May 5, 1961 as I watched Alan Shepard launch into space. The passion has never changed, only the focus. Mars looms ahead for us in the future. And while I will not be on the first mission to land astronauts there decades in the future, maybe one of the students I reach and inspire today will be making that journey.

My total time in space during my four flights was 43 days, 8 hours, 14 minutes, and 12 seconds and I enjoyed every day, hour, minute, and second of it.

Tom Henricks

Tom went on to serve as commander of STS-78, the Spacelab Life Sciences mission aboard space shuttle *Columbia* which launched

June 20, 1996, less than a year after STS-70. Kevin was his pilot once again on that flight. It was a bit unusual for NASA to assign the same commander and pilot on their next missions, but the science crew members of that flight had been assigned a year or more earlier and NASA needed to complete the crew by assigning a commander and pilot. Tom and Kevin, fresh from STS-70 with a proven record of working well together, were assigned to STS-78 on October 6, 1995. That mission turned out to be the second longest of the space shuttle program lasting 16 days, 21 hours, 48 minutes, 30 seconds. During the flight, Tom became the first person to log one thousand hours on the space shuttle as a pilot and commander.

An October 17, 1997 news release announced Tom's decision to retire from the Air Force after a stellar twenty-three year career, twelve years of which he spent at NASA. Tom had been asked by the Astronaut Office to be commander of another space shuttle mission, but delays in the program and a reduction in the number of shuttles that would be flying each year helped him to decide to leave the program. In November 1997, he joined the Timken Company in Canton, Ohio, where he worked on government programs. As a manufacturer of ball bearings, some of which were used in the wheels of the space shuttle, Tom still had his hands in the space business after leaving NASA.

Tom then was named vice president of Government Business Development at Textron, Inc., and went on to become the president of Aviation Week, where he worked until 2010. Today, Tom runs his own company, Henricks Enterprises, Inc., where he consults on various aerospace and media projects. He also is a partner in the Newport Board Group which aids small and medium-sized companies to improve their business strategies, operations, and capital markets.

He and Becky live on a small ranch midway between San Antonio and Austin and he still enjoys recreational flying.

Tom's total time in space during his four flights was 42 days, 18 hours, and 39 minutes.

Kevin Kregel

Kevin went on to fly as the pilot of STS-78, the Spacelab Life Sciences mission aboard space shuttle *Columbia* which launched June 20, 1996, less than a year after STS-70. During that mission the crew performed twenty-two life and microgravity sciences experiments which helped prepare the way for future ISS crews. Landing occurred on July 7, 1996.

Almost immediately following that flight, Kevin was assigned as commander of STS-87, the fourth United States Microgravity Payload (USMP) flight aboard *Columbia* which launched November 19, 1997. Kevin and his crew conducted many materials processing experiments and deployed and recaptured a Spartan satellite that studied the sun's outer atmospheric layers. Two spacewalks also were conducted during the flight which helped develop procedures and techniques that would be used during future spacewalks aboard the ISS.

Kevin went on to fly as commander of STS-99 aboard space shuttle *Endeavour* from February 11 to February 22, 2000. That eleven-day mission featured the Shuttle Radar Topography experiment which used a radar system to map more than four million miles of the Earth's surface. During that mission, Kevin finally got the opportunity to fly over his two home states—New York and Ohio—many, many times.

Kevin retired from NASA in December 2003, after thirteen years with the space agency and joined Southwest Airlines as one of their top pilots. Kevin absolutely loved to fly and it was no big surprise to me that he ended up flying for Southwest after he left NASA. So the next time you are on Southwest Airlines and have an extra-smooth flight, it may very well be famed 'Spaceman' Kevin Kregel at the controls.

Today he and Jeanne live in Seabrook, Texas, not far from the Johnson Space Center.

Kevin's total time in space during his four flights was 52 days, 18 hours, and 23 minutes.

Nancy Currie

Nancy went on to complete a life-long ambition and received her doctorate in industrial engineering from the University of Houston in 1997. Shortly afterward, she was assigned to the prestigious STS-88 flight aboard space shuttle *Endeavour*. That was the first construction mission of the International Space Station and Nancy and her crew successfully delivered and installed the United States docking node called *Unity* that connected the Russian-built portion of the ISS to the American-built segment. Launching on December 4, 1998, Nancy primarily was responsible for robotic arm operations and successfully used the arm to attach the new node to the Russian-built FGB module that had been launched by the Russians a month earlier. With that flight, assembly of the ISS began in earnest.

Nancy then was assigned to STS-109, the fourth Hubble Space Telescope servicing mission. Again Nancy's role was chief operator of the robotic arm which aided the spacewalking astronauts in replacing new solar arrays and the installation of a new power control unit and a new camera called the Advanced Camera for Surveys which was responsible for many of Hubble's most impressive deep space images.

After that flight, Nancy retired from the Astronaut Office to work in the Engineering Directorate at the Johnson Space Center as the deputy director of engineering until the space shuttle *Columbia* accident in 2003 when she was asked to lead the space shuttle program's Office of Safety and Mission Assurance. Nancy went on to serve as the chief engineer for the NASA Engineering and Safety Center (NESC) at the Johnson Space Center. In 2012, she accepted a temporary position at North Carolina State University in the Fitts Department of Industrial and Systems Engineering where she taught human factors, ergonomics, and safety engineering.

Today Nancy works as the Principal Engineer at the NESC at JSC and helps teach an aerospace systems engineering course at Rice University. In her spare time, she enjoys riding her horse, Harrison.

Nancy's total time in space during her four flights was 41 days, 15 hours, and 35 minutes.

Mary Ellen Weber

Mary Ellen went on to work extensively in technology commercialization and on the commercial development of bioreactors similar to the experiment we flew on STS-70. She was assigned to NASA Headquarters in Washington, D.C., where she worked closely with NASA Administrator Dan Goldin and was assigned to be NASA's legislative affairs liaison.

She flew her second and final space shuttle mission—STS-101—in 2000 as a mission specialist aboard space shuttle *Atlantis* which launched May 19, 2000. That mission took supplies and spare parts to the ISS. Mary Ellen's main responsibility was operation of the robotic arm that was used during two ISS construction and assembly spacewalks. She also was in charge of the coordination and transfer of three thousand pounds of equipment and supplies from the space shuttle to the ISS while the two were docked.

Mary Ellen received a master's in business administration degree from Southern Methodist University in 2002 and shortly thereafter resigned from NASA after a ten year career. She became vice president for Government Affairs and Policy at the University of Texas Southern Medical Center in Dallas.

In 2012, Mary Ellen founded her own consulting company, Stellar Strategies, LLC, where she works today. She continues to be an active skydiver and has logged nearly five thousand jumps. She has won thirteen silver and bronze medals at the US National Skydiving Championships over the years, and in 2002 was part of a world record freefall formation involving three hundred skydivers.

Mary Ellen and Jerry live in the Dallas area.

Mary Ellen's total time in space during her two flights was 18 days, 19 hours, and 30 minutes.

Woody Woodpecker

The plush, stuffed Woody Woodpecker that spent the time during our flight travelling from console to console in the new Mission Control room at the Johnson Space Center was presented to Tom by

Rob Kelso, the STS-70 lead flight director. Woody was considered the mission's unofficial mascot.

Tom took Woody with us during our postflight visit to the Lockheed Martin Assembly Plant in Michoud, Louisiana, where the space shuttle external tanks were all manufactured. During our visit, Tom presented Woody to the workers there as a reminder of the time when a woodpecker attacked one of their tanks.

A decade later during a tour of the facility with my family, I noticed Woody in a display case near the main offices for the engineers working at the plant. They told me it was the same one Tom had presented to them. They were still proud of it and kept it on display so it could keep an eye on the manufacture, assembly, and checkout of every external tank manufactured for the space shuttle program. With the end of the shuttle program, I assume Woody is still watching over the plant in hopes that it will be involved in the manufacture of the next generation of launch vehicles or spacecraft. Maybe one day he will make his way over to the National Air and Space Museum in Washington, D.C. to watch over space shuttle *Discovery* on permanent display there to help tell the story of the All-Ohio Space Shuttle Mission and the day a woodpecker attacked *Discovery*'s external tank.

The smaller Woody Woodpecker stuffed animal that flew with us aboard *Discovery* and helped us with our numerous experiments is still hanging out with our commander. Ever since we landed, Woody has been watching over Tom and helping him whenever needed.

The TDRS-G Satellite

After being deployed from *Discovery* on July 13, 1995 and after a series of burns using the two-stage Inertial Upper Stage and deployment of its communications antennas and large solar arrays, TRDS-G finally made its way to its proper geosynchronous orbit 22,300 miles above the Earth in what is known as the 'TDRS West' position at 171 degrees west longitude over the mid-Pacific Ocean. It was first powered up ten days later and after a few weeks to allow any residual gases to vaporize, technicians began the initial checkout and testing in early August 1995. After that was completed, it was powered down and put in storage for future use.

In early 1998, TDRS-G was powered up once again and put into service until April 2002, when the newer TDRS-8 satellite was launched aboard an expendable rocket and put into service. In August 2009, TDRS-G was moved from the 'TDRS West' position to its new location at 275 degrees west longitude, over the island of Guam. Its main function was to eliminate the old communications gap between TDRS-East and TDRS-West which was known as the Zone of Exclusion located over the middle of the Indian Ocean. Until that time, all astronauts aboard space shuttles or the International Space Station could count on anywhere from 5 to 11 minutes of total silence during each orbit while passing through the ZOE when communication was not possible. With TDRS-G positioned over Guam, NASA then enjoyed nearly 100 percent communications coverage for its orbiting spacecraft, including the ISS.

Though launched over eighteen years ago, TDRS-G continues to perform without any major problems. While most of the other 'first generation' TDRS satellites (TDRS A, C, D, E, and F) deployed from the space shuttles are all showing signs of age and slowly degrading (TDRS-D has totally failed and TDRS-C is on its last legs), TDRS-G remains unique among its group in its reliability and capability. Because it was the last one built in the first generation series, TDRS-G had a number of upgrades made to it during manufacture that made it one of the most reliable in the fleet. Those upgrades made TDRS-G a bit more expensive than the earlier versions which led some of the engineers to refer to it as 'the Cadillac of the First Generation' (TDRS-G is also referred to as 'TDRS-Goofy' or simply 'Goofy' because it is slightly different than the earlier TDRS A-F models).

The added upgrades and resulting expenses invested in TDRS-G clearly paid off for the program in its overall performance over the years. Nearly two decades after being launched, the engineers and technicians at the White Sands Space Complex characterize TDRS-G as "a pretty quiet bird" and as "their well-behaved child...not unruly like some of the others!"

TDRS-G will remain over the Guam Station for the foreseeable future supporting a wide range of missions beyond the International Space Station, the Hubble Space telescope, and other scientific satellites. On May 31, 2012, TDRS-G was instrumental in the successful return and splashdown of the new commercial Dragon capsule launched to the ISS by the Space Exploration Technologies Corp., better known as SpaceX. Engineers estimate the TDRS-G has sufficient maneuvering and pointing propellant to last an astonishing eighty years although other systems (such as its batteries, electronics, etc) are expected to fail long before that. Originally designed for a maximum ten-year lifetime, the TDRS satellites have experienced unmatched reliability and will continue to support human spaceflight and scientific missions for many more years to come.

Space Shuttle *Discovery*

STS-70 was the twenty-first flight of space shuttle *Discovery*. Immediately after the 'All-Ohio Space Shuttle Mission,' *Discovery* was flown to Palmdale, California, where it underwent a scheduled period of upgrades, inspections, and modifications that kept it grounded for nearly two years. Afterward, it went on to fly another eighteen missions before its retirement, including taking John Glenn back into space on the STS-95 mission in 1998. She flew two additional Hubble Space Telescope servicing missions—STS-82 in 1997 and STS-103 in 1999. She paved the way for International Space Station construction as the first orbiter to dock to the ISS during the STS-96 mission in 1999. In total, *Discovery* completed thirteen missions to the ISS, including the STS-114 mission commanded by Eileen Collins, the first female space shuttle commander.

Discovery's final flight was the STS-133 mission which launched February 24, 2011. NASA invited all former astronauts to attend one of the final three space shuttle launches and I chose to attend *Discovery*'s last launch. I planned to take Kai with me so we could finally see a launch together. I was really looking forward to sharing that experience with him. Unfortunately, flight delays and my work commitments conspired to keep me from attending. Simone took Kai to KSC for the February 24, 2011, event.

Delayed nearly four months due to hydrogen leaks and concerns about cracks in the external tank, *Discovery* was delayed one final time by a last-minute computer glitch. The problem was not with *Discovery*, but rather one of Air Force range safety office's computers. With only two seconds to spare, the Air Force was able to fix the problems and cleared *Discovery* for launch. "Get ready to witness the majesty and the power of *Discovery* as she lifts off one final time," Commander Steve Lindsey radioed.

And then at 4:53 p.m.—three minutes later than originally planned—the three main engines ignited followed by the twin solid rocket boosters. *Discovery* thundered into the clear blue sky, making, as George Diller, a KSC public affairs officer, put it, "one last reach for the stars."

Two days later, *Discovery* docked to the ISS for the last time and the crew began transferring thousands of pounds of equipment and supplies. Undocking for its return to Earth occurred March 7, 2011.

The next evening, with crystal clear skies in the Baltimore area and a thin crescent moon low in the west, Simone, Kai, and I walked to a big field down the street from our house and waited for a final look at *Discovery* doing what she did best. Simone saw her first, rising over the horizon. Dim at first, *Discovery* brightened quickly as it rose higher in the sky. About two minutes behind it was the ISS, also increasing in brightness as it rose. I glanced back at the ISS for a moment, but then firmly fixed my gaze on *Discovery*. As it passed nearly directly overhead, its speed seemed to increase as it glided smoothly and silently across the sky. It was just a mere point of light, but it represented so much more to me; I was seeing *Discovery* and its crew of six astronauts. But more than that, I saw the thirty-nine flights it had accomplished, including my own STS-70 mission.

As it moved toward the southeast, the light rapidly faded and then was gone—just like the soon-to-end space shuttle program. It was sad to realize I would never see *Discovery* in space again. It would be back on Earth in the next sixteen hours, destined for the Smithsonian's National Air and Space Museum's Steven F. Udvar-Hazy Center near Dulles International Airport, Chantilly, Virginia. It really was an incredible light that I watched move across the sky that night.

Discovery landed at KSC March 9, 2011. "For the final time, wheels stop," Commander Lindsey reported. Josh Byerly, the NASA Mission Control commentator, added, "To the ship that has led the way time and time again, we say 'Farewell *Discovery*.'" It brought a tear to my eye hearing that. *Discovery*'s flying days were over.

Later that afternoon, I along with the rest of my STS-70 crew mates received an incredibly touching email from Nancy Currie. She wrote that as she watched *Discovery* land for the final time, she reflected on *Discovery*'s superb performance during STS-70. And true to form, she noted, there were very few problems with *Discovery* on her final mission. Nancy thanked all of us for being such great crew mates on *Discovery*. "It was a wonderful mission, and I was fortunate to be part of the All-Ohio Crew," she wrote.

About eight months later, I was at the Kennedy Space Center and visited the Orbiter Processing Facility one last time to take a look at *Discovery* before she left Florida for her new home. It was great to see her again, but also a little sad considering her future.

I spent about forty-five minutes with *Discovery* that afternoon. I walked underneath, looking up at the black tiles that covered her belly. I kicked the tires. I climbed to the upper levels to see the flight deck. I saw the aft compartment from where they had removed all the plumbing lines for the main engines. I looked over the payload bay doors into the payload bay area and touched one of the doors for the last time. The next morning, those doors were scheduled to be closed for the final time. More than once, I sighed deeply and shook my head. It was difficult seeing *Discovery* being readied for retirement instead of for flight. *Discovery* was meant to soar majestically, not sit idle in a museum.

After a few more months, *Discovery*, riding atop a NASA 747, made her final flight on April 17, 2012, from the Kennedy Space Center to Dulles International Airport outside Washington. After takeoff she flew one final pass over KSC and along the Space Coast beaches before arriving later that morning in the Washington area. She flew low over such landmarks as the US Capitol Building, the White House, Washington Monument, and Lincoln Memorial. The entire city came to a halt as people looked skyward—just as others had done many times before for her launches and landings—to see *Discovery*. She then made her way to the Goddard Spaceflight Center in Greenbelt, Maryland, to give the NASA employees and contractors there a final glimpse. And then it was on to Dulles.

As she made her final landing approach about 11:00 a.m., traffic on the nearby roadways came to a standstill as motorists got out of their vehicles for the once-in-a-lifetime look. They ignored the AAA Mid-Atlantic warnings: "Don't let anyone or anything—even a space shuttle overhead—distract you" and "For safe shuttle-spotting, pull off the road and park your car."

After touchdown, the 747 taxied over to the Udvar-Hazy Center where she and her passenger were welcomed by numerous shuttle astronauts, many of whom, including John Glenn, had flown on *Discovery*. There were cheers and tears in about equal amounts.

I went to Udvar-Hazy Center two weeks later for a special dinner. Along with fellow shuttle astronauts Joe Allen, Rick Hauck, Tom Jones, Carl Walz, Frank Culbertson, Tammy Jernigan, and Charlie Bolden, who at the time was the NASA administrator, I waited for that familiar sound—a countdown. 10…9…8…7…6…5…4…3…2…1. A set of curtains parted and there, spectacularly illuminated, was *Discovery*. What an impressive and emotional moment. I just stood there motionless for a few minutes staring at that incredible vehicle. For some reason, I slowly looked up and spotted another incredible sight. Hanging directly above *Discovery*'s payload bay was a full-scale model of a Tracking and Data Relay Satellite with its solar panels and huge antennas fully deployed. *Discovery* had been used to launch and deploy three of those satellites, so it seemed only fitting that the two were back together again on permanent display.

I hope to be a frequent visitor to the Udvar-Hazy Center. I know that every time I step into the James S. McDonnell Space Hangar my thoughts will go immediately to the time I was able to spend aboard *Discovery* on the STS-70 mission. I'll be thinking about my crew mates—Tom, Kevin, Nancy, and Mary Ellen. And I'll relive the incredible adventure we all shared aboard her that July in 1995—one of the most incredible journeys of my life.

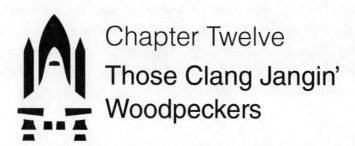 # Chapter Twelve
Those Clang Jangin' Woodpeckers

NASA's space shuttle program, which began April 12, 1981, and ended July 21, 2011, will be remembered both for its great triumphs and even greater tragedies. The loss of seven astronauts aboard the space shuttle *Challenger* in 1986 and the loss of seven more aboard *Columbia* seventeen years later will forever be etched in the annals of spaceflight.

So, too, will be the United States' development of a fleet of reusable spacecraft—*Columbia, Challenger, Discovery, Atlantis,* and *Endeavour*—that launched telescopes to give us exceptional views of faraway heavenly bodies and cosmic phenomenon, provided the opportunity for extensive microgravity scientific research in a myriad of disciplines, deployed satellites that explored our universe and beyond looking for the secrets to our existence, and facilitated the construction of a fully operational outpost that's manned year round—the International Space Station.

And then there was the human element—thousands upon thousands of people—from the astronauts who flew aboard the shuttles and the ground personnel at NASA centers and contractor employees who readied the missions, to the taxpayers who funded it all. We will never forget their commitment to space exploration.

It is my hope that the legacy of STS-70 is a mix of the substantive and the whimsical.

From the deployment of a Tracking and Data Relay Satellite that almost a generation later continues to provide virtually instantaneous communication between spacecraft and ground controllers to a cartoon character—Woody Woodpecker, to be precise—that embodied some of its pesky all-too-real brethren who pecked a hole—okay, over two hundred of them—in NASA's plans to launch a group of astronauts—all but one—from the same Midwestern state, Ohio.

But in many ways, as is often the case, the unusual overrides the normal, the humorous supersedes the serious in the collective memory. And so it probably will be for STS-70, the All-Ohio Space Shuttle Mission.

Todd Halvorson, veteran space reporter for the *Florida Today* newspaper which circulates in the Space Coast area that includes the Kennedy Space Center, decided that STS-70 experienced the weirdest launch delay in shuttle program history, "Set to launch on the historic one hundredth US human space flight, the mission was delayed after Yellow Shafted Flicker Woodpeckers drilled dozens of holes in external tank foam insulation. The shuttle was returned to its assembly building for repairs. The mission ultimately became the one hundred and first US human space flight," he wrote in an article marking the program's twenty-fifth anniversary.

Another veteran space reporter, Irene Klotz, writing for the Discovery Channel, compiled a list of the "Top Ten Reasons for Shuttle Launch Delays." While the attack of the woodpecker did not make the much coveted number one position, it did claim the number ten spot.

In a similar Discovery Space website ranking for the best space wake-up music ever, Klotz ranked the Woody Woodpecker theme song played for STS-70 as number two of all time and ranked it the top wake-up music of the entire space shuttle program.

And Roger Balettie, a long-time Mission Control employee who worked as the lead Flight Dynamics Officer, wrote in his blog that the STS-70 woodpecker incident was one of his most humorous experiences working for NASA. He even posted on his website a picture of himself posing with the stuffed Woody Woodpecker in Mission Control taken during our flight. He recalled how Woody was adopted as an unofficial mascot for that mission and that flight controllers took turns having their picture taken with Woody.

For many of the NASA workers at both the Kennedy Space Center and the Johnson Space Center who supported our mission, as well as those who worked on the TDRS we deployed, STS-70 will always be one of their more unusual missions. STS-70 clearly was as much fun for our ground team as it had been for all of us aboard *Discovery*—and it remains so to this day.

Woody was added to an unofficial version of the official STS-70 crew patch. That version, designed by JSC employees Paula Vargas and Andrew Parris, featured Woody Woodpecker standing behind the space shuttle in our official patch with arms outstretched as if to say "Here I am, folks!" The STS-70 crew received a few of those special patches after we landed, so unfortunately none of them ever flew aboard STS-70. But I think the entire crew got a kick out of the new design, which I would have preferred to wear on my launch and entry suit had they had been available before our flight.

Also after the mission, I received a polo shirt from the Kennedy Space Center with a unique embroidery honoring of our flight.

Those Clang Jangin' Woodpeckers

The shirts became a popular item at KSC for the STS-70 team that successfully launched us to space. I enjoyed seeing them whenever I returned to KSC.

And a song was written about the woodpecker attack a month after STS-70 landed. The lyrics, written by Charles W. Morley Jr., can be sung to the tune "Stonecutters Cut it on Stone," which was composed by Rodgers and Hammerstein for the musical *Carousel*. The new lyrics to the song titled "Those *?%@## (Clang Jangin') Woodpeckers" as published by the Space to Share Production Co. in 1995 are as follows:

> Woodpeckers peckin' on foam, Space Shuttle had to go home,
> Those yellow-shafted Flickers nearly punctured ET,
> And *Discovery* had to sit on the loam.
> Woodpeckers peckin' on steel, know exactly how they must feel
> Old Mother Nature stuck her hand in the scene,
> And *Discovery* did the V-A-B reel.
> Woodpeckers peckin' on foam, Space Shuttle's had to go home,
> God in his wisdom made a change in the plan,
> And *Atlantis* got its chance to roam.
> Space Shuttle knocked on *Mir*'s door, come in, have a seat on the floor,
> *Unity*'s light made the station shine bright,
> And we'll have to GO for many more.
> Woodpeckers peckin' on foam, Space Shuttle had to go home,
> History's made the whole planet bright,
> As *Atlantis* made the hundredth aplomb.
> Woodpeckers peckin' on foam, Space Shuttle had to go home,
> Cuttin' the budget makes our space program tight,
> And the light's out in the Capitol Dome.
> Bean counters counting their beans, blinder than I've ever seen,
> Cutting the growth crop before it can flower,
> A no brainer, if you know what I mean.

Woodpeckers peckin' on foam, finally peckin' through to the dome,
Multi-year funding: the prescription to fill,
To make Alpha, our Space Station Home.
Woodpeckers peckin' on foam, Space Shuttle had to go home,
It takes special people to make good on our dream,
And we need to give them room to roam.
Pad workers patched all the foam, made *Discovery* ready to roam,
TDRS and the Science came off slicker than slick,
As a great crew made the mission a poem.

Morley had worked for various NASA contractors since the Apollo program. Getting into the spirit and fun of the woodpecker attack, he penned the song and first performed it for us at the Rockwell Space Day Rally in Cape Canaveral after our flight. In his introduction, he jokingly congratulated our crew for conducting one of the most boring missions he had ever seen. "Your mission went so smoothly that if it hadn't been for our little feathered friends, this mission would have ho-hummed into the log books and been soon forgotten. But because of these pesky rascals, you've become a household name—the Woodpecker Crew."

Besides our notoriety for being the 'All-Ohio Space Shuttle Mission,' STS-70 will forever be known as the 'Woodpecker Flight.'

Most people no longer remember exactly what the STS-70 mission accomplished—even die-hard space enthusiasts. But the minute I refer to our flight as the woodpecker mission, big smiles appear and people say, "Oh, I remember that one." Totally unique among the 135 space shuttle missions—who could forget the Woodpecker Flight?

While none of the crew members went on to achieve great fame and fortune, we were immortalized on a postage stamp. In 2010, the Republic of Malawi, a small landlocked country in southeastern Africa, issued a series of stamps honoring all the space shuttle crews. One of the stamps in the series features the NASA portrait of the STS-70 crew. The stamp's monetary value is officially eighty Kwacha, or the equivalent of about fifty cents in the United States, just slightly above US first class postage.

And during a recent trip to the United Kingdom, someone presented me with a different embroidered patch that featured a woodpecker standing next to our official NASA version of the STS-70 patch. I laughed when I saw it and thought, *Our mission is still well known around the world.*

Discovery and the STS-70 mission were also immortalized in Toledo, Ohio, when the city named a street in our honor. Shortly after the flight, Tom visited Toledo to return a glass key to the city that flew on STS-70 (a glass key because Toledo is known as the Glass Capital of the World). While there, Tom also helped dedicate a new street—Discovery Way. It's located just outside of what at the time was known as COSI (Center of Science and Industry) in downtown Toledo along the Maumee River. Honoring both the STS-70 mission and the discoveries young people make every day as they explore the exhibits and displays at the science center—now known as Imagination Station, the street sign features the STS-70 crew patch and the traditional NASA meatball logo. The City of Toledo was kind enough to give each crew member one of the street signs, and mine still hangs proudly above my desk in my office at home, a great reminder of the All-Ohio Space Shuttle Mission.

The NASA Glenn Research Center in Cleveland, to signify its pride in Ohio astronauts, created a special poster featuring twenty-four of the state's astronauts and created a link on their website about them. Astronaut biographies are provided and they also have an 'Ohio Astronaut Quiz.' As part of the NASA display at the Great Lakes Science Center located next to the Rock and Roll Hall of Fame there is a special area highlighting Ohio's astronauts.

No other state has recognized and honored its astronauts quite like Ohio. If you look at the state quarters issued by the US Mint between 1999 and 2008, you will see cows, corn, buffaloes, bridges, and many national monuments. But only Ohio has an astronaut on its quarter. You can almost feel the envy from some of the other states when they jokingly ask, "What is it about your state that makes people want to flee the Earth?" As John Glenn pointed out in his foreword to this book, "Ohioans have a strong heritage of discovery, and exploration is in their blood."

So what *was* the STS-70 mission all about?

If you look it up in the NASA archives, it was about deploying the last Tracking and Data Relay Satellite from a space shuttle in 1995—a replacement communications satellite for the one lost during the *Challenger* accident in 1986. It was about bioreactors and exploring new technologies to potentially grow human tissue in space. It was about new techniques to incorporate position data into photographs taken from orbit. It was about growing larger protein crystals with greater purity than achievable on Earth to benefit the development of pharmaceutical drugs in the future. And it was about studying the Earth and improving life on our planet.

For the STS-70 crew and the many people who worked on the mission, it was about all those things and much more. For us the flight was about building NASA's infrastructure in space in order to support future missions—the Hubble Space Telescope, International Space Station, and countless other scientific satellites. The mission was about teamwork, dedication, solving problems, and overcoming obstacles. But it was also about the laughs created by a single woodpecker bringing the space shuttle program to a halt—a story of David and Goliath. Woodpecker vs. Shuttle.

The STS-70 mission was very much about the fun that we had during the space shuttle era. Everyone worked extremely hard every day on the program, but we had tremendous amounts of fun while we did. We loved our jobs and the work we were doing. We loved the mission. Was everyone frustrated by the delays caused by the woodpecker? You bet we were, but we didn't let it bother us too much.

In typical NASA fashion, we formed a team—the BIRD team—to investigate the situation, we identified the problem (a love-sick woodpecker during mating season), found a solution (placed owl figurines and predator balloons around the launch pad), and carried on with the program. And in a most strikingly human fashion, we laughed about what had happened and made fun of the situation. It took a few weeks after the attack, but soon most everyone appreciated the ridiculousness of what had occurred. Thanks to the folks at Universal Studios, Woody Woodpecker became our mission's unofficial mascot. A small Woody Woodpecker doll made it aboard *Discovery* and flew with us, while another ended up in the main Flight Control Room in Mission Control. Woodpecker wake-up music. Woodpecker cartoons. Woodpecker jokes.

For Tom, Nancy, Mary Ellen, and me, the mission also was about our pride in our home state and Ohio's rich history and tradition in aeronautics and aerospace. The Wright Brothers. John Glenn. Neil Armstrong. There wasn't one of us who hadn't been influenced and inspired by those legends. We were as surprised as most to realize how many of the astronauts had come from Ohio. Why? What could account for that? Was it the role models we had in Glenn, Armstrong, and others? Was it Ohio's education system, public schools, and outstanding colleges and universities? Was it simply part of the Midwest work ethic? Was it because, as one blogger has suggested, Ohio was just being true to her moniker as the Birthplace of Aviation? Or was it the water we drank from Lake Erie?

I don't think anyone has a definitive answer for why so many astronauts have come from Ohio, but one thing is sure: the Ohio astronauts are extremely proud of where they come from and are hopeful that the trend will continue. Why shouldn't the first person to set foot on Mars come from Ohio?

In August 2008 as part of NASA's fiftieth anniversary, the Glenn Research Center invited all of the Ohio astronauts to Cleveland for a black tie gala celebration. Seventeen of us attended along with two astronauts living and working in Ohio at the time. Included among the Ohio astronauts were John Glenn, Neil Armstrong, and Jim Lovell, along with fourteen who had participated in the space shuttle and International Space Station programs. Only Mike Good was unable to attend because he was busy training for one of the final space shuttle missions. It was the first ever Ohio astronaut reunion, and a special evening for everyone in attendance.

The five members of the STS-70 crew were there, including honorary Ohioan Kevin Kregel. On the first evening of the celebration, we all met in the hotel lobby with our spouses and shared stories and reminisced about the All-Ohio Space Shuttle Mission. Tom and Becky. Kevin and Jeanne. Nancy and Dave. Mary Ellen and Jerry. Simone and I. Everyone was there except Woody. As I sat there enjoying being with my former crew members, I knew what a special moment it was to have us all together. It was one of those evenings that you never want to end.

For me personally STS-70 was my most memorable and enjoyable flight, which is why I chose to write this book. While it's hard to top the thrill and exhilaration of your first mission to space, overall STS-70 still remains my favorite. I had more fun on STS-70 than on all my other three missions combined.

Now eighteen years after the STS-70 mission, I harbor no ill-will for woodpeckers. Quite the contrary, I have the utmost respect and admiration for them. Behind my house outside Baltimore, I hang special woodpecker suet to attract them to my yard and I enjoy watching them feed. I frequently see red-bellied woodpeckers, downey woodpeckers, occasionally a Woody look-alike pileated woodpecker, and just recently I spotted my first northern flicker. As I walk through the woods, I frequently hear the woodpeckers drumming as they search for insects in old trees. I enjoy the sounds of their rat-tat-tat-tatting. I recently purchased a new birdhouse designed especially for woodpeckers and look forward to having more woodpeckers for neighbors in the spring.

Woodpeckers played such an important role in the STS-70 story and I find my connection to them only growing over time. Maybe that is simply due to nostalgic longings for the good old days of the shuttle program and all the fun we had way back when now that the shuttle era has come to an end.

Whatever the reason, I know I will always have a warm spot for woodpeckers in my heart that never would have been there had it not been for the attack one of them perpetrated on my space shuttle in the late spring of 1995.

Appendix
The Twenty-Six Ohio Astronauts

Ohio Astronaut #1: John H. Glenn

Mercury-Atlas 6, *Friendship 7*
STS-95 (Space Shuttle *Discovery*)

First American to Orbit the Earth
First Ohioan in Space
Oldest Ohioan to Fly in Space
Longest Duration Between Spaceflights:
    36 years

John H. Glenn Jr. was born in Cambridge, Ohio, on July 18, 1921. At an early age, he moved with his parents to New Concord. He graduated from New Concord High School and attended Muskingum College. Shortly after Pearl Harbor was attacked, Glenn enlisted in the Naval Aviation Cadet Program and became a Marine pilot in 1943. He flew 59 combat missions in the South Pacific during World War II. When the Korean Conflict began, Glenn asked for combat duty and flew 63 missions. For his total of 149 missions during the two wars he received many decorations, including the Distinguished Flying Cross six times.

Following the Korean Conflict, Glenn attended test pilot school. He served as a test pilot for Naval and Marine aircraft. In 1957, he set a speed record, flying from Los Angeles to New York in three hours and twenty-three minutes.

In 1959, Glenn was selected to be one of the first seven astronauts in the US space program. On February 20, 1962, he became the first American to orbit the Earth. He piloted the *Friendship 7* spacecraft around the globe three times.

Glenn served as an advisor to the Apollo program until he resigned from the Manned Spacecraft Center in 1964. He was promoted to the rank of colonel the same year. He retired from the Marines in 1965. Following several years in private business, Glenn was elected to the US Senate in 1974, where he served many years.

On January 16, 1998, NASA announced that John Glenn had been assigned to his second spaceflight, this time as a payload specialist aboard a much roomier space shuttle on the STS-95 mission. Launching on October 29, 1998, and landing on November 7, this nine-day mission was a science research flight and also included the deployment of the *Spartan 201* solar-observing spacecraft. This mission was the twenty-fifth flight of space shuttle *Discovery* and the ninety-second flight of the space shuttle program. During the flight Glenn performed a number of the experiments, some of which investigated the aging process and how the absence of gravity affects balance and perception, immune system response, bone and muscle density, metabolism and blood flow, and sleep. Just after landing, Glenn radioed to Mission Control in Houston, "One g and I feel fine," a reference to one of his more famous quotes from his first mission back in 1962. Glenn holds the record for the oldest human to fly in space and for the longest period between flights (thirty-six years).

During his two missions in space, Glenn completed 137 orbits of the Earth, traveling 3.7 million miles. His total time in space is 218 hours, 39 minutes, and 23 seconds. John Glenn was the fifth person to fly in space.

Ohio Astronaut #2: James A. Lovell

Gemini 7
Gemini 12
Apollo 8
Apollo 13

First Ohioan to Orbit the Moon: Apollo 8
Only Ohioan to go to the Moon Twice: Apollo 8, 13

James Arthur Lovell Jr. was born in Cleveland, Ohio, on March 25, 1928. Lovell graduated from Juneau High School in Milwaukee, Wisconsin. He attended the University of Wisconsin for two years before receiving a bachelor of science degree from the US Naval Academy in 1952.

Lovell's military and flight career began in 1952 when he became an ensign in the Navy. While in the Navy he attended flight school and served as a flight instructor and test pilot. In September 1962, he was selected in NASA's second group of astronauts.

Lovell's first NASA duty was as the backup pilot for Gemini 4. He made his first space flight in 1965 on the Gemini 7 mission. That mission was the first time two manned spacecraft rendezvoused in space. In 1966, Lovell flew on the last space flight of the Gemini program, Gemini 12.

Lovell's third space flight was aboard Apollo 8 in 1968. He was command module pilot on man's first flight to the Moon. Launched aboard the Saturn V rocket, the crew of Apollo 8 became the first humans to leave Earth's gravitational field. They traveled 223,000 miles, circled the Moon ten times and broadcast live images of the Moon back to Earth.

Lovell's last space flight was as commander of the Apollo 13 mission in 1970. An explosion took place on board fifty-five hours into the flight which led Lovell to radio down to Mission Control, "Houston. We've had a problem here." The crew had to give up its original plan to walk on the Moon. Lovell and his crew worked closely with the Houston ground controllers to develop a rescue plan. After circling the Moon, the crew completed the first deep space emergency recovery in history. They returned safely to Earth on April 16, 1970.

Following his heroics aboard Apollo 13, Lovell attended the advanced management program at Harvard University. In 1971, he was named deputy director for science and applications at the Johnson Space Center. He served there until he retired from NASA in 1973.

During his four spaceflights, Lovell completed 267 orbits of the Earth and 10 orbits of the Moon, traveling 7.4 million miles. His total time in space is 715 hours, 4 minutes, and 57 seconds. Jim Lovell was the twenty-fourth person to fly in space.

Ohio Astronaut #3: Neil A. Armstrong

Gemini 8
Apollo 11

Performed the First Docking in Space
First Man on the Moon
The First and Only Ohioan to Walk on the Moon

Neil A. Armstrong was born in Wapakoneta, Ohio, on August 5, 1930. He graduated from high school there. He attended Purdue University for two years on a Navy scholarship. Armstrong's education was interrupted when he was called to active duty during the Korean conflict. While in Korea he flew seventy-eight combat missions. After the war, Armstrong returned to Purdue. He received a bachelor of science degree in engineering in 1955.

After graduating from Purdue, Armstrong went to work at NASA's Lewis Research Center as an engineer. Within a year, he transferred to the Flight Research Center at Edwards Air Force Base in California. While at Edwards, Armstrong tested high-speed aircraft, including the famous X-15.

Armstrong became an astronaut in 1962. On his first space mission he served as command pilot for the Gemini 8 mission in 1966. On that mission he performed the first successful docking of two vehicles in space. Armstrong also served as backup command pilot for Gemini 11.

Armstrong was chosen as commander of Apollo 11, the first manned lunar landing mission. On July 20, 1969, he became the first human being to set foot on the Moon.

Following the Apollo 11 flight, Armstrong held several jobs with NASA. He resigned in 1971 to become a professor of engineering at the University of Cincinnati. Armstrong served as vice chairman of the Rogers Commission, which investigated the space shuttle *Challenger* accident in 1986. In the early 1990s, he hosted a television documentary entitled *First Flights*.

Armstrong passed away on August 25, 2012.

During his two missions Armstrong completed 8 orbits of the Earth and 18 orbits of the Moon, traveling slightly over 1 million miles. His total time in space is 206 hours and 1 second. Neil Armstrong was the twenty-sixth person to fly in space.

Ohio Astronaut #4: Charles A. Bassett II

Assigned to Gemini 9 (not flown)

First Astronaut to attend Ohio State University

Charlie Bassett was born in Dayton, Ohio, on December 30, 1931, while his father was stationed at Wright Patterson Air Force Base. Bassett knew exactly what he wanted to do with his life...fly! He once told a reporter, "I remember when I was a kid five years old. I used to draw pictures of airplanes as they were taking off at Wright Field near Dayton, Ohio. I'd look out the window and draw pictures" (Burgess and Dooloan 2003, 50). Growing up in close proximity to Wright Patterson Air Force Base was clearly an influencing factor in his life. He moved around quite a bit as a child and eventually ended up in Berea outside Cleveland. In Cleveland, Bassett made his first solo flight in a private airplane on December 30, 1947, his sixteenth birthday. He spent a lot of his free time working around the airport helping grease and polish airplanes to pay for his flying lessons. He was active in the Boy Scouts where he attained the level of Life Scout. He graduated from Berea High School in 1950 and attended Ohio State University from 1950–52. He then joined the Air Force ROTC program at Ohio State and entered the Air Force as an aviation cadet in October 1952 and soon found himself in flight training at Bryan Air Force Base in Texas. He also attended Texas Technological College from 1958–60 and received a bachelor of science degree in electrical engineering with honors. He also did some graduate work at the University of Southern California. As an Air Force captain he attended the Aerospace Research Pilot School and the Air Force Experimental Pilot School. While in the Air Force he sought out advice from Colonel Chuck Yeager on getting into the astronaut program. Yeager told him to "Shoot for getting as high as you can go."

Bassett had over 3,600 hours of flying time, including over 2,900 hours in jet aircraft.

He was selected by NASA in the third astronaut group in October 1963. His first technical assignment within the astronaut office was to follow the training and simulators programs at the NASA Manned Spacecraft Center.

On November 8, 1965, he was assigned to be the pilot of the upcoming Gemini 9 mission along with Elliott See as commander. Only a few months later, on February 28, 1966, Bassett and See died in a crash of their T-38 jet as they approached the runway during a routine training session at McDonnell Aircraft, where their Gemini 9 spacecraft was being built. Their jet crashed into McDonnell Aircraft Building 101, known as the McDonnell Space Center, within five hundred feet of where their capsule was being built. A NASA investigative panel later concluded that pilot error, caused by poor visibility due to bad weather, was the principal cause of the accident. The panel concluded that See was flying too low to the ground during his approach, probably as a result of the poor visibility. Their backup crew of Tom Stafford and Gene Cernan were flying in a second T-38 and witnessed the crash and safely landed a few minutes later. Stafford and Cernan went on to fly the Gemini 9 mission, which is sometimes referred to as Gemini 9A.

Bassett was survived by his wife and two children. He is buried at Arlington National Cemetery. Directly below his name on his grave marker, it reads "OHIO" in recognition of his home state.

Had he flown the Gemini 9 mission as planned, Charlie Bassett would have been the twenty-eighth human in space and the fourth Ohio astronaut to orbit the Earth.

Ohio Astronaut #5: Donn Eisele

Apollo 7

First Ohioan to fly in the Apollo Program

Donn F. Eisele was born in Columbus, Ohio, on June 23, 1930. He graduated from West High School in Columbus in 1948 and went on to receive a bachelor of science degree from the US Naval Academy in 1952 and a master of science degree in astronautics in 1960 from the Air Force Institute of Technology at Wright Patterson Air Force Base. He was a member of the Freemasons, where he belonged to the Turner Lodge #732 in Columbus.

Eisele was selected in NASA's third group of astronauts (along with fellow Ohioan Charlie Bassett) in October 1963. He went on to fly as the command module pilot on the Apollo 7 mission along with Wally Schirra and Walt Cunningham. Launching on October 11, 1968, Apollo 7 was the first manned flight of the Apollo Program. Eisele participated in and executed numerous maneuvers that would be needed for the Apollo lunar landing missions. During the mission the crew also participated in the first television broadcast of onboard crew activities for which NASA later received an Emmy Award. Apollo 7 successfully landed in the Atlantic Ocean on October 22, 1968, some eight miles from the aircraft carrier *Essex*.

Following Apollo 7, Eisele went on to serve as the backup command module pilot for the Apollo 10 mission.

In 1972, Eisele retired from NASA and the US Air Force to become director of the US Peace Corps in Thailand. He later became sales manager for the Marion Power Shovel Company. He died of a heart attack on December 2, 1987, while on a business trip to Tokyo, Japan, where he was attending the opening of a new Space Camp patterned on the one at the US Space & Rocket Center in Huntsville, Alabama. He is buried at Arlington National Cemetery.

To honor Eisele, his hometown of Columbus named a street after him near Rickenbacker Air Force Base. Donn Eisele Street is right off John Glenn Avenue, and one block over from Judy Resnik Street.

During Apollo 7 Eisele completed 163 orbits of the Earth, traveling 4.5 million miles. His total time in space is 260 hours, 9 minutes, 3 seconds. Donn Eisele was the thirty-second person to fly in space.

Ohio Astronaut #6: Bob Overmyer

STS-5 (Space Shuttle *Columbia*)
STS-51B (Space Shuttle *Challenger*)

First Ohio Air Force Astronaut
First Ohioan to fly on the Space Shuttle

Robert F. Overmyer was born in Lorain, Ohio, on July 14, 1936, but considers Westlake as his hometown. He graduated from Westlake High School in 1954. Overmyer earned a bachelor of science degree in physics from Baldwin Wallace College in 1958. In 1964, he received a master of science degree in aeronautical engineering from the US Naval Postgraduate School. Overmyer later received an honorary doctor of philosophy degree from Baldwin Wallace College in 1982.

Overmyer's military career began in 1958 with the Marine Corps. During the next eleven years, he attended naval flight training, received his postgraduate degree, served in Japan and attended the Air Force Test Pilot School. In 1966, he was selected as an astronaut for the US Air Force's Manned Orbiting Laboratory (MOL) Program. When this program was canceled in 1969, Overmyer was transferred to NASA to join NASA's seventh group of astronauts, along with six other MOL astronauts.

During his early years with NASA, Overmyer worked on developing the Skylab Program. He also worked in Moscow on the Apollo-Soyuz Test Project. In addition, he worked on the Space Shuttle Approach and Landing Test Program. Overmyer's first space flight came in 1982 where he piloted space shuttle *Columbia* on STS-5, the first fully operational flight of the shuttle program. This was the first mission with a four-person crew and the first to launch large satellites from the space shuttle.

In 1985, Overmyer was commander of STS-51B, the Spacelab-3 mission. On that flight the astronauts conducted a wide range of scientific experiments while in space.

In 1986, Overmyer served on the Rogers Commission, which investigated the *Challenger* accident. He retired that year to pursue private business opportunities. In 1996, he was killed when a small plane he was test piloting crashed near Duluth, Minnesota. He is buried at Arlington National Cemetery.

During his two missions, Overmyer completed 192 orbits of the Earth, traveling 5 million miles. His total time in space is 290 hours, 23 minutes, and 12 seconds. Bob Overmyer was the 113th person to fly in space.

Ohio Astronaut #7: Judy Resnik

STS-41C (Space Shuttle *Discovery*)
STS-51L (Space Shuttle *Challenger*)

First Ohio Woman in Space

Judith A. Resnik was born in Akron, Ohio, on April 5, 1949. She was a product of Akron's public schools and attended Fairlawn Elementary, Simon Perkins Junior High, and Firestone High

School, graduating in 1966. Resnik earned a bachelor of science degree in electrical engineering from Carnegie Mellon University in 1970. Following college, Resnik worked at RCA in their missile and service radar divisions. She also attended graduate school. In 1974, she left RCA to become a staff scientist at the neurophysiology laboratory at the National Institutes of Health in Maryland. Resnik continued her schooling and earned a doctorate in electrical engineering from the University of Maryland in 1977.

In 1978, Resnik and five other women were selected to become America's first female astronauts in NASA's eighth group of astronauts. After completing the training program for shuttle mission specialists, Resnik held several roles at the Johnson Space Center. In 1984, Resnik became the second American woman in space when she flew on the first flight of space shuttle *Discovery*. During the mission she conducted experiments with a solar array and performed biomedical research.

Resnik was chosen as a mission specialist for shuttle mission STS-51L, aboard space shuttle *Challenger*. Tragically, she and six other astronauts were killed in an explosion shortly after lift-off on January 28, 1986.

A crater located within the Apollo impact basin on the far side of the Moon is named in her honor. Fairlawn Elementary School, the elementary she attended, was renamed Judith A. Resnik Community Learning Center when it reopened after renovations in December 2006.

During her two missions, Resnik completed 97 orbits of the Earth, traveling nearly 2.5 million miles. Her total time in space is 144 hours, 56 minutes, and 5 seconds. Judy Resnik was the 146th person to fly in space.

Ohio Astronaut #8: Karl Henize

STS-51F (Space Shuttle *Challenger*)

Ohio's First Science Astronaut

Karl Henize was born in Cincinnati, Ohio, on October 17, 1926. He grew up exploring the hills and valleys of Plainville and Marimont outside Cincinnati. His home was built on a hilltop above the Little Miami River and bordered the Taft Estate on one side.

With the United States' entry into World War II, Henize chose not to finish high school and volunteered to enter the Navy's V-12 College Training Program, which first took him to Dennison University in Cincinnati and then to the University of Virginia, where he received a bachelor of arts degree in mathematics in 1947 and a master of arts degree in astronomy in 1948. In 1954 he received his doctorate in astronomy from the University of Michigan.

After a distinguished career in astronomy, he was selected as a science astronaut in NASA's fourth group of astronauts in August 1967. This was the first selection of scientists into the astronaut program and of the six scientists chosen in this group, four went on to fly in space.

Nearly eighteen years after being selected as an astronaut, Henize flew into space on July 29, 1985, aboard space shuttle *Challenger* on the STS-51F mission. This dedicated science flight known as Spacelab-2, featured a number of astronomy and solar physics experiments, along with a number of life science and earth sciences experiments. After 126 orbits, space shuttle *Challenger* successfully landed at Edwards Air Force Base in California on August 6, 1985.

In early 1993, he was invited to join an expedition to climb the north face of Mount Everest. During his second day after reaching advanced base camp, he started showing symptoms of extreme high altitude sickness. A valiant effort was made by the other members of the expedition to save him, but they could not get him off the mountain in time. He died on October 5, 1993, only twelve days short of his sixty-seventh birthday, of high altitude pulmonary edema at eighteen thousand feet, and was buried nearby, above the Changste Glacier.

During STS-51F, Henize completed 126 orbits of the Earth, traveling 3.3 million miles. His total time in space is 190 hours, 45 minutes, and 26 seconds. Karl Henize was the 176th person to fly in space.

Ohio Astronaut #9: Robert C. Springer

STS-29 (Space Shuttle *Discovery*)
STS-38 (Space Shuttle *Atlantis*)

Bob Springer was born in St. Louis, Missouri, on May 21, 1942, but considers Ashland, Ohio, to be his hometown. In 1960, he graduated from Ashland High School and went on to attend the US Naval Academy, where he received a bachelor of science degree in naval science in 1964. In 1971, he received a master of science degree in operations research and systems analysis from the US Naval Postgraduate School.

In 1975, he graduated from the US Navy Test Pilot School at Patuxent River, Maryland. He was selected by NASA to be an astronaut in 1980, where he joined twenty others in NASA's ninth group of astronauts.

On March 13, 1989, Springer launched aboard *Discovery* on the STS-29 mission, where they deployed the fourth Tracking and Data Relay Satellite (TDRS). During the mission, the crew also tested out a new heat-pipe design for transferring heat in space for future use on the ISS and performed a number of other experiments. On March 18, 1989, STS-29 came to a successful conclusion with a smooth landing at Edwards Air Force Base in California.

Almost immediately, Springer was assigned to his second mission, STS-38, aboard space shuttle *Atlantis*. Launching on November 15, 1990, Springer and his other four crew members conducted a secret

Department of Defense mission. On November 20, 1990, Springer and his crew successfully landed at the Kennedy Space Center in Florida. Following this mission, Springer retired from NASA and joined Boeing as the director of quality systems.

During his two missions Springer completed 160 orbits of the Earth, traveling 4 million miles. His total time in space is 237 hours, 33 minutes, and 23 seconds. Bob Springer was the 215th person to fly in space.

Ohio Astronaut #10: David Low

STS-32 (Space Shuttle *Columbia*)
STS-43 (Space Shuttle *Atlantis*)
STS-57 (Space Shuttle *Endeavour*)

First Ohioan to do a Spacewalk from the Space Shuttle

G. David Low was born in Cleveland, Ohio, on February 19, 1956, while his father, George Low, was starting his NASA career at the Lewis (now Glenn) Research Center in Cleveland. A few years later his dad was transferred to work with the Space Task Group at the NASA Langley Research Center in Langley, Virginia. David Low graduated from Langley High School in McLean, Virginia, in 1974. He attended Washington & Lee University, where he earned a bachelor of science degree in physics-engineering in 1978. Low received a second bachelor of science degree in mechanical engineering from Cornell University in 1980. He also received a master of science degree in aeronautics and astronautics from Stanford University in 1983.

Low was selected to become an astronaut in 1984 in NASA's tenth astronaut group, affectionately called *The Maggots*. During his training Low held many jobs with NASA, including spacecraft communicator (Capcom) in the Mission Control Center during the STS-26, 27, 29 and 30 missions.

Low flew on three space flights. In 1990, he served as a mission specialist on STS-32, launching aboard space shuttle *Columbia* on January 9, 1990. The main objective of this mission was to retrieve the Long Duration Exposure Facility (LDEF), a materials exposure satellite that had been deployed years earlier in 1984 during STS-41C.

Low next served as the flight engineer aboard space shuttle *Atlantis* on STS-43 in 1991. Launching on August 2, 1991, the crew deployed the fifth Tracking and Data Relay Satellite (TRDS) from the space shuttle.

On his last flight, Low served as payload commander for STS-57, launching aboard space shuttle *Endeavour* on June 21, 1993. This mission was the first flight of a new commercial science research facility called Spacehab. During the mission the crew retrieved the European Retrievable Carrier satellite (EURECA) using the space shuttle robotic arm and conducted numerous experiments. During STS-57, Low and fellow crewmate Jeff Wisoff performed a spacewalk in the payload bay for five hours and fifty minutes, evaluating special tools to be used on future missions.

Low left NASA in 1996 to pursue a private career with Orbital Sciences Corporation's Launch Systems Group in Dulles, Virginia. On March 15, 2008, he passed away in Reston, Virginia, after a battle with cancer.

During his three missions Low completed 470 orbits of the Earth, traveling 12.3 million miles. His total time in space is 714 hours, 6 minutes, and 55 seconds. David Low was the 225th person to fly in space.

Ohio Astronaut #11: Ronald A. Parise

STS-35 (Space Shuttle *Columbia*)
STS-67 (Space Shuttle *Endeavour*)

First Payload Specialist from Ohio

Ron Parise was born in Warren, Ohio, on May 24, 1951. As a young boy, Parise had a strong interest in amateur radio and became a licensed operator at the age of eleven. He also had a keen interest in astronomy and aviation and received his pilot's license when he was still in his teens. He graduated from Western Reserve High School in Warren in 1969 and then attended Youngstown State University, where he received a bachelor of science degree in physics with minors in mathematics, astronomy, and geology in 1973. He then went on to the University of Florida where he received a master of science and doctor of philosophy degrees in astronomy in 1977 and 1979 respectively.

From there he started working at Operations Research Inc., and then Computer Sciences Corporation, supporting several different NASA science satellite missions, such as the International Ultraviolet Explorer. In 1981, he began working on a new Spacelab experiment called the Ultraviolet Imaging Telescope and in 1984, he was selected by NASA as a payload specialist astronaut in support of the newly formed Astro series Spacelab missions.

Parise was originally scheduled to make his first flight on STS-61E, which had been scheduled to launch in March 1986, but only thirty-five days before his scheduled launch the *Challenger* accident occurred, forcing NASA to cancel all shuttle missions. It would be nearly five years before he would get his chance to fly in space on the rescheduled mission known as STS-35/Astro-1.

On December 2, 1990, he launched into space aboard space shuttle *Columbia* on the STS-35/Astro-1 mission, which carried three unique telescopes which could simultaneously record spectral data, polarimetric data, and imagery of faint astronomical objects in the far-ultraviolet. After nearly nine days in space they landed at Edwards Air Force Base in California on December 10, 1990.

Parise also flew aboard space shuttle *Endeavour* on the STS-67/Astro-2 mission, the second flight of the Astro observatory. Launch was on March 2, 1995 from pad 39A at the Kennedy Space Center in Florida. Similar to the Astro-1 mission, this flight focused on far ultraviolet observations of faint astronomical objects such as hot stars and distant galaxies. During their record-setting flight—which lasted 16 days, 15 hours, and 8 minutes, the longest of the space shuttle program up to that time—the crew observed a different region of the sky than previously studied on the Astro-1 mission, which allowed them to make significantly more observations during the flight.

On May 9, 2008, Parise passed away at the age of fifty-six after a long battle with cancer.

During his two shuttle missions Parise completed 406 orbits of the Earth, traveling 10.6 million miles. His total time in space is 614 hours, 13 minutes, and 56 seconds. Ron Parise was the 238th person to fly in space.

Ohio Astronaut #12: Kenneth D. Cameron

STS-37 (Space Shuttle *Atlantis*)
STS-56 (Space Shuttle *Discovery*)
STS-74 (Space Shuttle *Atlantis*)

First Space Shuttle Mission to Carry a Space Station (*Mir*) Module

Ken Cameron was born in Cleveland, Ohio, on November 29, 1949. He graduated from Rocky River High School in 1967 and then went on to the Massachusetts Institute of Technology, where he received a bachelor of science degree in astronautics and aeronautics in 1978. He went on to receive a master of science degree from MIT in 1979. In 1983, he graduated from the US Navy Test Pilot School and received a master of business administration degree from Michigan State University in 2002.

Prior to joining NASA Cameron had a distinguished career with the US Marine Corp, where he did extensive flying. He has over four thousand hours of flying time in forty-eight different types of aircraft.

In 1984, he was selected as a pilot in NASA's tenth astronaut group (*The Maggots*) along with fellow-Ohioan David Low. This was the first astronaut class that had two Ohioans (two Clevelanders no less!).

Cameron flew his first mission as pilot of space shuttle *Atlantis* on the STS-37 mission. Launching on April 5, 1991, Cameron and his crew deployed the Gamma Ray Observatory, which studied gamma ray sources throughout the universe. After a one day delay for weather, STS-37 landed on April 11, 1991, after a nearly six-day mission.

Cameron's next flight was as commander of the STS-56/Atlas-2 mission, aboard space shuttle *Discovery*. Atlas-2 (Atmospheric Laboratory for Applications and Science-2) was designed to collect data on the relationship between the sun's energy output and Earth's middle atmosphere and how these factors affect the ozone layer. STS-56 launched on April 8, 1993, and landed on April 17, 1993, at the Kennedy Space Center in Florida.

Cameron's third mission was as commander of the STS-74 mission aboard space shuttle *Atlantis*. This mission was the fourth in a series of US/Russian joint missions. It was the second docking of the space shuttle to the Russian space station *Mir* and was the first mission that used the space shuttle to launch and assemble a module to a space station (*Mir*). STS-74 delivered the Russian-built *Mir* Docking Module along with a new pair of solar arrays to help increase power generation aboard *Mir*. The delivered *Mir* Docking Module would be used by all subsequent space shuttles that docked with *Mir* in the Shuttle-*Mir* Program. STS-74 launched on November 12, 1995, and landed at the Kennedy Space Center in Florida on November 20, 1995.

Cameron left NASA on August 5, 1996, to join Hughes Training, Inc., a subsidiary of General Motors Corporation, as executive director of Houston Operations. In October 2003, Cameron returned to the space program, taking a position as principal engineer in the NASA Engineering & Safety Center, based at the NASA Langley Research Center, in Hampton, Virginia.

During his three spaceflights, Cameron completed 369 orbits of the Earth, traveling 9.8 million miles. His total time is space is 562 hours, 12 minutes, and 50 seconds. Ken Cameron was the 241st person to fly in space.

Ohio Astronaut #13: Gregory J. Harbaugh

STS-39 (Space Shuttle *Discovery*)
STS-54 (Space Shuttle *Endeavour*)
STS-71 (Space Shuttle *Atlantis*)
STS-82 (Space Shuttle *Discovery*)

First Ohioan to Visit the Russian Space Station *Mir*
First Ohioan to Fly to the Hubble Space Telescope
Member of the Crew for the United State's 100th Manned Spaceflight

Greg Harbaugh was born in Cleveland, Ohio, on April 15, 1956, and grew up in Willoughby, Ohio. He graduated from Willoughby South High School in 1974 and went on to receive a bachelor of science degree in astronautics and aeronautical engineering from Purdue University in 1978. He received a master of science degree in physical science from the University of Houston-Clear Lake in 1986.

Harbaugh joined NASA in 1978 and worked as a flight controller in Mission Control until he was selected to be a mission specialist astronaut in NASA's twelfth group in June 1987.

Harbaugh's first flight was aboard space shuttle *Discovery* on the STS-39 mission. This was an unclassified Department of Defense mission that involved research for the Strategic Defense Initiative (SDI). Harbaugh was responsible for operation of the Remote Manipulator System (robotic arm) on the space shuttle during the mission. Launching on April 28, 1991, they landed eight days later on May 6, 1991.

Harbaugh's second flight was aboard space shuttle *Endeavour*, which was a six-day mission that deployed the sixth Tracking and Data Relay Satellite (TDRS). Launching on January 13, 1993, Harbaugh also performed his first spacewalk of 4.5 hours before landing on January 19, 1993.

Harbaugh's third mission was aboard space shuttle *Atlantis* on the STS-71 mission. This flight was the first docking of a space shuttle to the Russian space station and involved an exchange of crews. Harbaugh was responsible for the docking system that allowed the space shuttle and *Mir* to dock.

Harbaugh's fourth and final flight was aboard space shuttle *Discovery* on the STS-82 mission, which was the second Hubble Space Telescope servicing mission. During this mission Harbaugh made two of the five spacewalks that replaced instruments and made repairs to the Hubble Space Telescope. His total time for spacewalks during this mission was 14 hours and 1 minute. STS-82 launched on February 11, 1997, and returned on February 21, 1997. Both launch and landing occurred at night.

During his four missions, Harbaugh completed 532 orbits of the Earth, traveling 16.6 million miles. His total time in space is 818 hours and 2 minutes. Greg Harbaugh was the 245th person to fly in space.

Ohio Astronaut #14: Terence T. 'Tom' Henricks

STS-44 (Space Shuttle *Atlantis*)
STS-55 (Space Shuttle *Columbia*)
STS-70 (Space Shuttle *Discovery*)
STS-78 (Space Shuttle *Columbia*)

First Ohioan to Fly on Three Different
  Space Shuttles
First Spaceflight with Two Ohioans Onboard:
  STS-44 (with Tom Hennen)
Member of the All-Ohio Space Shuttle Mission:
  STS-70
First Spaceflight with Five Ohioans Onboard:
  STS-70

Tom Henricks was born on July 5, 1952 in Bryan, Ohio, but considers Woodville, Ohio to be his hometown. He is considered 'Astronaut #1' in the city of Toledo. Henricks graduated from Woodmore High School in 1970 and then entered the US Air Force Academy, where he received a bachelor of science degree in civil engineering in 1974.

Additionally, he holds a master of science degree in public administration from Golden Gate University. In 1993, he was awarded an honorary doctor of science degree from Defiance College located in Defiance, Ohio.

Henricks was selected in the eleventh astronaut group in June 1985, the only Ohioan selected in this class. His first flight was as the pilot on space shuttle *Atlantis* for the STS-44 mission. Launching at night on November 24, 1991, the main objective of this mission was to deploy the Defense Support Program (DSP) satellite with an Inertial Upper Stage. After 110 orbits of the Earth, *Atlantis* landed at Edwards Air Force base in California on December 1, 1991.

Henrick's second flight was pilot of space shuttle *Columbia* for the STS-55 German D-2 Spacelab mission. Carrying a compliment of eighty-nine German and other international experiments, STS-55 launched on April 26, 1993. After their ten-day mission, they successfully landed at Edwards Air Force base in California on May 6, 1993.

STS-70 was Henricks' third flight and was his first as space shuttle commander. Launching aboard space shuttle *Discovery* on July 13, 1995, Henricks and his 'All-Ohio' crew successfully deployed the final Tracking and Data Relay Satellite in the space shuttle program. After 142 orbits, Henricks landed *Discovery* on July 22, 1995, at the Kennedy Space Center in Florida, bringing great pride to the Great State of Ohio for the accomplishments of this mission.

Henricks' fourth and final flight was commander of the STS-78 Life and Microgravity Sciences Spacelab mission aboard space shuttle *Columbia*. Launching on June 20, 1996, Henricks and his crew completed experiments from ten different nations before landing at the Kennedy Space Center in Florida on July 7, 1996. The pilot for this flight was once again one of NASA's more famous shuttle pilots, Honorary Ohioan Kevin Kregel. This mission will go down as the second longest mission of the space shuttle program, lasting 16 days, 21 hours, and 48 minutes. It served as a model for future studies performed onboard the International Space Station years later.

During his four flights, Henricks completed 684 orbits of the Earth, traveling 17.8 million miles. His total time in space is 1026 hours, 39 minutes, and 18 seconds. Henricks became the first person to log over 1,000 hours as a space shuttle pilot/commander. Tom Henricks was the 259th person to fly in space.

Ohio Astronaut #15: Thomas J. Hennen

STS-44 (Space Shuttle *Atlantis*)

First Spaceflight With Two Ohioans Onboard: STS-44 (with Tom Henricks)
First and Only Warrant Officer (US Army) to Fly in Space

Tom Hennen was born on August 17, 1952, in Albany, Georgia, but considers Columbus, Ohio, to be his hometown. He graduated from Groveport-Madison High School in Groveport, Ohio, in 1970 and then attended Urbana College in Urbana, Ohio, from 1970–72, before joining the US Army.

Hennen was selected as a payload specialist astronaut in August 1989 and assigned to the Terra Scout experiment that was scheduled to fly on STS-44. Launching aboard space shuttle *Atlantis* on November 24, 1991, along with fellow-Ohioan Tom Henricks, Hennen became the first warrant officer to fly in space. Landing took place on December 1, 1991.

Hennen retired from the US Army in December 1995 and is cofounder and executive director of the Atlantis Foundation, a nonprofit organization that is an advocate and service provider for people with various developmental disabilities.

During the STS-44 mission, Hennen completed 110 orbits of the Earth, traveling 2.9 million miles. His total time in space is 166 hours, 50 minutes, and 44 seconds. Tom Hennen was the 262nd person to fly in space.

Ohio Astronaut #16: Nancy J. Currie

STS-57 (Space Shuttle *Endeavour*)
STS-70 (Space Shuttle *Discovery*)
STS-88 (Space Shuttle *Endeavour*)
STS-109 (Space Shuttle *Columbia*)

Member of the All-Ohio Space Shuttle Mission: STS-70
First Spaceflight with Five Ohioans Onboard: STS-70
First Ohioan to Visit the ISS

Nancy J. Currie was born in Wilmington, Delaware, on December 29, 1958, but proudly considers Troy, Ohio, to be her hometown. She graduated from Troy High School in 1977. Currie earned a bachelor of arts degree in biological science from Ohio State University in 1980. Currie also holds a master of science degree in safety engineering from the University of Southern California. In 1997, Currie received a doctorate degree in industrial engineering from the University of Houston.

After her graduation from Ohio State University, Currie served as a neuropathology research assistant at the OSU College of Medicine. In 1981, she became a second lieutenant in the US Army and attended Army Aviation School. Following her flight training, Currie became a helicopter flight instructor and a senior Army aviator. She has logged over four thousand flying hours in a variety of rotary-wing and fixed-wing aircraft.

Currie went to work at the Johnson Space Center in 1987 as an engineer on the flight simulator for the Shuttle Training Aircraft. She was selected in NASA's thirteenth group of astronauts in January 1990, along with fellow-Ohioans Ron Sega, Carl Walz, and Don Thomas. This was the first astronaut group selected with four Ohioans.

Currie's first space flight was STS-57 aboard space shuttle *Endeavour*, where she was assigned with fellow Ohio astronaut David Low. Launch occurred on June 21, 1993, and the main objective of the mission was the retrieval of the European Retrievable Carrier Satellite (EURECA). The mission also featured the first flight of the commercial science module called Spacehab, where Currie took part in many experiments in materials and life sciences. Currie was also the prime operator for the shuttle robotic arm that was used extensively during the flight. Landing took place on July 1, 1993.

Currie's second flight was aboard space shuttle *Discovery* as part of the historic STS-70 All-Ohio Shuttle Mission. STS-70 launched on July 13, 1995, and the main objective of the mission was to deploy the final Tracking and Data Relay Satellite (TDRS) from the space shuttle. The TDRS deployed on STS-70 was the replacement satellite for the one lost on space shuttle *Challenger* during the STS-51L mission in January 1986. The STS-70 crew successfully deployed the satellite and landed nine days later on July 22, 1995.

Currie's third flight was aboard space shuttle *Endeavour* on the STS-88 mission. This was the first of many space shuttle construction mission for the International Space Station. STS-88 launched on December 4, 1998, carrying the first space station node called *Unity*, where Currie was responsible for operation of the shuttle robotic arm that successfully joined the US Node to the Russian Functional Cargo Block module. After 185 orbits Currie and her crew landed back at the Kennedy Space Center on December 15, 1998.

Currie's final mission was on space shuttle *Columbia* on the STS-109 mission, which was the fourth servicing mission of the Hubble Space Telescope. Launch occurred on March 1, 2002. During the mission the crew replaced both solar arrays, the primary power control unit, and installed the Advanced Camera for Surveys and a scientific instrument cooling system. Currie's main role in the mission was primary operator of the shuttle robotic arm that assisted her spacewalking crew members in making all the necessary repairs and upgrades. After 165 orbits, *Columbia* landed on March 12, 2002.

During STS-109, Nancy Currie and fellow astronaut and Ohio State University alumnus Rick Linnehan carried four new Ohio state quarters depicting the Wright brothers' plane and Neil Armstrong on the Moon, which are now on display at the US Air Force Museum in Dayton, Ohio.

Years later while reflecting on her missions, Nancy emphasized that, "I am a tried and true Buckeye. I'm a product of the public school education in Ohio and I'm extremely proud of that" (Sims 2008).

During Currie's four missions she completed 648 orbits of the Earth, traveling 16.4 million miles. Her total time in space is 999 hours, 34 minutes, and 55 seconds. Nancy Currie was the 294th person to fly in space.

Ohio Astronaut #17: Carl E. Walz

STS-51 (Space Shuttle *Discovery*)
STS-65 (Space Shuttle *Columbia*)
STS-79 (Space Shuttle *Atlantis*)

ISS Expedition 4 (International Space Station)
 Launched on STS-108/
  Space Shuttle *Endeavour*
 Landed on STS-111/
  Space Shuttle *Endeavour*
First Spaceflight with two Clevelanders
 Onboard: STS-65 (with Don Thomas)
Only Ohioan to Fly on Six Different
 Spacecraft—*Discovery, Columbia, Atlantis, Endeavour,* ISS, and *Soyuz*
First Ohioan to Live on the International
 Space Station
Longest Duration Spaceflight of any Ohioan:
 195 days, 12 hours

Carl Walz was born on September 6, 1955, in Cleveland, Ohio. In 1973, he graduated from Brush High School in Lyndhurst, Ohio. He received his bachelor of science degree in physics from Kent State University in 1977 and a master of science degree in solid state physics in 1979 from John Carroll University in Cleveland.

Walz was selected in NASA's thirteenth group of astronauts in January 1990, along with fellow-Clevelanders Ron Sega and Don Thomas, and fellow-Ohioan Nancy Currie.

Walz's first flight was aboard space shuttle *Discovery* on the STS-51 mission launching on September 12, 1993. During the mission he helped deploy the Advanced Communications Technology Satellite (ACTS) that had been developed by the NASA Glenn Research Center. They also deployed a Shuttle Pallet Satellite (SPAS) that had both NASA and German investigations aboard. During the flight Walz made his first spacewalk which lasted seven hours. During the spacewalk he helped evaluate tools for future Hubble Space Telescope serving missions. STS-51 landed on September 22, 1993.

Walz's second flight was aboard space shuttle *Columbia* on the STS-65 mission, which flew the second International Microgravity Laboratory Spacelab module. Launching along with fellow-Clevelander Don Thomas on July 8, 1994, the crew performed nearly eighty different experiments from around the world, focusing on materials and life sciences. *Columbia* landed on July 23, 1994, and set a new flight duration record for the space shuttle (14 days, 17 hours, and 55 minutes).

Walz's third flight was aboard space shuttle *Endeavour* on the STS-79 mission. This mission was the fourth space shuttle mission to dock with the Russian space station *Mir*. During the mission, the *Atlantis/Mir* complex set a record for docked mass in space, a record that would last until surpassed by the International Space Station years later. STS-79 returned to Earth on September 26, 1996.

Walz's fourth and final flight was as a member of the Expedition 4 crew to the International Space Station. Launching aboard space shuttle *Endeavour* on the STS-108 mission on December 5, 2001, two days later he began his stay on the ISS, returning to Earth aboard space shuttle *Endeavour* once again on June 19, 2002. During Expedition 4 Watz completed two different spacewalks (one in a Russian spacesuit and the second in a US spacesuit). He also participated in a *Soyuz* relocation maneuver, separating from the ISS and flying from one docking port to another to make room for the arrival of a new *Soyuz* capsule. This maneuver lead to a record six different spacecraft that Walz had flown in space. On June 19, 2002, he returned to Earth on STS-111, landing at Edwards Air Force Base in California, shattering all previous flight duration records held by Ohio astronauts.

One of Walz's most memorable moments from his missions was during a spacewalk he performed outside the ISS during Expedition 4 on February 20, 2002 (the fortieth anniversary of Glenn's flight aboard *Friendship 7*), when he got to talk with one of his childhood heroes.

> We stopped in the middle of the spacewalk to do a telephone conversation with John Glenn to congratulate him on his anniversary. It was really cool to be outside the space station working and then have the chance to talk to John Glenn and reflect on his accomplishments and how the early pioneers had led to where we were when we were building the International Space Station. (Sims 2008)

During Walz's four flights he completed 3,620 orbits of the Earth, traveling 94.9 million miles. His total time in space is 5,525 hours, 3 minutes, and 52 seconds. Carl Walz was the 301st person to fly in space.

Ohio Astronaut #18: Ron Sega

STS-60 (Space Shuttle *Discovery*)
STS-76 (Space Shuttle *Atlantis*)

First Ohioan to Fly with a Russian Cosmonaut:
    STS-60 (Sergei Krikalov)
The First Astronaut of Slovenian Heritage to
    Fly in Space

Ron Sega was born in Cleveland, Ohio, on December 4, 1952, and graduated from Nordonia High School in Macedonia in 1970. He received a bachelor of science degree in mathematics and physics from the US Air Force Academy in 1974 and a master of science degree in physics from Ohio State University in 1975. In 1982 he received a doctorate in electrical engineering from the University of Colorado.

During his career, Sega logged over four thousand hours flying jet aircraft in the Air Force, Air Force Reserves, and with NASA.

Sega was selected in NASA's thirteenth group of astronauts in January 1990, along with fellow-Clevelanders Don Thomas and Carl Walz, and fellow-Ohioan Nancy Currie.

Sega's first flight was aboard space shuttle *Discovery* on the STS-60 mission. This mission was the first joint US/Russian space shuttle mission, with cosmonaut Sergei Krikalev a member of the STS-60 crew. It was the second flight of the commercial Spacehab science module and the first flight of the Wake Shield Facility (WSF), designed to investigate the formation of thin films under ultra-high vacuum conditions. Launching on February 3, 1994, the crew orbited the Earth 130 times before landing on February 11, 1994.

Sega's second flight was aboard space shuttle *Atlantis* on the STS-76 mission, which featured the third docking of the space shuttle with the Russian space station *Mir*. During STS-76, Sega operated many experiments within the Spacehab module, including numerous operations with the Biorack facility for fundamental biology experiments. Sega and the rest of the crew successfully transferred 4,800 pounds of science and mission hardware, food, water, and air to *Mir* and returned nearly 1,100 pounds of European science and Russian hardware. After 144 orbits, STS-76 landed on March 31, 1996.

Following his years at NASA, Sega went on to become an Air Force major general and the under secretary of the Air Force.

During his two missions Sega completed 275 orbits of the Earth, traveling 7.2 million miles. His total time in space is 420 hours, 26 minutes, and 10 seconds. Ron Sega was the 307th person to fly in space.

Ohio Astronaut #19: Donald A. Thomas

STS-65 (Space Shuttle *Columbia*)
STS-70 (Space Shuttle *Discovery*)
STS-83 (Space Shuttle *Columbia*)
STS-94 (Space Shuttle *Columbia*)

First Spaceflight with two Clevelanders
  Onboard: STS-65 (with Carl Walz)
Member of the All-Ohio Space Shuttle Mission:
  STS-70
First Spaceflight with Five Ohioans Onboard:
  STS-70
Shortest Duration Between Spaceflights:
  88 days
Most Spaceflights Flown in the Shortest Time:
  4 flights in 3.1 years

Don Thomas was born on May 6, 1955, in Cleveland, Ohio. He grew up in Independence and Cleveland Heights and graduated from Cleveland Heights High School in 1973. He received a bachelor of science degree in physics from Case Western Reserve University in 1977 and a master of science and doctorate in materials science and engineering from Cornell University in 1980 and 1982 respectively.

From 1982–87, Thomas was a senior member of the technical staff at AT&T Bell Laboratories in Princeton, New Jersey. He then joined the NASA Johnson Space Center in Houston, working as a materials engineer for the space shuttle program.

Thomas was selected in NASA's thirteenth group of astronauts in January 1990, along with fellow-Clevelanders Ron Sega and Carl Walz, and fellow-Ohioan Nancy Currie.

Thomas's first mission was aboard space shuttle *Columbia* on STS-65, which flew the second International Microgravity Laboratory Spacelab module. Launching along with fellow-Clevelander Carl Walz on July 8, 1994, the crew performed nearly eighty different experiments from around the world, focusing on materials and life sciences. *Columbia* landed on July 23, 1994, and set a new flight duration record for the space shuttle (14 days, 17 hours, and 55 minutes).

Thomas's second flight was aboard space shuttle *Discovery* as part of the historic STS-70 All-Ohio Shuttle Mission. STS-70 launched on July 13, 1995, and deployed the final Tracking and Data Relay Satellite (TDRS) from the space shuttle. Landing took place at KSC nine days later on July 22, 1995.

Thomas's third flight was aboard space shuttle *Columbia* on the STS-83 mission. This dedicated Spacelab science mission, called Microgravity Science Laboratory, featured a wide range of microgravity experiments from around the world. Launching on April 4, 1997, along with fellow-Ohioan Mike Gernhardt, a problem with one of *Columbia*'s three fuel cells developed very soon into the mission. With uncertainty about whether it might catch fire or explode, Mission Control instructed the crew to shut it down and to start preparing for an early return to Earth. A safe landing was made on April 8, 1997, after only four days in space.

Almost immediately after landing, NASA started to analyze the fuel cell problem and planned to refly space shuttle *Columbia* and the STS-83 crew once again as soon as feasible. The mission was renamed STS-94 and was nearly identical to the originally planned STS-83 mission. After a record turn-around of *Columbia*, STS-94 launched Thomas on his fourth flight on July 1, 1997. This mission flew eleven different experiments and facilities that were developed at the NASA Glenn Research Center. After 251 orbits of the Earth, STS-94 landed on July 17, 1997, at the Kennedy Space Center in Florida. The eighty-eight days between the STS-83 landing and the STS-94 launch remains the record for the shortest period between flights in space for this seven member crew.

During Thomas's four missions he completed 692 orbits of the Earth, traveling 17.5 million miles. His total time in space is 1040 hours, 14 minutes, and 12 seconds. He has flown with more Ohioans (six) than any other person in the history of space travel and flew with fellow Ohio astronauts on every one of his four flights. He was the first astronaut to take the famous Stadium Mustard into space and the first to fly Tony Packo's hot dog sauce, a local Toledo delicacy. He is also credited with taking the first pizza (pepperoni) into space on STS-83. Don Thomas was the 312th person to fly in space.

Ohio Astronaut #20: Kevin Kregel

STS-70 (Space Shuttle *Discovery*)
STS-78 (Space Shuttle *Columbia*)
STS-87 (Space Shuttle *Columbia*)
STS-99 (Space Shuttle *Endeavour*)

First Astronaut to Become an Honorary Ohioan Member of the All-Ohio Space Shuttle Mission: STS-70
First Spaceflight with Five Ohioans Onboard: STS-70

Kevin Kregel was born in Amityville, New York, (home of the Amityville Horror) on September 16, 1956, and became an honorary Ohioan on May 5, 1995, by official proclamation of Ohio Governor George Voinovich. Kregel graduated from Amityville Memorial High School in 1974 and then attended the US Air Force Academy, where he received a bachelor of science degree in astronautical engineering in 1978. In 1988, he received a master's degree in public administration from Troy State University.

In 1990, Kregel joined the NASA Johnson Space Center in Houston, Texas, as an aerospace engineer and instructor pilot for the NASA astronauts. He has logged more than five thousand flight hours in more than thirty different aircraft.

Kregel was selected in NASA's fourteenth group of astronauts in 1992, along with Ohioans Mary Ellen Weber and Mike Gernhardt.

Kregel's first flight was as pilot on the notable STS-70 mission, the All-Ohio Space Shuttle Mission. Launching on July 13, 1995, the

STS-70 crew successfully deployed the final Tracking and Data Relay Satellite from the shuttle before landing nine days later on July 22, 1995.

Kregel's second flight was aboard space shuttle *Columbia*, where he was paired up once again with Commander Tom Henricks, a fellow-Ohioan, on STS-78. This flight was a dedicated Spacelab science mission called the Life and Microgravity Sciences mission. Launching on June 20, 1996, Kregel and the crew completed experiments from ten different nations before landing at the Kennedy Space Center in Florida on July 7, 1995. This mission will go down as the second longest mission of the space shuttle program, lasting 16 days, 21 hours, and 48 minutes. It served as a model for conducting scientific research onboard the International Space Station years later.

Kregels' third flight was as commander of the STS-87 mission aboard space shuttle *Columbia*, and launched on November 19, 1997. This flight was the fourth US Microgravity Payload flight, involving a large group of materials science experiments that were mounted on a pallet in the payload bay of the shuttle. A solar observation satellite called Spartan was also released and later retrieved on this flight. Two members of the crew conducted a spacewalk during which they captured the Spartan satellite by hand and tested out tools for future use on the International Space Station. After 252 orbits they landed on December 5, 1997.

Kregel's fourth and final flight was as commander aboard space shuttle *Endeavour* on the STS-99 mission. This was the Shuttle Radar Topography mission and the objective of the flight was to use radar to map the Earth's surface. Launching on February 11, 2000, the crew collected nearly one trillion measurements of Earth's topography before landing on February 22, 2000.

During Kregel's four missions he completed 847 orbits of the Earth, traveling 21.3 million miles. His total time in space is 1266 hours, 23 minutes, and 17 seconds. Kevin Kregel was the 328th person to fly in space.

Ohio Astronaut #21: Mary Ellen Weber

STS-70 (Space Shuttle *Discovery*)
STS-101 (Space Shuttle *Atlantis*)

Member of the All-Ohio Space Shuttle Mission: STS-70
First Spaceflight with Five Ohioans Onboard: STS-70

Mary Ellen Weber was born in Cleveland, Ohio, on August 24, 1962, but considers Bedford, Ohio, as her hometown. She graduated from Bedford High School in 1980 and then went on to Purdue University, where she received a bachelor of science degree in chemical engineering in 1984. In 1988, she received a doctorate degree in physical chemistry from the University of California. She also received a master of business administration from Southern Methodist University in 2002.

Weber was selected in NASA's fourteenth group of astronauts in 1992, along with fellow-Ohioans Mike Gernhardt and Kevin Kregel.

Weber's first flight was on the notable STS-70 mission, the All-Ohio Space Shuttle Mission, aboard space shuttle *Discovery*. Launching on July 13, 1995, the STS-70 crew successfully deployed the final Tracking and Data Relay Satellite from the shuttle before landing nine days later on July 22, 1995.

Weber's second and final flight was aboard space shuttle *Atlantis* on the STS-101 mission, which launched on May 19, 2000. This flight was the third space shuttle mission devoted to construction of the International Space Station. Weber operated the shuttle robotic arm during a spacewalk and helped transfer over three thousand pounds of supplies to the ISS. After 155 orbits, STS-101 landed on May 29, 2000.

Weber is an avid skydiver and has logged over five thousand jumps. She is a thirteen-time medalist at the US National Skydiving Championships, and was on a world record skydive for the largest freefall formation with three hundred skydivers in 2002.

During Weber's two missions she completed 298 orbits of the Earth, traveling 7.8 million miles. Her total time in space is 451 hours, 30 minutes, and 15 seconds. Mary Ellen Weber was the 329th person to fly in space.

Ohio Astronaut #22: Michael L. Gernhardt

STS-69 (Space Shuttle *Endeavour*)
STS-83 (Space Shuttle *Columbia*)
STS-94 (Space Shuttle *Columbia*)
STS-104 (Space Shuttle *Atlantis*)

First US Spacewalk from the ISS

Mike Gernhardt was born in Mansfield, Ohio, on May 4, 1956. He graduated from Malabar High School in Mansfield in 1974. He received a bachelor of science degree in physics from Vanderbilt University in 1978 and a master of science degree and a doctorate in bioengineering from the University of Pennsylvania in 1983 and 1991, respectively.

Gernhardt was selected in NASA's fourteenth group of astronauts in 1992, along with fellow Ohioans Mary Ellen Weber and Kevin Kregel.

Gernhardt's first flight was aboard space shuttle *Endeavour* on the STS-69 mission. Launching on September 7, 1995, the main objectives of the mission were to deploy and later retrieve a Spartan satellite and the Wake Shield Facility (WSF), which was designed to process ultra-pure semiconductor materials under ultra-vacuum conditions. Gernhardt performed his first spacewalk on this flight, lasting 6 hours and 46 minutes. During the spacewalk he helped evaluate future space station tools and hardware. Landing occurred on September 18, 1995.

Gernhardt's second flight was aboard space shuttle *Columbia*, along with fellow-Ohioan Don Thomas, on the STS-83 mission. This dedicated Spacelab science mission called Microgravity Science Laboratory featured a wide range of microgravity experiments from around the world. Launching on April 4, 1997, a problem with one of *Columbia*'s three fuel cells developed very soon into the mission. With uncertainty about whether it might catch fire or explode, Mission Control instructed the crew to shut it down and to start preparing for an early return to Earth. A safe landing was made on April 8, 1997, after only four days in space.

Almost immediately after landing, NASA started to analyze the fuel cell problem and planned to refly space shuttle *Columbia* and the STS-83 crew once again as soon as feasible. The mission was renamed STS-94 and was nearly identical to the originally planned STS-83 mission. After a record turn-around of *Columbia*, STS-94 launched Gernhardt on his fourth flight on July 1, 1997. This mission flew eleven different experiments and facilities that were developed at the NASA Glenn Research Center. After 251 orbits of the Earth, STS-94 landed on July 17, 1997, at the Kennedy Space Center in Florida. The eighty-eight days between the STS-83 landing and the STS-94 launch remains the record for the shortest period between flights in space for this seven member crew.

Gernhardt's fourth flight was aboard space shuttle *Atlantis* on the STS-104 mission. Launching on July 12, 2001, STS-104 was the tenth space shuttle mission to the International Space Station. One of the main objectives of this mission was to deliver and install a joint airlock capable of servicing both US and Russian spacesuits. During the mission Gernhardt led three spacewalks to install the airlock, named Quest, including the first US spacewalk from the ISS. After completing two hundred orbits, space shuttle *Atlantis* and the STS-104 crew successfully landed on July 24, 2001.

Gernhardt is an experienced deep sea diver and has logged over seven hundred deep sea dives.

During his four missions Gernhardt completed 685 orbits of the Earth, traveling 17.5 million miles. His total time in space is 1039 hours, 6 minutes, and 42 seconds. Mike Gernhardt was the 332nd person to fly in space.

Ohio Astronaut #23: Sunita Williams

ISS Expedition 14/15
  Launched on STS-116 (Space Shuttle *Discovery*)
  Landed on STS-117 (Space Shuttle *Atlantis*)
ISS Expedition 32/33
  Launched/Landed on *Soyuz* TMA-05M

Most Time in Space by an Ohioan: 322 days
Greatest Distance Traveled in Space by an Ohioan: 131 million miles
Greatest Number of Orbits Around the Earth by an Ohioan: 5,147
Greatest Number of Spacewalks by an Ohioan: 7
Most Time Spacewalking by a Female: 50 hours 40 minutes
Longest Duration in Space for a Female Ohioan: 194 days 18 hours
Completed First Marathon in Space (2007)
Completed First Triathlon in Space (2012)

Sunita Williams was born in Euclid, Ohio, on September 19, 1965, and moved to Needham, Massachusetts, at a very young age. There she attended Hillside Elementary and Newman Middle Schools and graduated from Needham High School in 1983.

After an eleven-year career in the US Navy, Williams was selected in NASA's seventeenth astronaut group on June 4, 1998, along with fellow-Ohioans Mike Foreman and Greg Johnson.

Williams's first flight was as the flight engineer aboard the International Space Station, overlapping Expeditions 14 and 15. She launched to the ISS aboard space shuttle *Discovery* on the STS-116 mission on December 9, 2006, and arrived at the ISS two days later.

As a member of the Expedition 14 crew Williams established a world record for females with four spacewalks, totaling 29 hours and 17 minutes outside the ISS. An accomplished marathoner, she was the first astronaut to ever run the Boston Marathon from space. Wearing official bib #14,000, she was among 24,000 runners participating in the 2007 event. She circled the Earth nearly three times during the marathon, averaging nearly 17,503 miles per hour in the process. Traveling 5 miles per second aboard the ISS, she traveled a total distance of 76,000 miles during the race, while her Earth-based competition covered only the traditional 26.2 miles. Her official time for the marathon was 4 hours, 24 minutes.

During her first mission to the International Space Station, Williams completed 3,116 orbits of the Earth, traveling 77.9 million miles. During this mission she broke the previous world record for a female in space, spending nearly 195 days aboard the ISS.

On July 14, 2012 Williams launched from the Baikonur Cosmodrome for her second long duration stay aboard the ISS, during which she conducted numerous experiments and performed three spacewalks. She now holds the record for total cumulative time performing spacewalks by a female astronaut at 50 hours and 40 minutes.

During her second flight in September 2012, Williams became the first person to do a triathlon in space, the Nautica Malibu Triathlon. During the triathlon she ran on the treadmill and biked on the ergometer aboard ISS. For the swimming portion, she used the Advanced Resistive Exercise Device to approximate swimming in microgravity. Her time for the event was 1 hour, 48 minutes, and 33 seconds.

During her two missions Williams completed 5,147 orbits of the Earth, traveling 131.3 million miles. Her total time in space is 7,721 hours, 15 minutes, and 23 seconds. She ranks sixth among all US astronauts for longest time in space and second among all women. She was the second woman of Indian descent to fly in space. Sunita Williams was the 454th person to fly in space.

Ohio Astronaut #24: Gregory H. Johnson

STS-123 (Space Shuttle *Endeavour*)
STS-134 (Space Shuttle *Endeavour*)

Last Ohioan to fly on the Space Shuttle: STS-134

Greg Johnson was born in South Ruislip, United Kingdom, but considers Fairborn, Ohio, and Traverse City, Michigan, as his hometowns. He graduated from Park Hills High School in Fairborn, Ohio, in 1980. He received his bachelor of science degree in aeronautical engineering from the US Air Force Academy in 1984. He received a master of science degree in flight structures engineering from Columbia University in 1985, and went on to receive a master of business administration degree from the University of Texas in 2005.

After eighteen years flying in the US Air Force, where he flew in both Operation Desert Storm and Operation Southern Watch, he was selected in NASA's seventeenth class of astronauts in 1998, along with fellow-Ohioans Mike Foreman and Sunita Williams.

Johnson is an experienced Air Force pilot with over five thousand flight hours in more than fifty different aircraft. He retired from the Air Force in 2009.

Johnson's first flight was as the pilot of *Endeavour* on the STS-123 mission, along with fellow Ohioan Mike Foreman. Launching on March 11, 2008, Johnson and his crew delivered the Japanese Experiment Logistics Module, the pressurized portion of Japan's Kibo Laboratory, to the ISS. Johnson was the primary robotic arm operator during the flight, which saw a record five spacewalks while docked to the ISS. Landing occurred on March 26, 2008. During the flight he completed 250 orbits of the Earth, traveling 6.5 million miles in 15 days, 18 hours, 12 minutes, and 27 seconds.

Johnson's second flight was also as pilot aboard space shuttle *Endeavour*, on the STS-134 mission, launching on May 16, 2011. This mission to the ISS was the final flight of *Endeavour* and the second last of the space shuttle program. During the mission, Johnson and his crew mates delivered and installed the Alpha Magnetic Spectrometer experiment to the ISS, where once again he was the lead robotics arm operator. The mission included four spacewalks before landing on June 1, 2011. On this flight he completed 248 orbits of the Earth, traveling 6.5 million miles in 15 days, 17 hours, 38 minutes, and 23 seconds.

During Johnson's two missions he completed 498 orbits of the Earth, traveling over 13 million miles. His total time in space is 755 hours, 50 minutes, and 50 seconds. Greg Johnson was the 471st person to fly in space and was the last Ohioan ever to have flown on the space shuttle.

Ohio Astronaut #25: Michael J. Foreman

STS-123 (Space Shuttle *Endeavour*)
STS-129 (Space Shuttle *Atlantis*)

Mike Foreman was born on March 29, 1957, in Columbus, Ohio, and considers Wadsworth, Ohio, to be his hometown. He graduated from Wadsworth High School in 1975, and then received a bachelor of science degree in aerospace engineering from the US Naval Academy in 1979. He also received a master of science degree in aeronautical engineering from the US Naval Postgraduate School in 1986.

An experienced naval aviator, he has logged more than five thousand hours in more than fifty different aircraft.

Foreman was selected in NASA's seventeenth astronaut group on June 4, 1998, along with fellow-Ohioans Sunita Williams and Greg Johnson.

Foreman's first flight was aboard space shuttle *Endeavour* on the STS-123 mission. Launched at night on March 11, 2008, this was the twenty-fifth space shuttle ISS assembly mission. During the mission they delivered and installed the Japanese Experiment Logistics

Module, the first pressurized component of the Japanese Space Agency's Kibo science laboratory, one of the ISS's major science laboratories. During the mission, Foreman performed three spacewalks for a total of 19 hours and 34 minutes. Landing occurred at night on March 26, 2008.

Foreman's second flight was aboard space shuttle *Atlantis* on the STS-129 mission, which was the thirty-first space shuttle ISS assembly mission. During the flight they delivered nearly thirty thousand pounds of equipment and hardware to the ISS. During the mission Foreman performed two spacewalks, for a total of 12 hours and 45 minutes.

During Foreman's two missions he completed 421 orbits of the Earth, traveling 11.1 million miles. His total time in space is 637 hours, 28 minutes, and 40 seconds. Mike Foreman was the 473rd person to fly in space.

Ohio Astronaut #26: Michael T. Good

STS-125 (Space Shuttle *Atlantis*)
STS-132 (Space Shuttle *Atlantis*)

Mike Good was born in Parma, Ohio, on October 13, 1962, but considers Broadview Heights, Ohio, to be his hometown. He graduated from Brecksville-Broadview Heights High School in 1980, and then received a bachelor of science degree and master of science degree in aeronautical engineering from the University of Notre Dame in 1984 and 1986, respectively.

Good is an experience Air Force aviator with over 2,650 flight hours in thirty different aircraft.

Good was selected in NASA's eighteenth group of astronauts on July 26, 2000.

Good's first flight was aboard space shuttle *Atlantis* on the STS-125 mission, which was the fifth and final Hubble Space Telescope servicing mission. Adding new instruments, batteries, gyroscopes, computer, and outer thermal blankets were the main tasks of this mission, which were accomplished during five different spacewalks. Good performed two of the spacewalks, with a total time outside *Atlantis* of 15 hours and 58 minutes, becoming the ninth Ohioan to perform a spacewalk. After nearly thirteen days in space, STS-125 landed on May 24, 2009, completing 197 orbits of the Earth and traveling 5.2 million miles.

Good's second flight was aboard space shuttle *Atlantis* on the STS-132 mission. Launching on May 14, 2010, Good and his crew delivered an Integrated Cargo Carrier and a Russian-built Mini Research Module to the International Space Station. During seven days of docked operations, three spacewalks were conducted with Good logging 13 hours and 55 minutes during his two spacewalks. On the second spacewalk, Good and astronaut Stephen Bowen replaced batteries on the P6 Truss that stores solar energy. On the final spacewalk of the mission, Good and astronaut Garrett Reisman replaced the last of the P6 Truss batteries and retrieved a power data grapple fixture for installation at a later date. The STS-132 mission was completed in 186 orbits, traveling 4.8 million miles in 11 days, 18 hours, 28 minutes and 2 seconds.

During Good's two missions he completed 421 orbits of the Earth, traveling 11.1 million miles. His total time in space is 592 hours, 6 minutes, and 18 seconds. Mike Good was the 495th person to fly in space.

References

Burgess, Colin and Kate Doolan. 2003. *Fallen Astronauts*. Lincoln, NE: University of Nebraska Press.

Sims, Damon. 2008. "Astronauts with Ohio ties speak at NASA 50th anniversary." *Plain Dealer*, August 29. http://blog.cleveland.com/metro/2008/08/theyre_a_little_awestruck_by.html.

Index

A
Abbey, George, 63, 65, 113, 137, 314
airglow, 241–42
airlock, 172
Aldrin, Buzz, 49–50, 217
Alexander, Jean, 133, 148
Allen, Joe, 346
All-Ohio Crew, 1–2, 4, 7, 109–12
All-Ohio Space Shuttle Mission, 111–12, 126–27, 209, 349, 353. *See also* STS-70 mission
Amityville, 96, 111–12, 319–20
Apollo 7, 25
Apollo 8, 49
Apollo 11, 49–50, 216, 217
Apollo 13, 84
APUs. *See* auxiliary power units
Armstrong, Carol, 331–32
Armstrong, Neil, 49–50, 95, 126, 217, 303, 331–32, 355, 360–61
ASP. *See* astronaut support person
Astronaut Class of 1990, 75
Astronaut Crew Quarters, 8, 26–29, 62
astronauts. *See also* crew members; specific astronauts
    fifty-mile mark and, 89–90
    from Ohio, 1–2, 4, 7, 109–12, 259, 353–54, 355–56, 357–96
    quarantine of, 24–29
    selection of, 54–64, 71–73
    training of, 8–11, 23–24, 30, 75–78, 87–88, 114–17, 121–22, 128–34, 329–30

astronaut ice cream, 195–96
Astronaut Selection Office, 59, 62, 71
astronaut selection process, 59–64
astronaut support person (ASP), 133, 148, 290
AT&T Bell Laboratories, 55, 65, 67
Atlantis, 11, 96, 109, 125, 335, 340, 348
auxiliary power units (APUs), 70, 257, 278, 288

B
Baker, Ellen, 298
Balettie, Roger, 350
basal energy expenditure (BEE), 195
Bassett, Charles A., II, 361–63
bathing in space, 188–89
BDS. *See* Bioreactor Development System
beach house, 34–37, 130–31
BEE. *See* basal energy expenditure
Bioreactor Development System (BDS), 220–21, 224, 251
Bird Investigation Review and Deterrent (BIRD) team, 17–18, 354
Bird Scare, 17
Bird-X, 17
Blackman, Buddy, 20
Block I engine, 106–7
Boeing Space Center, 120
Bolden, Charlie, 346

Borman, Frank, 303
Brandenstein, Dan, 60, 70–71
Brown, Curt, 159, 164, 261, 269, 275, 276, 277
Brown, Dave, 333–34
Burge, Allen, 114, 302
Bursch, Dan, 149, 298
buzzards at KSC, 150–51

C

Cabana, Bob, 87, 89–91, 113
Cameron, Kenneth D., 100, 372–73
Campbell, Don, 111
Capcoms, 80, 84–86
Cape Canaveral, 169, 231, 275, 284, 314–17
Carpenter, Scott, 192
Case Western Reserve University, 53–54
Castner, Bud, 66, 68, 69, 71
Cenkar, Bob, 59
Chaffee, Roger, 48–49
Challenger, 102–3
Challenger accident, 59, 67, 103, 104, 141, 218, 324, 348
Chawla, K. C., 333
Chen, Phil, 298
Chiao, Leroy, 36, 86, 87, 93, 233–34
children of STS-70 crew, 149–50
CICs. *See* Crew Interface Coordinators
cities, from space, 240, 244
Clark, Laurel, 334
Claussen, Harland, 210–12
Cleveland, Ohio, 47–53, 100–101
Cleveland Air Show, 319
Cleveland Browns, 321
Cleveland Heights High School, 52–53
Cockerell, Ken, 328–29
Coleman, Cady, 330, 331
Collins, Eileen, 75, 329, 343
Collins, Michael, 49
Columbia, 54, 86, 88–91, 96, 177, 216, 292–93, 329, 332, 334, 336–38, 348
Comet Hale-Bopp, 332

communications satellite, 1, 102–4, 118–19, 217, 341–43
Cooper, Gordon, 192
Copeland, Al, 65–66
Cornell University, 54–55
cosmic ray flashes, 174
Crew Interface Coordinators (CICs), 85
crew notebooks, 29
crew patch, 112–14, 350
crew quarters, 8, 26–29, 62
Crew Rescue Sphere, 61–62
Crew Transfer Vehicle, 289, 292
Crippen, Bob, 292
Cronkite, Walter, 49
Crouch, Roger, 329
Cuda, Tony, 304
Culbertson, Frank, 346
Currie, Dave, 122
Currie, Nancy, 1, 3, 7, 75, 94, 121, 128, 165–66, 251, 285
  background of, 98–100
  career of, 378–80
  crew patch and, 112–13
  at first crew meeting, 107
  inspiration for being astronaut, 217–18
  interview with, 213, 224–25
  post-STS-70 mission, 339
  wedding of, 122
Cutler, Linda, 322

D

Dark Side of the Moon (Pink Floyd), 242–43, 256
Davis, Jim, 148
Davis, Trudy, 295
Dawson, Mike, 110
Decker, Walter, 217–18
deorbit burn, 275–83
Diller, George, 155, 344
Discovery, 1–5, 109, 348. *See also* STS-70 mission
  Block I engine, 106–7
  final flight of, 343–47
  move to launch pad, 123–25
  night viewing of, 40

Index 399

postlanding inspections, 323–24
prelaunch, 145–47, 149, 150
repairs and upgrades to, 325–26
rollout of, 19–24
STS-26 mission, 67
windows on, 247–48
woodpecker attack on, 8–18, 149, 164, 354–56, 349–50
Discovery Way, 353
Dragon capsule, 343
Dunbar, Bonnie, 70–71
Dwoskin, David, 319

E

Earth
　viewed at night, 239–44
　view from space, 227–45, 271
Eglin Air Force Base, 31, 32
Egypt, 237–38
EI. *See* entry interface
Eisele, Donn, 363–64
Elkind, Jerry, 101
El Lago, Texas, 303–4
Ellington Field, 300, 301
Endeavour, 81, 99, 125, 259, 325, 338, 339, 348
entry interface (EI), 280
ET. *See* external tank
EVA. *See* extra-vehicular activity
EVA training, 121–22
exercising, in space (ergometer), 187–89
Expedition 6, 334–35
external tank (ET), 13–16, 21, 81, 124, 134
extra-vehicular activity (EVA), 107, 121–22

F

family escorts for STS-70, 149
FDF. *See* Flight Data File
fifty-mile mark, 90–91, 159–60
Final Inspection Team, 149
fixed-base simulator, 135
Flight Control Room, 170, 257
Flight Data File (FDF), 191–93
flight lessons, 57–58

Florinef, 268–69, 271–72, 294–95, 309
Foale, Mike, 241
food, in space, 193–200, 226
Foreman, Michael J., 394–95
Freedom Star, 204
fresh food in space, 199–200
Friendship 7, 45
Friends of Students, 72–73

G

Gagarin, Yuri, 43
Gagarin Cosmonaut Training Center, 334
Gee, Gordon, 312
Gernhardt, Michael L., 329, 389–90
Gibson, Hoot, 92
Glenn, John, 44–46, 95, 118, 126, 127, 136–40, 192, 193, 207, 217, 228, 298, 343, 346, 354, 355, 357–58
Glenn Research Center, 353, 355
　See also Lewis Research Center
Global Tracking Network, 102
Goddard Space Flight Center, 104
gold astronaut pin, 314
Goldin, Dan, 113, 138, 340
Good, Michael T., 355, 395–96
Grantham, Rebecca, 96
Graubard, Robert, 114, 302
Great Barrier Reef, 235
Great Wall of China, 233–34
Grissom, Betty, 192
Grissom, Gus, 48–49, 192
Gross, John, 119
ground crew, 289–91
g-suit, 265, 279
Guantanamo Bay, Cuba, 230

H

Hackemack, Gene, 252
Hackerman Academy of Mathematics and Science, 336
Halsell, Jim, 87, 91, 329
Halvorson, Todd, 124, 127, 349
Harbaugh, Gregory J., 100, 374–75
Harrington, Jim, 131, 149, 293

Hart, Terry, 59, 65
Hartsfield, Hank, 303
Hauck, Rick, 346
Hawaiian Islands, 235–36
Hawley, Steve, 143, 145, 293
Heading Alignment Circle, 284
Health Stabilization Program, 25–26
Henize, Karl, 367–68
Hennen, Thomas J., 377
Henricks, Terence T. 'Tom', 1, 2, 4, 7, 82, 117, 119
   background of, 94–96
   career of, 375–77
   chat with veteran by, 210–12
   interviews with, 126–27, 214, 217, 219
   post-landing press conference, 299
   post-STS-70 mission, 336–37
   Toledo and, 322, 353
HERCULES camera, 184, 207–8, 224, 256
Herschitz, Roman, 59
Hickam, Homer, 85
Hieb, Rick, 86, 87, 89, 93
Holliman, John, 224–26
homecoming ceremony, 301–2
Honeycutt, Jay, 38, 131, 316
Houston, Texas, 231–32, 275, 300–305
Hubble Space Telescope, 1, 104, 105, 209, 339, 343
Huntoon, Carolyn, 39, 301–2, 314
hydrazine, 288

I

Ibarra, Ralph, 118
IMAX films, 200, 228
IML-2. *See* International Microgravity Laboratory
Inertial Upper Stage (IUS) rocket, 104–5, 120–21, 167
International Microgravity Laboratory (IML-2), 87, 90
International Space Station, 1, 69, 105, 249, 334–36, 339, 343, 348

Israel, 238
IUS. *See* Inertial Upper Stage rocket
Ivins, Marsha, 301

J

Jackson, Rich, 288
Jernigan, Tammy, 346
jet contrails, 236–37, 283
Jet Propulsion Laboratory, 79
Johnson, Gregory H., 393–94
Johnson Space Center (JSC), 8, 169
   crew quarters, 26–29
   debriefing, 310
   Glenn visit to, 136–40
   Spouse Day at, 135–36
   working at, 66–73
Jones, Tom, 167, 202, 249, 302, 346
JSC. *See* Johnson Space Center
Juvenile Court Volunteers, 72–73

K

Kandler, Max, 133, 144, 148
KC-135 zero-gravity aircraft, 81–83
Kelso, Rob, 176, 202–3, 263, 341
Kennedy Space Center, 1–3, 7, 12, 80, 231, 275
Klotz, Irene, 349
Kregel, Frances, 150
Kregel, Jeanne, 98
Kregel, Kevin, 1, 2, 7, 93–94, 115, 119, 266–67, 284, 356
   background of, 96–98
   career of, 386–87
   as honorary Ohioan, 109–11, 127, 246–47
   inspiration for being astronaut, 217
   interviews with, 217, 219
   during liftoff, 159–60
   post-STS-70 mission, 337, 338

L

Launch and Entry Suits (LES), 131–32, 134, 264–67
Launch Control Center (LCC), 133, 144, 149
launch-day cake, 142, 297

Index

launch invitations, 117–18
LCC. *See* Launch Control Center
LCVG. *See* liquid cooling ventilation garments
Leestma, Dave, 86, 113
Lehmann, Simone, 82–84, 93, 116–17, 136, 214–15, 223–24, 255, 295
LES. *See* Launch and Entry Suits
Letterman, David, 16
Lewis Research Center, 329. *See also* Glenn Research Center
Liberty Star, 204
Life magazine, 46, 47
Lindsey, Steve, 344, 345
Linteris, Greg, 329
liquid cooling ventilation garments (LCVG), 265
liquid hydrogen, 13, 155
liquid nitrogen, 149
liquid oxygen, 13, 149, 155
Littleton, Carolyn, 302, 304–5
Littleton, Frank, 302, 304–5
Lockheed Engineering and Science, 66–69
Lockheed Martin, 11, 341
Lovell, James A., 355, 359–60
Low, David, 100, 369–70

M

M-113 armored personnel carrier, 130
Magellanic Clouds, 242, 271
Manned Flight Awareness Program, 301, 315, 319–20
Manned Spacecraft Center, 169. *See also* Johnson Space Center
Marshall Space Flight Center, 55, 85, 88, 329
McCully, Mike, 293
meals
during quarantine, 27–28
in space, 200–201
Meals Ready to Eat (MREs), 197, 198
media, 77–78, 209–17, 224–26, 259, 288–89, 298–99

medical exams, 61–62, 71, 295–96, 308–9
medical experiments, 268–69, 271–72, 285
Meinert, Chris, 148
meteorites, 243
Microencapsulation in Space (MIS) experiment, 249–50, 256
Microgravity Disturbances Experiment, 71, 74, 333
Mike and Maty Show, 212–14
Mir space station, 11, 109, 209, 241
Mission Control, 74, 80, 84–86, 133, 166, 169–70, 187
Moon
landing, 49–50, 216, 217
orbit of, 49
return to, 217
Morley, Charles, W., Jr., 351–52
morning routine, in space, 179–83
motion-based simulator, 30, 135–38
Mount Everest, 234
Mount Rainier, 120
Mount St. Helens, 119–20
MREs. *See* Meals Ready to Eat
Mukai, Chiaki, 86, 87, 93
muscle atrophy, 187, 286–87
Musgrave, Story, 177, 216, 247, 253
music
on STS-70, 189–91, 242–43, 251, 256
wake-up, 177–78, 207, 216, 223, 246–47, 252–53, 256–57, 262–63, 349

N

NASA
application to, 55–56, 59–64
astronaut selection by, 54–64, 71–73
public affairs team, 209, 217
working at, 66–73
Nelson, Pinky, 64, 65
news media. *See* media
New Ulm, Minnesota, 321
New York, 298
night-before-launch parties, 39–40

nightscapes, 239–44
No Rinse Shampoo, 189

O

Oates, Jay, 330, 331
October Sky (Hickam), 85
Office of Safety and Mission
    Assurance, 339
Ohio
    astronauts from, 259, 353–56,
        357–96
    postflight visit to, 317–19
Ohio State Fair, 4, 297, 298–99,
    310–13
OMS. *See* Orbital Maneuvering
    System engines
one hundredth manned spaceflight,
    7, 11, 125
OPF. *See* Orbiter Processing Facility
Orbital Maneuvering System
    (OMS) engines, 275–76
Orbiter Processing Facility (OPF),
    123, 323
Orbiter Recovery Team, 277, 285,
    286, 287–88
orbiter test conductor (OTC), 152
O-ring problem, 298, 324–25
orthostatic intolerance, 268–69,
    295–96
Oswald, Steve, 269, 277–78
OTC. *See* orbiter test conductor
Overmyer, Bob, 364–65

P

Pad 39B, 39, 40
pantry, 199
Parise, Ronald A., 370–72
Pavlik, Emil, 111, 320
Payloads Branch, 328, 333–34
PCC. *See* Prime Crew Contact
Pepper, Larry, 259, 290
Personal Rescue System, 61–62
Peterson, Don, 303
Petty, Tom, 190–91
PFC. *See* Private Family Conference
Pickering, John, 72–73, 230
Pifer, Mark, 45

Pilot Science experiments, 249–50
pilot's license, 58
plasma layer, 279
plasma trail, 282
postflight appearances, 310–22
postflight crew report, 309–10
postflight period, 306–22
postflight presentation, 313–14
postlanding crew party, 307–8
postsleep activities, 179–83
practice countdown, 128–34
Predator Eyes, 17
preflight press conference, 125–28
prelaunch activities, 30–41, 141–54
press conferences, 125–28, 186–87,
    216–17, 259, 288–89, 298–99
Preston, Bill, 114, 302
prime crew, 31
Prime Crew Contact (PCC), 25
Private Family Conference (PFC),
    223–24
Probst, Ray, 152
P.S. I Listened to Your Heartbeat
    (Glenn), 139, 192, 193
public affairs team, 209, 217
Puddy, Don, 74
Put-in-Bay, 317–18

Q

quarantine, 7, 24–29, 33–34

R

radiation, 335
radio blackout, 279
rainforests, 232–33
Ramon, Illan, 334
range safety officer, 153
rat experiment, 184–85
recovery ships, 204–5
reentry procedures, 264–69,
    273–74, 278
Resnik, Judy, 55, 103, 356–66
Return to Launch Site (RTLS), 138
Rhodes, James, 312
Ride, Sally, 218
Rock and Roll Hall of Fame, 318–19
Rockwell Aerospace, 314–15

Index

Rockwell International facility, 79, 81
roll program, 156
RTLS. *See* Return to Launch Site
Runco, Mario, 298
Ruppucci, George, 118
Russia, 334–35

S

SAREX. *See* Shuttle Amateur Radio Experiment
SAS. *See* Space Adaptation Syndrome
satellites, 1, 102–4, 118–19, 217, 241, 341–43
Saturn V rockets, 20
Schirra, Wally, 49, 192
Schmidt, Harrison, 54
science astronauts, 52, 54
scientific experiments. *See* secondary experiments
"Script Ohio," 127–28
secondary experiments, 71, 74, 183–87, 220–21, 224, 249–50, 256, 268–69, 271–72, 333
Sega, Ron, 76, 100, 302, 382–83
shaving, 182
Shaw, Brewster, 288
Shepard, Alan B., 42–44, 192, 217, 336
Shepherd, Beth, 330–31
Shepherd, Bill, 85, 331
ship wakes, 238–39
shooting stars, 240–41
shrimp cocktail, 194–95
Shuttle Amateur Radio Experiment (SAREX), 105, 185–86, 208, 225, 250–51, 254–55, 258
Shuttle Landing Facility, 23, 32, 134, 256, 277, 284, 315
Shuttle Radar Topography, 338
shuttle rollout, 19–24
Shuttle Safety Review Panel (SSRP), 80–81
Shuttle Training Aircraft (STA), 115, 277–78
shuttle transporter, 20–21

Sieck, Bob, 38
Silver Snoopy Award, 315, 317
single system trainer (SST), 114
Skid Strip runway, 284, 300
Slayton, Deke, 192
sleeping, in space, 170–75
sleeping bags, 171
Smith, Steve, 38, 133, 148, 159, 290
solid rocket boosters (SRBs), 20, 123–24, 154–55, 158, 324–25
sonic boom, 282, 283
Southern Cross, 242, 271
Space Adaptation Syndrome (SAS), 162–63, 200, 202, 223
Space Day Rally, 314–17
Space Flight Medals, 313–14
Space Frontier Foundation, 16
space junk, 247–49
Spacelab, 84–85, 90, 96, 314, 336–38
space motion sickness, 162–63, 200, 202, 223. *See also* Space Adaptation Syndrome
space shuttle food, 193–200
space shuttle main engines (SSMEs), 81
space shuttle toilet, 179–80 *See also* waste containment system
space stations
    International Space Station, 1, 69, 105, 249, 334–36, 339, 343, 348
    Mir, 11, 109, 209, 241
    Skylab, 52, 53
Space Tissue Loss (STL) experiment, 224
spacewalk, 107, 168
SpaceX, 343
Special Flight Data File, 191–93
Spirit Lake, 119–20
Spouse Day, 135–36
Springer, Robert C., 368–69
SRBs. *See* solid rocket boosters
SRB separation, 158
SSMEs. *See* space shuttle main engines
SSRP. *See* Shuttle Safety Review Panel

SST. *See* single system trainer
STA. *See* Shuttle Training Aircraft
Stadium Mustard, 197, 200, 226, 247, 311, 319
stamp, commemorative, 352–53
Star City, 334–35
Stepaniak, Phil, 260
Still, Susan, 329
Stone, Randy, 219, 248, 263
STS-6 mission, 102–3
STS-26 mission, 67, 103
STS-29 mission, 103
STS-32 mission, 70–71, 74
STS-43 mission, 103
STS-47 mission, 84–85
STS-52 mission, 85–86
STS-53 mission, 87
STS-54 mission, 103
STS-55 mission, 82, 83–84
STS-57 mission, 99
STS-61C mission, 59
STS-65 mission, 19, 86–91, 143, 156
STS-69 mission, 125, 259, 325
STS-70 Execute Package, 178
STS-70 mission, 1–5
　as All-Ohio Space Shuttle Mission, 111–12, 126–27, 209, 349, 353
　artwork, 150
　assignment to, 92–94
　Block I engine, 106–7
　countdown to launch, 151–54
　crew members, 1–2, 4, 94–102, 109–12
　crew patch, 112–14
　Day 1, 161–75
　Day 2, 176–78, 202
　Day 3, 206–8
　Day 4, 216–22
　Day 5, 222–26
　Day 6, 246–51
　Day 7, 251–56
　Day 8, 256–61
　emergency training, 134
　end of, 321–22
　extra day on, 269–73
　food aboard, 193–200
　Landing Day, 262–69, 273–305
　Launch Day, 141–54
　launch delay, 8–12, 14
　launch invitations, 117–18
　legacy of, 349–56
　liftoff, 154–60
　mealtimes, 200–201
　morning routine, 179–83
　music on, 189–91, 242–43, 251, 256
　postflight period, 306–22
　practice countdown, 128–34
　preflight press conference, 125–28
　prelaunch activities, 141–54
　primary payload of, 102–5
　secondary experiments, 102–5, 183–87, 220–21, 224, 249–50, 256, 268–69, 271–72
　settling in, 183–87
　sleeping during, 170–75
　training for, 114–17, 121–22
　TSRS deployment, 164–68
　woodpecker jokes on, 176–78
STS-71 mission, 11, 125, 126, 209, 298, 324–25
STS-78 mission, 314, 336–37, 338
STS-82 mission, 343
STS-83 mission, 328–33
STS-87 mission, 338
STS-88 mission, 339
STS-94 mission, 291, 333
STS-95 mission, 343
STS-99 mission, 338
STS-101 mission, 340
STS-103 mission, 343
STS-107 mission, 334
STS-109 mission, 339
STS-113 mission, 335
STS-114 mission, 343
STS-133 mission, 343–47
Stucky, Mark 'Forger', 31, 33
suiting up process, 143, 264–67, 274
Super Bowl party, 115–16

Index 405

T

T-38s, 30–33, 76, 81, 119–20
Tanner, Joe, 290
TCDT. *See* Terminal Countdown Demonstration Test
TDRS. *See* Tracking and Data Relay Satellite
TDRS-G satellite, 104, 341–43. *See also* Tracking and Data Relay Satellite
teeth brushing, 181
Tepoorten, John M., 16–17
Terminal Countdown Demonstration Test (TCDT), 128–34, 143
Terror Eyes, 17
Terry, Tim, 30, 115, 138, 307
Thomas, Dennis, 37, 42, 48, 51
Thomas, Don, 1, 384–86
   application to NASA, 59–64
   broken ankle of, 330–31
   Capcom assignment, 84–86
   career at JSC, 66–73
   childhood, 42–53
   college years, 53–55
   marriage to Simone, 84
   parents divorce and, 47–48, 51
   postflight life for, 328, 333–34
   post-NASA career, 336
   rejection by NASA, 64–68
   selection of, as astronaut, 74–78
   STS-65 mission, 19, 86–91, 143, 156
   STS-83 mission, 328–33
   STS-94 mission, 291, 333
   STS-113 mission, 335
   TSRS deployment and, 164–68
   volunteer work, 71–73
Thomas, Kai, 116–17, 120, 214–15, 259–60, 295, 301, 328, 335–36
thunderstorms, 240
thyroid cancer, 78–81, 335
Todd, Bill, 131–32, 144, 291
Toledo, Ohio, 322, 353
Tony Packo's, 322
touchdown, 283–89
Towson University, 336
Tracking and Data Relay Satellite (TDRS), 1, 8, 92, 103–5, 118–19, 164–68, 183–84, 218, 288, 341–43, 349, 354
tracking stations, 102
training sessions, 8–11, 23–24, 30, 75–78, 87–88, 114–17, 121–22, 329–30
   emergency training, 134
   EVA, 121–22
   practice countdown, 128–34
Tropical Storm Chantal, 202, 208, 226
twang, 154

U

Udvar-Hazy Center, 346–47
United States Microgravity Payload (USMP), 338
Unity, 339
USMP. *See* United States Microgravity Payload

V

VAB. *See* Vehicle Assembly Building
Vargas, Paula, 350
Vehicle Assembly Building (VAB), 10, 11, 19, 123, 231, 284, 323
Vehicle Integration and Test Team (VITT), 20, 33, 131
vehicle walk-around, 292–95
VITT. *See* Vehicle Integration and Test Team
Voinovich, George, 2–5, 7, 109–10, 111, 297–99, 310–13, 319
Voinovich, Janet, 2–5, 297, 310–13, 319
volunteer work, 71–73
Vomit Comet, 82
Voss, Janice, 329

W

wake-up music, 177–78, 207, 216, 223, 246–47, 252–53, 256–57, 262–63, 349

walk-around, 292–95
Walker, Dave, 259
Walz, Carl, 36, 87, 100, 149, 346, 380–82
waste containment system (WCS), 163–64, 179–80
WCS. *See* waste containment system
Weber, Mary Ellen, 1, 7–9, 94, 161–62
    background of, 100–102
    career of, 388–89
    at first crew meeting, 107–8
    inspiration for being astronaut, 218
    interviews with, 128, 213, 218–19, 299
    liftoff and, 159–60
    post-STS 70 mission, 340
    space sickness, 162–63, 202, 223
    training, 114–15, 121
    TSRS deployment and, 164–68
Weightless Environmental Test Facility (WETF), 76–77, 116, 121–22
welcome home ceremony, 301–2
WETF. *See* Weightless Environmental Test Facility
Wetherbee, Jim, 298
White, Ed, 47, 48–49, 303
White Room, 151
White Sands Missile Range, 115, 166
Williams, Sunita, 391–92
WINDEX experiment, 184, 224, 256
woodpecker attack, 8–18, 149, 164, 349–50, 351–52, 354–56
woodpecker jokes, 176–78, 221–22, 316, 317, 350–52, 354–55
Woods, Ron, 331–32
Woody Woodpecker, 144, 150, 176–77, 192, 202, 212, 221, 249, 316, 317, 340–41, 349, 350, 355

Y

Young, Ginny, 114, 302
Young, John, 60, 138, 292, 303

Z

zero-gravity aircraft, 81–83
ZOE. *See* Zone of Exclusion
Zone of Exclusion (ZOE), 103, 276, 342